Vegetation–Climate Interaction

How Plants Make the Global Environment

(Second Edition)

Jonathan Adams

Vegetation–Climate Interaction

How Plants Make the Global Environment

(Second Edition)

Springer

Published in association with
Praxis Publishing
Chichester, UK

PRAXIS

Dr Jonathan Adams
Assistant Professor in Biological Sciences
Seoul National University
Gwanak-Ro
Seoul
Republic of Korea

SPRINGER–PRAXIS BOOKS IN ENVIRONMENTAL SCIENCES
SUBJECT *ADVISORY EDITOR*: John Mason, M.B.E., B.Sc., M.Sc., Ph.D.

ISBN 978-3-642-26905-9 Springer Berlin Heidelberg New York

Springer is part of Springer-Science + Business Media (springer.com)

Second Edition published 2010
First Edition published 2007
Reprinted in 2008 twice
Softcover reprint of the hardcover 2nd edition 2009

Cover design: Jim Wilkie
Project management: OPS Ltd, Gt Yarmouth, Norfolk, UK

Printed on acid-free paper

Contents

Preface to the Second Edition xi

Preface to the First Edition xiii

Foreword xv

List of figures xvii

List of tables xxiii

List of abbreviations and acronyms xxv

About the author xxvii

1 The climate system 1
- 1.1 Why does climate vary from one place to another? 2
  - 1.1.1 Why mountains are colder 4
- 1.2 Winds and currents: the atmosphere and oceans 6
- 1.3 The ocean circulation 9
  - 1.3.1 Ocean gyres and the "Roaring Forties" (or Furious Fifties) 9
  - 1.3.2 Winds and ocean currents push against one another . . . 10
- 1.4 The thermohaline circulation 10
- 1.5 The great heat-transporting machine 15
  - 1.5.1 The "continental" climate 17
  - 1.5.2 Patterns of precipitation 18

2 From climate to vegetation 27
- 2.1 Biomes: the broad vegetation types of the world 27
- 2.2 An example of a biome or broad-scale vegetation type: tropical rainforest 28

2.3 The world's major vegetation types 31
2.4 Understanding the patterns 37
2.5 What favors forest vegetation 39
2.5.1 Why trees need more warmth 39
2.5.2 Why trees need more water. 40
2.6 Deciduous or evergreen: the adaptive choices that plants make. . 43
2.7 Cold-climate evergreenness 48
2.8 The latitudinal bands of evergreen and deciduous forest 50
2.9 Nutrients and evergreenness. 50
2.10 Other trends in forest with climate 52
2.11 Non-forest biomes 53
2.12 Scrub biomes 53
2.13 Grasslands 53
2.14 Deserts 54
2.15 Biomes are to some extent subjective. 54
2.16 Humans altering the natural vegetation, shifting biomes 55
2.17 "Predicting" where vegetation types will occur 55
2.18 Species distributions and climate 59
2.18.1 Patterns in species richness 60

3 Plants on the move 67
3.1 Vegetation can move as the climate shifts. 67
3.2 The Quaternary: the last 2.4 million years 67
3.3 Biomes in the distant past. 75
3.3.1 Sudden changes in climate, and how vegetation responds 75
3.4 The increasing greenhouse effect, and future vegetation change. . 81
3.5 Response of vegetation to the present warming of climate 81
3.6 Seasons as well as vegetation distribution are changing 86
3.7 What will happen as the warming continues? 88
3.7.1 Movement of biomes under greenhouse effect warming . 92

4 Microclimates and vegetation 97
4.1 What causes microclimates? 97
4.1.1 At the soil surface and below 98
4.1.2 Above the surface: the boundary layer and wind speed . 99
4.1.3 Roughness and turbulence 102
4.1.4 Microclimates of a forest canopy 103
4.1.5 Under the canopy 106
4.1.6 Big plants "make" the microclimates of smaller plants . . 108
4.1.7 The importance of sun angle. 110
4.1.8 Bumps and hollows in the landscape have their own microclimate 112
4.1.9 Life within rocks: endolithic lichens and algae 114
4.1.10 Plants creating their own microclimate 115
4.1.11 Dark colors 115

4.1.12 Protection against freezing ... 115
4.1.13 Internal heating ... 115
4.1.14 Volatiles from leaves ... 116
4.1.15 Utilization of microclimates in agriculture ... 116
4.2 From microclimates to macroclimates ... 117

5 The desert makes the desert: Climate feedbacks from the vegetation of arid zones ... 121
5.1 Geography makes deserts ... 121
5.2 But deserts make themselves ... 122
5.2.1 The Sahel and vegetation feedbacks ... 127
5.2.2 Have humans really caused the Sahelian droughts? ... 132
5.3 Could the Sahara be made green? ... 132
5.4 A human effect on climate? The grasslands of the Great Plains in the USA ... 136
5.5 The Green Sahara of the past ... 139
5.6 Could other arid regions show the same amplification of change by vegetation cover? ... 143
5.7 Dust ... 145
5.7.1 Sudden climate switches and dust ... 149
5.8 The future ... 150

6 Forests ... 153
6.1 Finding out what forests really do to climate ... 155
6.2 What deforestation does to climate within a region ... 161
6.3 Re-afforestation ... 169
6.4 The remote effects of deforestation ... 169
6.5 The role of forest feedback in broad swings in climate ... 170
6.5.1 Deforestation and the Little Ice Age ... 170
6.5.2 Deforestation around the Mediterranean and drying in North Africa ... 173
6.5.3 Forest feedbacks during the Quaternary ... 173
6.6 Volatile organic compounds and climate ... 176
6.7 Forest–climate feedbacks in the greenhouse world ... 177

7 Plants and the carbon cycle ... 181
7.1 The ocean ... 183
7.2 Plants as a control on CO_2 and O_2 ... 185
7.3 Methane: the other carbon gas ... 187
7.3.1 Carbon and the history of the earth's temperature ... 188
7.3.2 Plants, weathering and CO_2 ... 189
7.3.3 Plants, CO_2 and ice ages ... 193
7.4 Humans and the carbon store of plants ... 198

7.5 The present increase in CO_2 201
7.5.1 The oceans as a carbon sink 204
7.5.2 Seasonal and year-to-year wiggles in CO_2 level 205
7.6 The signal in the atmosphere 210
7.7 The strength of the seasonal "wiggle" in CO_2 212
7.8 Accounting errors: the missing sink 213
7.9 Watching forests take up carbon 215
7.9.1 Predicting changes in global carbon balance under global warming 217

8 The direct carbon dioxide effect on plants 221
8.1 The two direct effects of CO_2 on plants: photosynthesis and water balance 221
8.2 Increased CO_2 effects at the scale of a leaf 222
8.3 Modeling direct CO_2 effects 223
8.4 What models predict for increasing CO_2 and global vegetation 224
8.5 Adding climate change to the CO_2 fertilization effect 225
8.6 Experiments with raised CO_2 and whole plants 227
8.6.1 The sort of results that are found in CO_2 enrichment experiments 230
8.6.2 A decline in response with time 233
8.7 Temperature and CO_2 responses interacting 233
8.8 A few examples of what is found in FACE experiments 234
8.8.1 Forests 234
8.8.2 Semi-desert and dry grassland vegetation 236
8.8.3 Will C_4 plants lose out in an increased CO_2 world? 237
8.9 Other FACE experiments 242
8.9.1 FACE studies on agricultural systems 242
8.10 Some conclusions about FACE experiments 244
8.10.1 Will a high CO_2 world favor C_3 species over C_4 species? 244
8.10.2 What factors tend to decrease plant responses to CO_2 fertilization? 245
8.11 There are other effects of enhanced CO_2 on plants apart from growth rate 245
8.12 CO_2 fertilization and soils 246
8.13 CO_2 fertilization effects across trophic levels 247
8.13.1 Looking for signs of a CO_2 fertilization effect in agriculture 248
8.13.2 Looking for signs of a CO_2 fertilization effect in natural plant communities 249
8.13.3 The changing seasonal amplitude of CO_2 252
8.14 CO_2 levels and stomata out in nature 253
8.15 Direct CO_2 effects and the ecology of the past 253
8.15.1 Direct CO_2 effects on longer geological timescales 256
8.15.2 Ancient moist climates or high CO_2 effects? 257

8.16 Other direct CO_2 effects: in the oceans 258
8.17 The future direct CO_2 effect: a good or a bad thing for the natural world? 259
8.18 Conclusion: The limits to what we can know 260

Bibliography 261

Index 265

Preface to the Second Edition

In the two years since the First Edition of this book, the study of climate and vegetation has continued to advance rapidly, with many new and interesting things to write about. I have also been able to benefit from the comments and suggestions of colleagues who have read the book. And, furthermore, I have thought up some new—and hopefully better—ways to explain things, including various new diagrams and photographs. All these seem like good reasons to attempt a new edition, which I hope will continue to serve as an up-to-date review of the complex role of vegetation in our planet's environment. Once again, my wife Mei Ling and my daughters have been a great help in encouraging my efforts, trying to improve my limited photographic skills, and politely listening as I try to explain earth system science to them.

Jonathan Adams
Seoul, Republic of Korea, 2009

Preface to the First Edition

I had wanted to write something like this book for many years, but would probably never have dared to attempt it unless I had been asked to by Clive Horwood at Praxis Publishing. As it is, this has been a rewarding experience for me personally, something which has forced me to read literature that I would not otherwise have read, and to clarify things in my head that would have remained muddled.

What I have set out to do here is provide an accessible textbook for university students, and a generalized source of current scientific information and opinion for both academics and the interested lay reader. I have myself often found it frustrating that there have been no accessible textbooks on most of the subjects dealt with here, and I hope that this book will fill the gap.

My friends and colleagues have provided valuable comment, amongst them David Schwartzman, Axel Kleidon, Alex Guenther, Ellen Thomas, Tyler Volk, Ning Zeng, Hans Renssen, Mary Killilea, Charlie Zender, Rich Norby, Christian Koerner and Roger Pielke Sr. I could not stop myself from adding to the manuscript even after they had sent me their careful advice, and any embarrassing errors that have slipped through are of course a result of my doing this. I am also very grateful to everyone who has generously given me permission to use their own photographs as illustrations in this book, and I have named each one in the photo caption. Lastly but very importantly, Mei Ling Lee has provided the encouragement to show that what I have been writing is of interest to somebody, somewhere.

Thanks in particular to Neil Cobb for providing the photo of a mountain scene, used on the cover of this book.

Jonathan Adams
Newark, New Jersey, 2007

Foreword

This book has been written with the aim of providing an accessible introduction to the many ways in which plants respond to and form the environment of our planet. As an academic scientist, and yet as a teacher, I have tried to balance conflicting needs between something which can be trusted and useful to my colleagues, and something which can enthuse newcomers to the subject. For too long, I feel, earth system science has been a closed door to students because of its jargon, its mathematics and its emphasis on meticulous but rather tedious explanations of concepts. I hate to think how many good potential scientists we have lost because of all this, and how many students who could have understood how the living earth worked have gone away bored or baffled. At a time when we may be facing one of the greatest challenges to our well-being in recent history, from global warming, it is essential that we recruit all the good researchers that we can. If we want the public, business people and politicians to understand the problems they are facing, we need to disseminate knowledge of earth system processes as widely as possibly.

In line with the aims of Praxis—and with my own aims too—I have not attempted a complete referenced literature review in this book. Instead, selected papers of authors named in the text are listed in a bibliography, to provide the reader with some useful leads into the literature. Many important studies are not directly referenced even if their findings are mentioned in the text, and I hope that the authors of these studies will not feel snubbed (because my selection of papers to reference was often fairly arbitrary). The text is written in an informal way, reflecting my own dislike of pomposity in academia. Jargon in science gives precision, but it also takes away understanding if newcomers to the subject are driven away by it. As part of my balancing act, I have tried to keep jargon to a minimum. I have also used some homey and traditional categories such as "plants" to apply to all photosynthesizers, bacterial or eukaryotic (I regard being a plant as a lifestyle, not a birthright), and somehow I could not bear to keep throwing the word "archaea" around when I could just call them "bacteria".

Dedicated to the irreverent and brilliant
Hugues Faure (1928–2003)

Figures

(asterisks indicate color images)

1.1	Why the tropics are colder than the poles	3
1.2	How the tilt of the earth's axis affects the angle of the sun, giving the seasons	4
1.3	Why the upper parts of mountains are colder	5
1.4	How mid-altitude warm belts form	5
1.5	The general position of the ITCZ along the equator	6
1.6	The intertropical convergence zone, a belt of rising air heated by the equatorial sun	7
1.7	The Coriolis effect, and the Ekman spiral	8
1.8	Thermohaline circulation in the Atlantic	11
1.9	Ocean gyres	13
1.10	How the greenhouse effect works	14
1.11	Antarctica is cut off by a continuous belt of winds and currents	16
1.12	The annual temperature cycle of an oceanic and a continental location compared	17
1.13	How the rain-making machine of the tropics works	18
1.14	How the monsoon rains move north then south of the equator during the year, following the zone where the sun is directly overhead	19
1.15	Two seasonal rainfall peaks at the equator	20
1.16	The passage of peak rains from south to north of the equator with the seasons	20
1.17*	A satellite image of the density of vegetation across northernmost South America (upper part of image can also be found in the colour section)	22
1.18	The Mediterranean climate. In winter, moist westerly winds cross southern Europe and bring rain. In summer, rain-bearing winds are pushed away by descending air from the "desert belt" of North Africa	23
1.19	Where cold seawater wells up off the coast, air cools and then is warmed as it passes over land; and the way in which coastal deserts form	24
1.20*	A view off the coast of Peru	25
1.21	The rainshadow effect	25
2.1*	(a) Map of major biome distributions	28
2.1*	(b) Areas of the most intense human alteration of vegetation	29

2.2* Buttress roots in a tropical rainforest tree 30
2.3* Drip tips on leaves of a rainforest tree shortly after a thunderstorm 31
2.4* An epiphyte growing on a tropical rainforest tree. 32
2.5 General form of vegetation 33
2.6* Tropical rainforest, Malaysia 34
2.7* Cold climate conifer forest, mountains of California. 34
2.8* Semi-deciduous forest during the dry season, with some trees leaf-less, Malaysia, near Thai border 35
2.9* Tropical montane forest with a cloud layer blanketing the upper parts of the forest, Sabah, Borneo 35
2.10* Deciduous oak scrub, western Iran 36
2.11* Grassland, California. 36
2.12* Tundra, above tree line in the Andes, Chile 37
2.13* Semi-desert, Mohave Desert, Arizona 38
2.14* Semi-desert, Iran 38
2.15 A tree's large area of leaves places a high demand in terms of evaporation, and a shrub loses less water by evaporation 41
2.16* Tree line on a mountain. 43
2.17* Autumn leaves in a northern temperate deciduous tree, Norway maple (*Acer platanoides*). 45
2.18 The relationship between January temperature and leafing out date in a range of North American trees 47
2.19* Toothed or lobed leaves are far more prevalent in cooler climate forests . . . 48
2.20 Typical leaf of red maple (*Acer rubrum*) population in Canada compared with one from Florida. 49
2.21 The proportion of species of trees with "entire" leaves depends closely on the warmth of the climate 49
2.22 Latitudinal bands of alternating evergreen and deciduous forest 51
2.23 Map of eastern Asia with zones of evergreen vs. deciduous forest 51
2.24 Holdridge's predictive scheme for relating biomes to climate. 58
2.25a* The cross-leaved heath (*Erica tetralix*), a plant with an oceanic distribution . 61
2.25b The cross-leaved heath (*Erica tetralix*) shows a typical "oceanic" range, along the western side of Europe next to the Atlantic Ocean 61
2.26* *Rhododendron ponticum* 62
2.27 Tree species richness map of parts of eastern Asia 63
2.28 Wild tree species richness for North America. 64
3.1 The global temperate history of the last 450,000 years showing a sawtooth pattern which appeared by 700,000 years ago 68
3.2 Distribution of forest vs desert 70
3.3 Biome distributions of Europe, North America at the present day and last glacial maximum 72
3.4* Temperature zones in the USA for the last glacial maximum 20,000 years ago and present day compared 74
3.5 Maps of migration rate of trees 76
3.6 Temperature history of the late glacial 77
3.7* The hazel tree (*Corylus avelana*) 79
3.8 Increase in pollen abundance of Scots pine (*Pinus sylestris*) in eastern England after the end of the ice age. 80
3.9* The greening trend around the Arctic from satellite data 82

3.10 Instrumental temperature record of the last 120 years. 83
3.11 Arctic shrub cover change in northern Canada 84
3.12 The heat-trapping effect of large cities alters the time of arrival of spring. . . 88
3.13 Sugar maple extends from southeastern Canada to the south–central USA. . 89
3.14* A bluebell (*Hyacinthoides non-scripta*) woodland in spring, southwestern England 95
4.1 The boundary layer over a surface 99
4.2 Shrubs trap more heat amongst their branches than trees do, because the wind cannot blow between the tightly packed branches of a shrub. 101
4.3* An alpine cushion plant. 102
4.4* The lumpy, uneven tree crowns of tropical forest create turbulence in the air that flows over them 103
4.5* The eery gloom of tropical montane forest shrouded by cloud. 105
4.6 During the day, heating of the canopy leaves gives an "inversion layer" of warm air floating within it; after the sun goes down, the canopy loses heat to the night sky, and the air floating just above it is left relatively warm, as it fails to cool down as fast. 107
4.7* This species of *Begonia* lives in the understory of mountain rainforests in South-East Asia 109
4.8 Distribution of temperatures on a sunny summer's day on a hill in Massachusetts, USA 111
4.9 Temperature profile against height on a cold spring morning in a Pennsylvania valley that acts as a frost hollow. 113
4.10 The daisyworld model of Lovelock illustrates how the microclimate effects of plants could scale up to global climates. 117
5.1 If we take two equally moist surfaces with different albedo values, the darker low-albedo one is more likely to give rain 124
5.2 How positive feedback affects the slope of a response. 126
5.3 A metastable system has multiple states. 127
5.4 The Sahel, at the southern border of the Sahara desert. 128
5.5 Record of rainfall in the Sahel since 1900 129
5.6 How a GCM works and how a mesoscale model slots into it 134
5.7 Temperature map for a warm day in northeastern Colorado. 137
5.8* The distribution of vegetation zones of the present-day and the Holocene "Green Sahara" 141
5.9 In the Sahara, during the last 9,000 years, the summer solar energy input changes slowly 144
5.10 The opposing effects of dust in the atmosphere on temperature at ground level. 147
6.1 Forest loss in tropical lands 156–8
6.2 Some of the ways in which forests modify temperature. 159
6.3 As the leaves come out, the progressive warming into spring halts for a few days because of the latent heat taken up by evaporation from the leaves 160
6.4 In tropical rainforest, loss of latent heat uptake and roughness dominates and deforestation is predicted to result in a regional temperature increase; in boreal forest the albedo effect dominates upon forest clearance, producing cooling . 162
6.5 In Georgia, USA, models suggest that after the lowlands were deforested, the maximum rainfall area shifted away from the mountains towards the boundary between forested land and cropland 165

6.6 In Costa Rica, before deforestation, evaporation from lowland forests supplied abundant clouds shrouding mountaintops and supporting cloud forest; now, lowland deforestation has resulted in less moisture supply to the atmosphere from the lowlands, so clouds are sparser and higher in the mountains 167
6.7 Having small open areas in a mainly forested landscape can actually increase rainfall by providing focal points for rising air which can be carried high and condense out rain clouds 168
6.8* Global temperature history of the last 2,000 years from several sources of tree ring data 171
6.9 Scene from a frozen river in Holland, 1608 172
7.1 Some basic components of the carbon cycle. 182
7.2 A huge amount of CO_2 is stored in the form of both bicarbonate and dissolved CO_2 in the ocean. 184
7.3 Estimated CO_2 concentrations in the atmosphere over the last several hundred million years. 186
7.4 One of the thousands of species of lichens—symbiotic combinations of a fungus and alga 190
7.5 Results of an experiment that compared the amounts of salts turning up in rainwater that had run off lichen-covered rocks vs. bare rock surfaces 191
7.6 History of temperature and atmospheric CO_2, deduced from polar ice cores 194
7.7 How plankton activity may have decreased the CO_2 concentration during glacials. 195
7.8 The distribution of forest and desert in the present natural world and the LGM 197
7.9 How the land reservoir of carbon may help keep up CO_2 concentrations in the atmosphere when the oceans are dragging carbon down 198
7.10 Ice core record of atmospheric CO_2 since 1000 AD, Law Dome, Antarctica ice cores 201
7.11* Annual net flux of carbon to the atmosphere from land use change: 1850–2000 203
7.12 When an area of land is allowed to return to forest, there is an initial accumulation of carbon 203
7.13 The record of atmospheric CO_2 increase since the 1950s, measured directly at the Mauna Loa Observatory in Hawaii 204
7.14 The seasonal cycle in CO_2 concentration varies with latitude. 206
7.15 "Lightening" of the isotope composition of atmospheric CO_2 over time. . . . 208
7.16 A carbon isotope shift around 7 million years ago indicates that C_4 plants suddenly became much more common 209
7.17* Map showing correlation between temperature and global CO_2 increment each year, and map showing correlation between the amount of rainfall and the size of the global CO_2 increment around the world. 211
7.18 The strength of the seasonal CO_2 wiggle is strongly related to the state of the NAO, and variation in the carbon isotope composition of CO_2. 213
7.19 The eddy flux covariance method picks up CO_2 215
7.20 Model results with and without the "gushing out" of carbon that would result from warming affecting the carbon balance of forests 218
8.1 Key steps in photosynthesis which are altered by CO_2 concentrations 223
8.2 The three types of increased CO_2 experiment: closed-chamber, open-top chamber, free air release 228
8.3 The Tennessee FACE site showing the towers used to release CO_2 into the forest 230
8.4 Aerial view of the Tennessee FACE experiment showing rings of towers . . . 231

8.5 The Swiss FACE site on mature mixed temperate forest uses a network of tubes twisting up the branches to deliver CO_2 in the right places and right quantities to simulate a higher CO_2 atmosphere 231

8.6* Scientists at the Swiss FACE site inspect the forest canopy for direct CO_2 effects using a crane 232

8.7 The sequence of reactions in a C_4 leaf. 239

8.8* Thermal infrared image taken looking down on a wheat field fertilized by a ring of pipes releasing CO_2 243

8.9 Stomatal index change. 246

8.10 The shift in ^{13}C in sediments in North America, South Asia, and in the global ocean indicating a "take-over" by C_4 plants 258

Tables

5.1 Typical values for various land surface types 123
5.2 Climate history of northwestern China over the last 10,000 years 145

Abbreviations and acronyms

CAM	Crassulacean Acid Metabolism
CDIAC	Carbon Dioxide Information and Analysis Center
CSIRO	Commonwealth Scientific and Industrial Research Organization
FACE	Free Air CO_2 Experiment
GCM	General Circulation Model
IPCC	Intergovernmental Panel on Climate Change
ITCZ	Inter-Tropical Convergence Zone
LAI	Leaf Area Index
LGM	Last Glacial Maximum
NCAR	National Center for Atmospheric Research
NCEP	National Centers for Environmental Prediction
NOAA	National Oceanic and Aerospace Administration
NPP	Net Primary Production
UV	UltraViolet
VOC	Volatile Organic Compound

About the author

Jonathan Adams was born in England and studied Botany at St Catherine's College of the University of Oxford. His PhD was in Geology from the University of Aix-Marseilles II, France, where his mentor was the distinguished Quaternary geologist Hugues Faure.

After postdoctoral studies at Cambridge University and at Oak Ridge National Laboratory (Tennessee), Jonathan Adams has taught at the University of Adelaide (Australia), Rutgers University (New Jersey), and latterly at Seoul National University (South Korea).

About the Author

1

The climate system

Though few people stop to think of it, much of the character of a place comes from its covering of plants. Southern France, with scented hard-leaved scrublands, has an entirely different feel about it from the tropical rainforest of Brazil, or the conifer forests of Canada. Vegetation is as important a part of the landscape as topography and the architecture of buildings, and yet it is an accepted and almost subconscious part of the order of things.

Even fewer people ever ask themselves "why" vegetation should be any different from one place to another. Why do conifers dominate in some parts of the world, but not others? Why are there broadleaved trees that drop their leaves in winter some places, while elsewhere they keep them all year round? Why are some places covered in grasslands and not forest? As with almost everything in nature, there is a combination of reasons why things are the way they are. Most important in the case of vegetation are two factors: humans, and climate.

In some cases, the landscape we see is almost completely a product of what mankind is currently doing. Humans have cleared away much of the world's natural plant cover, and replaced it with fields and buildings, or forest plantations of trees from other parts of the world. Yet, even in such heavily modified areas, fragments of the original vegetation often survive. In other instances the vegetation is a sort of hybrid of human influence and nature; battered by fires or by grazing animals, and yet still distinctive to its region. Most of the landscapes of Europe (including, for example, southern France) are like this, produced by the combination of climate, local flora and rural land use patterns.

However, over large areas the vegetation is still much as it was before humans dominated the planet. This original cover tends to survive in the areas where the landscape is too mountainous to farm, or the climate or soils are in other ways unsuitable for cultivation. Most of Siberia, Canada, the Himalayan Plateau and the Amazon Basin are like this, and scattered areas of protected wilderness survive in hilly or marginal areas in most countries. If we concentrate on these most natural areas in particular, there are clear trends in the look of vegetation which tend to correlate with climate. Such relationships between vegetation and climate first became apparent when explorers, traders and colonialists began to voyage around the world during the last few centuries. The tradition of natural history that grew out of these early explorations has tried to make sense of it all. Vegetation takes on a myriad of forms, which can be difficult to push into orderly boxes for classification. Yet there is no doubt that there is a lot of predictability about it.

Variation in climate, then, is a major factor that determines the way vegetation varies around the world. But why does the climate itself vary so much between different regions? The basic processes that make climate are important not just in understanding why vegetation types occur where they do, but also in understanding the complex feedbacks explored in the later chapters of this book. As we shall see, not only is the vegetation made by the climate, but the climate itself is also made by vegetation!

1.1 WHY DOES CLIMATE VARY FROM ONE PLACE TO ANOTHER?

Essentially, there are two main reasons that climate varies from place to place; first, the amount of energy arriving from the sun, and second the circulation of the atmosphere and oceans which carry heat and moisture from one place to another.

One of the major factors determining the relative warmth of a climate is the angle of the sun in the sky. The sun shines almost straight at the earth's equator, because the equator sits in the direct plane of the sun within the solar system. So, if you stand on the equator during the middle part of the day, the sun passes straight overhead. At higher latitudes, such as in Europe or North America, you would be standing a little way around the curve of the earth and so the sun always stays lower in the sky. The farther away from the equator you go, the lower the sun stays until at the poles it is really only barely above the horizon during the day.

Having the sun directly overhead gives a lot more energy to the surface than if the sun is at an angle. It is rather like shining a flashlight down onto a table. Hold the flashlight pointing straight down at the table and you have an intense beam on the surface. But hold it at an angle and the light is spread out across the table top and much weaker. If the sun is high in the sky, a lot of light energy hits each square kilometer of the earth's surface and warms the air above. If the sun is low in the sky, the energy is splurged out across the land; so there is less energy falling on the same unit area (Figure 1.1a). This tends to make the poles colder than the tropics, because they are getting less heat from sunlight.

A second factor relating to sun angle, which helps make the high latitudes cooler, is the depth of atmosphere that the sun's rays must pass through on the way to the earth's surface (Figure 1.1b). Because at high latitudes the sun is lower in the sky, it shines through the atmosphere on a slanting path. At this angle, the light must pass a longer distance through more gases, dust and haze. This keeps more of the sun's energy away from the surface, and what is absorbed high in the atmosphere is quickly lost again up into space. Think how weak the sun is around sunset just before it sinks

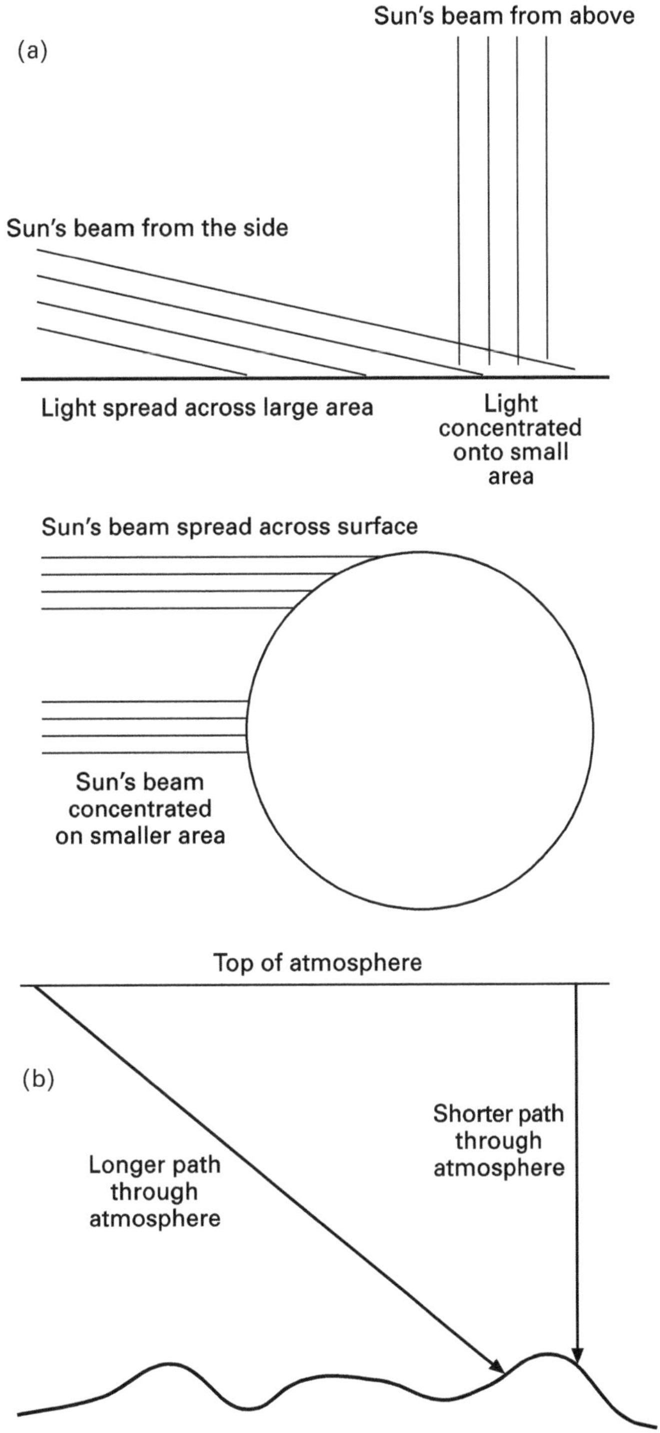

Figure 1.1. Why the tropics are colder than the poles. (a) A direct beam gives more energy than an angled beam. (b) Passing through greater depth of atmosphere absorbs more energy before it can hit the earth.

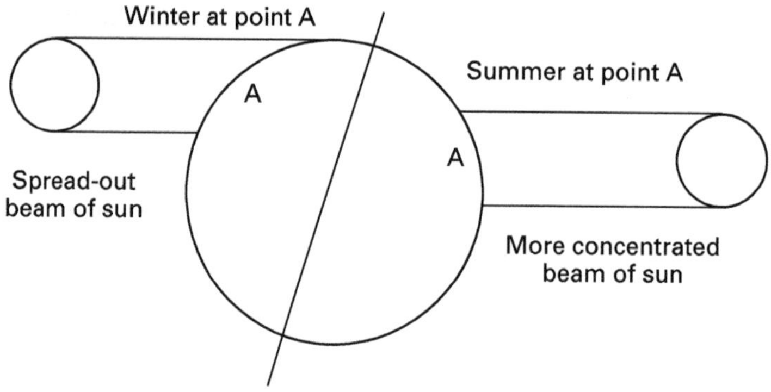

Figure 1.2. How the tilt of the earth's axis affects the angle of the sun, giving the seasons.

below the horizon—so weak that you can stare straight into it. The dimness of the setting sun is an example of the effect of it having to shine through a longer path of atmosphere, which absorbs and scatters the sun's light before it can reach the surface. So, the lower in the sky the sun is, the longer is its path through the atmosphere, and the less energy reaches the ground.

Only in the tropics is the sun right overhead throughout the year, giving the maximum amount of energy. This then is the key to why the poles are cooler than the tropics.

The seasons of the year are also basically the result of the same sun angle effects (Figure 1.2). The earth is rotating on its axis at a slight angle to the sun, and at one part of its yearly orbit the northern hemisphere is tilted so the sun is higher in the sky; it gets more energy. This time of year will be the northern summer. At the same time, the southern hemisphere is getting less energy due to the sun being lower. During the other half of the year, the southern hemisphere gets favored and this is the southern summer. Adding to these effects of sun angle is day length; the "winter" hemisphere is in night more of the time because the lower sun spends more time below the horizon. This adds to the coldness—the warming effect of the sun during the day lasts less time, because the days are shorter.

1.1.1 Why mountains are colder

If you climb up a mountain, the air usually gets colder. The temperature tends to decline by about 0.5°C for every hundred meters ascended, although this does vary. The rate of decrease of temperature with altitude is called the "lapse rate". Lapse rate tends to be less if the air is moist, and more if the air is dry. Generally, every 10 meters higher up a mountain is the climatic equivalent of traveling about 15 km towards the poles. Unlike the decline in temperature with latitude, sun angle does not explain why higher altitudes are generally colder. The relative coldness of mountains is a by-product of the way that the atmosphere acts as a blanket, letting the sun's light in but preventing heat from being lost into space (see Box Section 1.1 on the greenhouse effect). Because they protrude up into the atmosphere, mountain tops have less of this blanket above them, so they are colder (Figure 1.3).

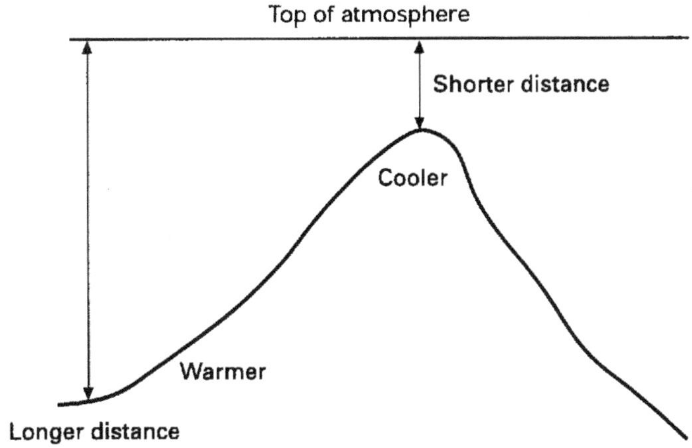

Figure 1.3. Why the upper parts of mountains are colder. A thinner layer of greenhouse gases causes them to lose heat rapidly.

There are however some exceptions to this pattern of temperature decline with altitude: places where the mid-altitudes of a mountain are warmer on average than the lowest altitudes. This occurs where there are enclosed valleys between mountains, where there is not much wind. At night, cold air from the upper mountain slopes tends to drain as a fluid into the valley below, and accumulate. Just above the level that this cold draining air tops up to, there is a warm mid-altitude belt that can have warmer-climate plants than the valley below (Figure 1.4). Mid-altitude warm belts like this often occur in the Austrian Alps, for example.

The general pattern of cooler temperatures at higher altitudes occurs not only on mountains, but through the atmosphere in general, essentially because of the same factor—a thinner blanket of greenhouse gases higher up. If air is rising up from the surface due to the sun's heating, it will tend to cool as it rises due to this same factor. Another thing that will tend to make it cool is that it expands as it rises into the thinner upper atmosphere—an expanding gas always takes up heat. If the rising air is moist, the cooling may cause it to condense out water droplets as cloud, and then perhaps rain drops which will fall back down to earth.

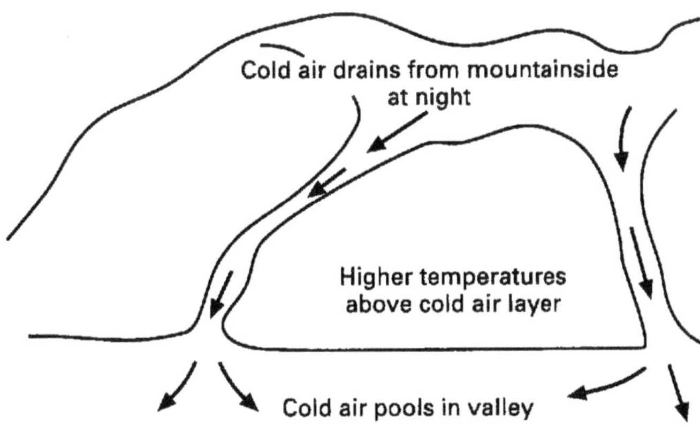

Figure 1.4. How mid-altitude warm belts form. Cold air drains down as "rivers" from the upper slopes of the mountain, and fills up the valley below. Just above the top of the accumulated cold air, temperatures are warmer.

1.2 WINDS AND CURRENTS: THE ATMOSPHERE AND OCEANS

Differences in the amount of the sun's energy received by the surface drive a powerful global circulation pattern of winds and water currents. The most basic feature of this circulation, and a major driving force for almost everything else, is a broad belt of rising air along the equator (Figures 1.5 and 1.6). This is known as the intertropical convergence zone, or ITCZ for short. The air within the ITCZ is rising by a process known as convection; intense tropical sunlight heats the land and ocean surface and the air above it warms and expands. Along most of this long belt, the expanding air rises up into the atmosphere as a plume, sucking in air sideways from near ground level to replace the air that has already risen up. Essentially the same process of convection occurs within a saucepan full of soup heated on a hot plate, or air warmed by a heater within a room; any fluid whether air or water can show convection if it is heated from below. The difference with the ITCZ, though, is that it is convection occurring on an enormous scale. Because air is being sucked away upwards, this means that the air pressure at ground level is reduced—so the ITCZ is a zone of low air pressure in the sense that it would be measured by a barometer at ground level.

What goes up has to come down, and the air that rises along the equator ends up cooling and sinking several hundred kilometers to the north or south of the equator. These two belts of sinking air press down on the ground from above, imposing higher pressure at the surface as they push downwards.

The air that sinks down in these outer tropical high-pressure belts gets sucked back at ground level towards the equator, to replace the air that is rising up from being heated by the sun. It would be easiest for these winds blowing back to the equator to take a simple north–south path; this after all is the shortest distance. But

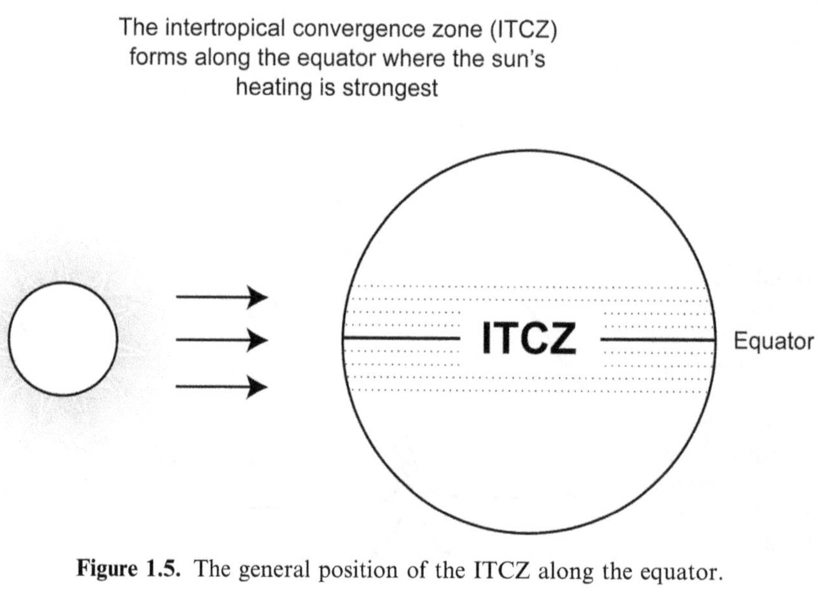

Figure 1.5. The general position of the ITCZ along the equator.

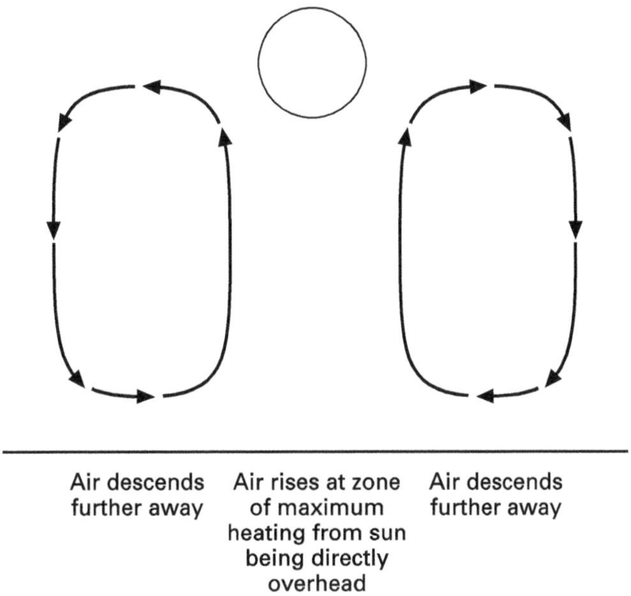

Figure 1.6. The intertropical convergence zone, a belt of rising air heated by the equatorial sun.

the earth is rotating, and in every 24 hour rotation the equator has a lot farther to travel round than the poles. So, the closer you are to the equator, the faster you are traveling as the earth turns. When wind comes from a slightly higher latitude, it comes from a part of the earth that is rotating more slowly. As it nears the equator, it gets "left behind"—and the closer to the equator it gets, the more it lags behind. So, because it is getting left behind the wind follows a curving path sideways. This lagging effect of differences in the earth's rotation speed with latitude is known as the "Coriolis effect", and any wind or ocean current that moves between different latitudes will be affected by it. It also explains, for example, why hurricanes rotate.

Although it has been moving towards the equator, much of this wind does not get there because the Coriolis effect turns it sideways. It ends up blowing westwards as two parallel belts of winds, one belt either side of the equator (Figure 1.7a). These are the trade winds, so-called because in the days of sail, merchant vessels could rely on these winds to carry them straight across an ocean.

There is another related effect—the "Ekman spiral"—when a wind bent by the Coriolis effect blows over the rough surface of the earth, the friction of the earth's surface—which remember is rotating underneath it at a different speed—will drag the wind along with the rotating earth, canceling out the Coriolis effect (Figure 1.7b). This causes the wind direction to change near the earth's surface, and is part of the reason why winds by the ground can be blowing in one direction, while the clouds up above are being blown in a different direction. Between the air nearest the ground and the air way above, the wind will be blowing at an intermediate angle; it is "bent" around slightly. The closer it gets to the surface the more bent off course it gets.

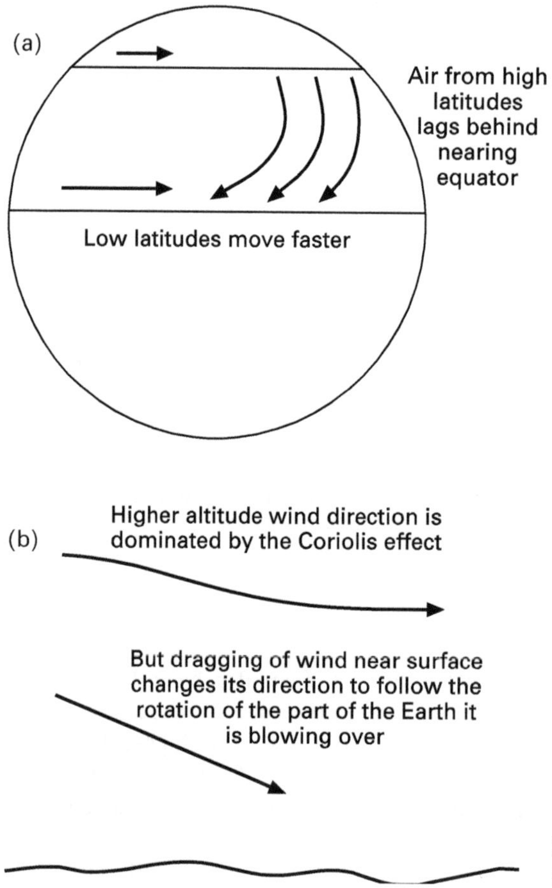

Figure 1.7. (a) The Coriolis effect. (b) The Ekman spiral.

There are many other aspects to the circulation pattern of the world's atmosphere, too many to properly describe here in a book that is mainly about vegetation. For instance, there is another convection cell of rising and sinking air just to the north of the outer tropical belt, and driven like a cog wheel by pushing against the cooling air that sinks back down there. A third convection cell sits over each of the poles.

Outside the tropics, air tends to move mostly in the form of huge "blobs" hundreds of miles across. These are known as "air masses". An air mass is formed when air stays still for days or weeks over a particular region, cooling off or heating up, and only later starts to drift away from where it formed. You might regard an air mass as resembling a big drop of syrup poured into a pan of water. It tends to spread out sideways, and also mix sideways with what is around it. The collision zone between an air mass and the air that it is moving into is known as a "front". When a front passes over, you get a change in the weather, and often rain.

In a sense, the detailed patterns of moving individual air masses are controlled by thin belts of higher altitude winds (at between 9 and 12 km altitude) in the atmosphere at the edge of the polar regions, and also at lower latitudes where the air from the ITCZ starts descending.

These eastward-trending winds are the jet streams. They "push around" the lower-level air masses like chess pieces. There is the subtropical jet stream and the polar jetstream in each hemisphere. That makes four jet streams in all. The jet streams are fed by air rising up into them moving in a polewards direction, and they are propelled east by the Coriolis force because the air comes from the faster-rotating lower latitudes.

1.3 THE OCEAN CIRCULATION

Just as winds move through the atmosphere, there are currents in the oceans. These too transport an immense amount of heat from the equator towards the higher latitudes. For the most part, ocean currents only exist because winds blow them along, pushing the water by friction. But part of the reason winds blow is that there are temperature differences at the surface, and ocean currents sometimes bring about such contrasts in temperature (especially if there is upwelling of cool water from below). So the water moves because the wind blows across it, yet the wind may blow because of the very same temperature contrasts that are brought about by the water moving!

Wind skimming across the surface will drive the top layer of water as a current in a particular direction, and if it moves towards or away from the equator the current will eventually get bent round by the Coriolis effect. So, for example, in each of the world's main ocean basins there are eastward-curving currents that travel out from the equator because of this mechanism (see below). But below the surface of a current being bent by the Coriolis effect, the deeper part of the current is being dragged by contact with the still waters below it. That dragging tends to move it along in the direction that the earth is rotating locally. So because of this dragging, this deeper water in the ocean ends up traveling in a slightly different direction. The deeper you go, the more the angle of the current is diverted by dragging against water below, and different layers in the ocean can be traveling in quite different directions. This is the same Ekman spiral effect as occurs in the atmosphere.

Winds blow fast but per volume of air they don't carry very much heat. The heat-carrying capacity of ocean water is much greater, but the ocean currents move much more slowly than the winds. In fact, both ocean currents and winds are important in transporting heat around the earth's surface.

1.3.1 Ocean gyres and the "Roaring Forties" (or Furious Fifties)

The most prominent feature of the world's ocean circulation are currents that run in big loops, known as gyres. They start off in the tropics moving west, and curve round eastwards in the higher latitude parts of each ocean basin, eventually coming back down to the tropics and completing a circle.

These gyres originate from the powerful trade winds that blow towards the west in the outer tropics. The winds push against the surface of the ocean producing these currents. But why does an ocean gyre eventually turn around and flow eastwards? It happens because the ocean currents are slammed against the shores on the west sides of ocean basins by the trade winds that blow west along the equator. Both the winds and the currents bounce off the western side of the basin, and start to head away from the equator. Because they are traveling with the same rotation speed as the equatorial zone, the Coriolis effect bends them off towards the east, diagonally across the ocean towards higher latitudes.

The winds that follow the outer parts of these ocean gyres, and help drive them, are powered by the big contrast in temperature created as the ocean currents move polewards and cool off. In the southern hemisphere these winds are known as the Roaring Forties, blowing west-to-east just south of South Africa and Tasmania, and hitting the southern tip of South America with a glancing blow. The nickname that generations of sailors have given these winds comes from their unrelenting power and their tendency to carry storms, and the fact that they stay within the 40s latitudes. In the northern hemisphere, the equivalent belt of winds is located more in the fifties and low sixties, hitting Iceland, the British Isles and the southwest Norwegian coast. These winds, even stormier, are known as the "Furious Fifties".

1.3.2 Winds and ocean currents push against one another

As I've implied above, surface ocean currents are driven by winds, but to some extent the winds are responding to pressure and temperature differences created by ocean currents beneath them. So it is a rather complex circular chicken-and-egg situation.

Actually, there is something peculiar about the North Atlantic circulation, beyond just the push of equatorial trade winds, which partly explains why it is strong enough to produce the Furious Fifties. As well as being pushed, it is also pulled along by another mechanism, the thermohaline circulation.

1.4 THE THERMOHALINE CIRCULATION

Ocean currents do not just move around on the surface. In some places, the upper ocean waters sink down into the deep ocean. This happens for example in the North Atlantic off Greenland, Iceland and Norway. Where the surface water sinks, this sends a "river" of surface water down into deep ocean. A similar sinking process happens off Antarctica, and in a small patch of the Mediterranean Sea (just south of Marseilles, France) in winter.

The reason these waters sink is that they are denser than the surrounding ocean. But why are they denser? It is mostly due to their higher salt content. Pour a dense brine solution into a bowl of fresh water and it will sink straight down to the bottom, and the same principle applies here. These denser, saltier ocean waters are derived from areas that undergo a lot of evaporation, because the climate is hot. Evaporation of water leaves a more concentrated salt solution behind, and this is the key to the whole mechanism. So, for example, the waters in the north Atlantic gyre are derived from the Gulf Stream that comes up from the Caribbean. Heated by the tropical sun, it has lost a fair amount of water by evaporation. After water vapor is transported away, the remaining seawater is left saltier and denser as it leaves on its path northwards across the surface of the Atlantic (Figures 1.8a and 1.9). But the water is not yet dense enough to sink because the Gulf Stream is still warm as it is transported northwards. Warm water tends to be less dense than cold water. Even though it is saltier, its extra warmth is keeping its density quite low and it can still float over the less salty but cold water below.

Only when it reaches northern latitudes does the Gulf Stream water cool off drastically, giving up its heat to the winds that blow east over Europe. Because it has cooled, the Gulf Stream water is now left heavier than the surrounding waters and it finally sinks, as "pipes" of descending water about a kilometer across that lead down to the ocean floor. These pipes tend to form in the spaces between sea ice floes when a cold wind skips across the surface. On reaching the bottom, the sunken waters pan out to form a discrete layer that eventually spreads through all the world's ocean basins (Figure 1.8b).

There are several different sinking regions that feed water down into the deep (the North Atlantic being just one of them), and they each produce their own mass of water. These different waters sit above one another in a sort of "layer cake" arrangement, that shows up in a cross section down through the ocean. Each layer has its own density, temperature/salinity balance, chemistry and is travelling in its own particular direction!

Just about all the world's deep ocean waters—those below about 300 metres—are cold (about 2 to 4°C), even though most of the ocean surface area is warmer. Even in the tropics, where surface water temperatures may be 32°C, the water below 300 m depth is about as cold as it would be in a domestic refrigerator. Why then are these

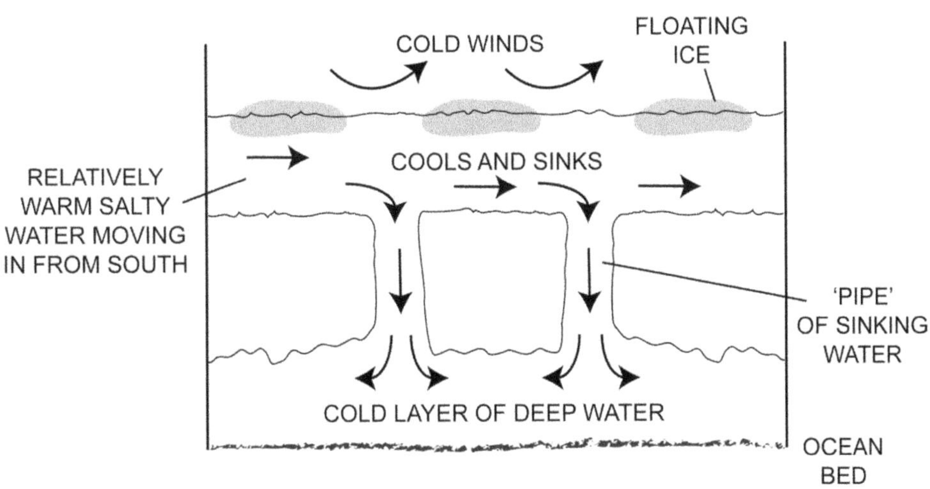

Figure 1.8. Thermohaline circulation in the Atlantic.

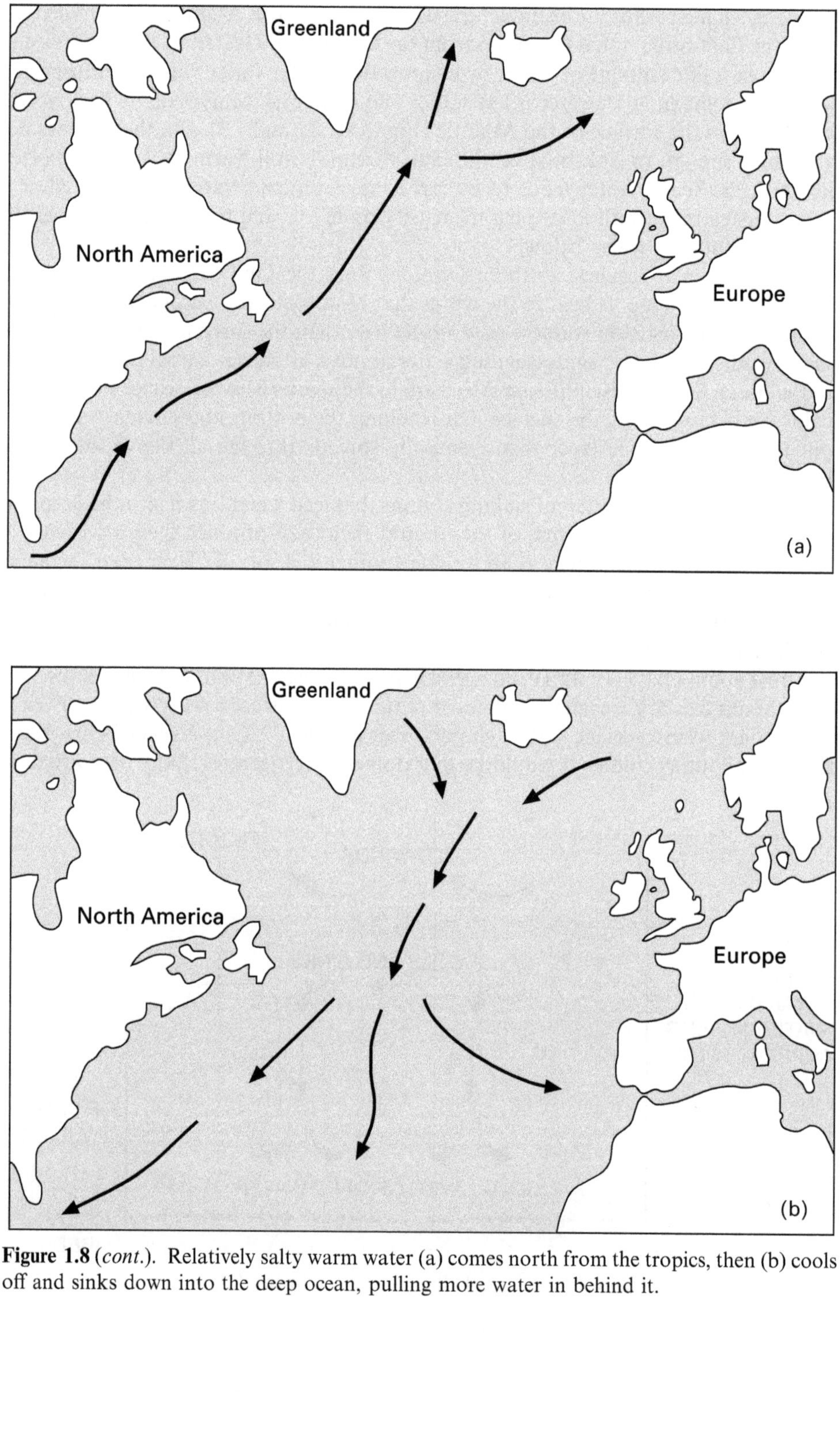

Figure 1.8 (*cont.*). Relatively salty warm water (a) comes north from the tropics, then (b) cools off and sinks down into the deep ocean, pulling more water in behind it.

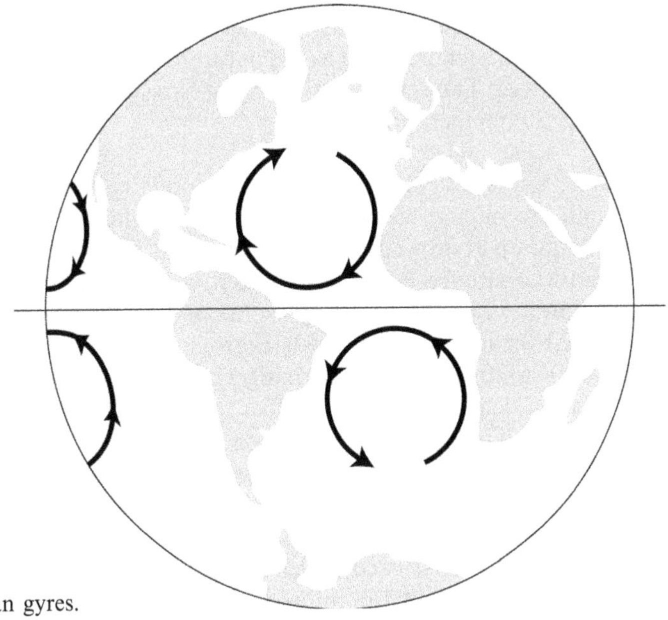

Figure 1.9. Ocean gyres.

deeper waters so cold? Because they originate as water that sinks in winter in the high latitudes, when the sea surface is cold. If other warmer waters at other temperatures had instead been filling the deep ocean, the mass of ocean water would reflect their particular temperature instead.

In fact, at other times in past (e.g., the early Eocene period, around 55 million years ago) the whole deep ocean was a mild 12°C instead of about 3°C at present. Why? Perhaps because the "feeding" of sinking water was occurring not only in chilly sub-polar seas but down in tropical latitudes, from places similar to the Arabian Gulf at present where warm but salty water (concentrated by evaporation) spills out into the Indian Ocean. What did this opposite circulation system do to climate? The climate scientists have no idea, really. But it could perhaps help explain the warmer world at such times, a world that, for example, had palm trees and crocodiles living near the poles.

Box 1.1 The greenhouse effect

The atmosphere tends to trap heat, through a process known as the "greenhouse effect". The gases in the atmosphere are mostly transparent to visible light, which is the main form in which the sun's energy arrives on earth. But many of these same gases tend to strongly absorb the invisible infra-red light that the earth's surface radiates to loose heat back to space. Some of the infra-red captured by the gas molecules in the atmosphere is sent back down to earth (as infra-red again) where it is absorbed by the surface once more and helps keep it warm. This is known as the "greenhouse effect".

If it were not for the combined greenhouse effect of naturally occurring gases in the atmosphere, the earth's temperature would naturally be somewhere around −20°C to −30°C on average. Thus this extra warming is very important in keeping the earth at a moderate temperature for life.

At present there is a lot of concern about an ongoing increase in the atmospheric levels of certain greenhouse gases due to human activities. For instance, carbon dioxide is building up at around 1% a year due to it being released by fossil fuel burning and forest clearance around the world (Chapter 7). It is set to reach double the concentration it was at 250 years ago some time during the mid-21st century. The worry is that the increase over the background level of this and other greenhouse gases will lead to major climate changes around the world over the coming centuries. Already, detectable warming does seem to be occurring and the likelihood is that this will intensify. Since plants are strongly affected by temperature, it is likely that global warming will change the distribution of biomes (see Chapters 2 and 3). Shifts in rainfall that result from the changing heat balance and circulation of the atmosphere may also turn out to be important. And because of the many "feedbacks" discussed in the later chapters of this book, a change in vegetation may in itself amplify an initial change in climate, resulting in a bigger change than would otherwise have occurred (Figure 1.10).

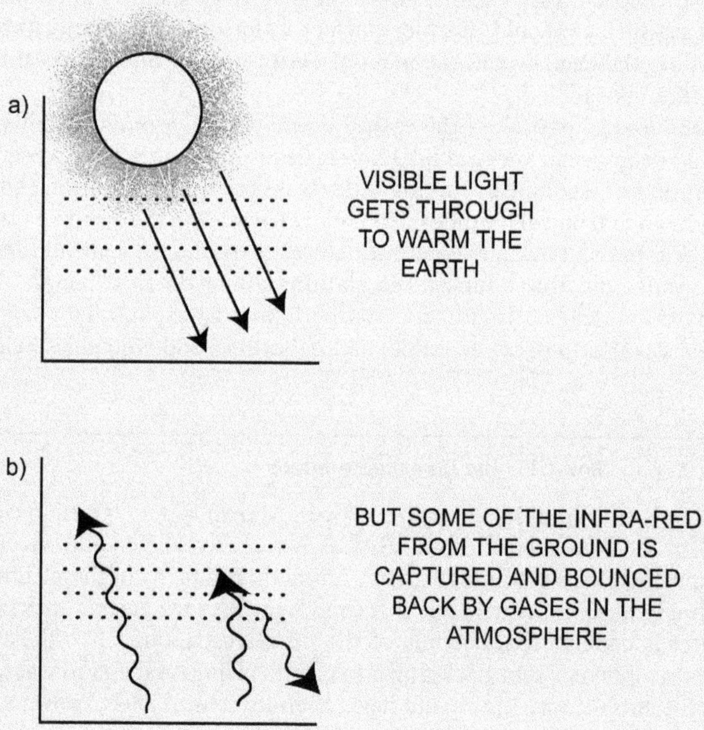

Figure 1.10. How the greenhouse effect works. (a) Visible light from the sun can get through the atmosphere. (b) Some of the infra-red leaving the earth's surface is absorbed and sent back down to earth.

1.5 THE GREAT HEAT-TRANSPORTING MACHINE

The decrease in temperature towards the poles forms the basic pattern of the earth's climates. But this pattern is greatly altered by the global circulation of two fluids: air and water. Factoring in the circulation of air and water enables us to understand the present-day patterns of climate in more detail.

One useful way to think of the world's climate circulation is to view it essentially as a heat-transporting machine that takes heat from the tropics and moves it to higher latitudes. It operates by movement of warm ocean currents, and also movement of winds and air masses (those great "blobs" of air) that move across the surface.

Heat is transported not just as the temperature that one can easily measure (known as "sensible" heat, because it can easily be "sensed"), but also in the form of "latent heat". This latent heat is hidden energy that comes out only if you try to lower the temperature of moist air until a fog of water droplets appears. As you attempt to cool it, the air temperature drops, but nowhere near as fast as you would expect, because the water vapor condensing out as droplets gives off heat that keeps the air warm.

If it were not for this movement of heat in air masses and ocean currents, the high latitudes would be far colder than they actually are. Heat transport from the tropics "subsidizes" the higher latitudes, by as much as 40–60% more than the heat that they get from the sun (and the higher the latitude, the more important this heat subsidy is). This draining of heat away from the tropics also makes them cooler than they would otherwise be.

Places in the high latitudes that are close to the oceans, and receive especially strong ocean currents from the tropics, can be a lot warmer than places that do not. A warm current known as the Gulf Stream (mentioned above) crosses the Atlantic up from the Caribbean, and across to northwestern Europe. Largely because of the Gulf Stream, Britain has a much warmer climate on average than Nova Scotia, the eastern tip of Canada which is at the same latitude on the west side of Atlantic. In England, grass stays green in January and palmettos can be grown outdoors because the winters are so mild. In Nova Scotia the snow lies deep all winter long, and temperatures can dip to −40°C. As mentioned above, part of the reason that the Gulf Stream flows so strongly northwards and carries so much heat is that it is essentially "sucked" northwards by the sinking water of the thermohaline circulation in the north Atlantic.

There is a similar "gulf stream" reaching the western side of North America (e.g., on rainy Vancouver Island) which has a very mild climate compared with the harsh winters of Sakhalin/northern Japan at the same latitude on the western side of the Pacific. However, because there is no strong sinking zone in the ocean to pull it in, its effect on climates is not as strong as in the north Atlantic.

High-latitude places that are isolated from tropical air masses and warm sea currents tend to be especially cold for most of the year. The most extreme example is Antarctica. It is cut off from the rest of world by the belt of swirling currents and winds known as the "Roaring Forties". This prevents much heat transfer from lower

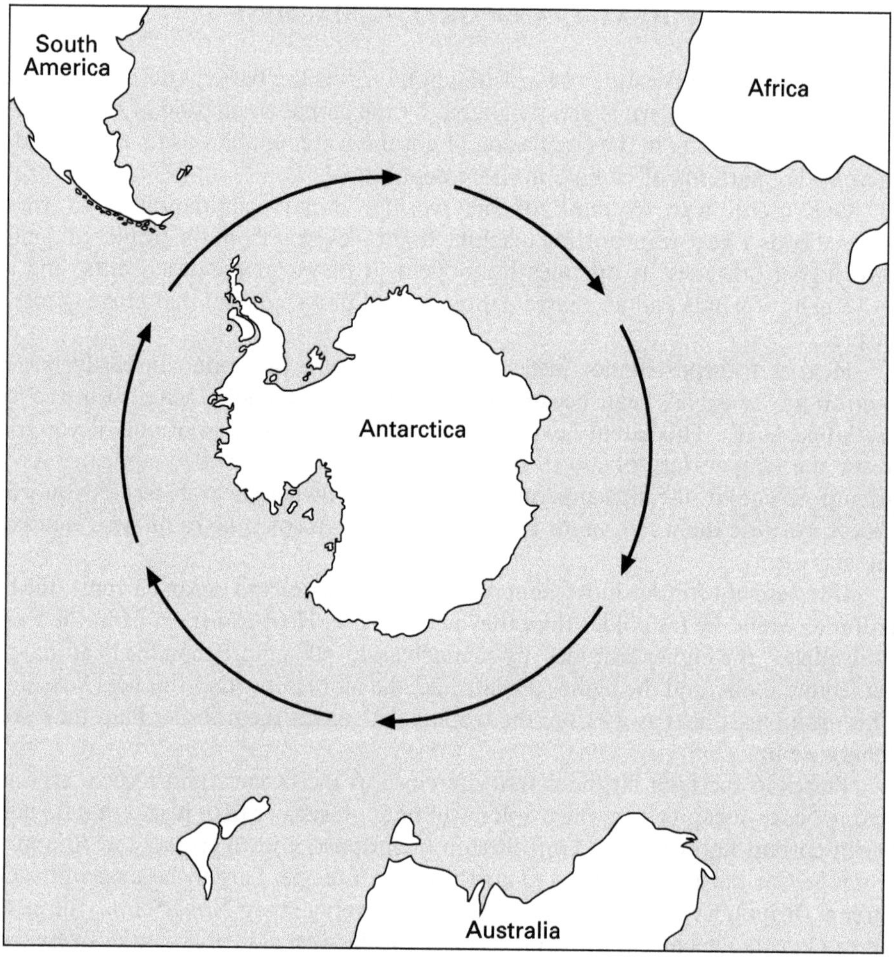

Figure 1.11. Antarctica is cut off by a continuous belt of winds and currents.

latitudes, so Antarctica is colder than the North Pole region which receives air masses and warmer ocean currents from low latitudes (Figure 1.11).

In some places an especially cold area of ocean just off the coast makes a difference to the climate inland. Although Nova Scotia is at a disadvantage for heat because it does not receive the Gulf Stream, the frigidity of its climate is added to by a cold sea current that comes down along the west side of Greenland, bringing water straight down from near the North Pole. Across the other side of North America, the remarkable climate of San Francisco in California, which almost never gets hot—and almost never has frost either—is caused by a zone of upwelling of cool deep ocean water just off the coast. A similar cool upwelling zone occurs off the coast of Peru, where it brings about the extreme aridity of the Atacama Desert (see below).

1.5.1 The "continental" climate

Areas far inland in the higher latitudes tend to experience wide seasonal swings in temperature, because they are cut off from the moderating influence of the oceans. Seas have a very high capacity to store up heat—so their temperature does not vary so much during the year (Figure 1.12a). In contrast, the land cools down or heats up far more quickly. An area far inland gets less oceanic influence and is more at the mercy of the amount of heat received from the differing sun angle and day length at different times of year. Hence in such places the seasonal differences in temperature can be extreme (Figure 1.12b). The coldest winters on earth outside Antarctica occur not at the North Pole but in the interior of northeastern Siberia, because of its isolation from the oceans. This is known as a "continental" climate, receiving little heat from the distant oceans, and not much warming water vapor in the atmosphere to release heat. The coldest temperature ever recorded in northeastern Siberia in winter was a bone-chilling −68°C. Yet, paradoxically, this same part of Siberia has warm summers too; temperatures can exceed 30°C. The summer warmth is the result of the same factor—isolation from cooling sea winds, which do not reach the interior of Siberia from the seas around its edges.

To a lesser extent, continental climates with wide seasonal temperature swings are found in central Canada and the USA, eastern Europe and central Asia.

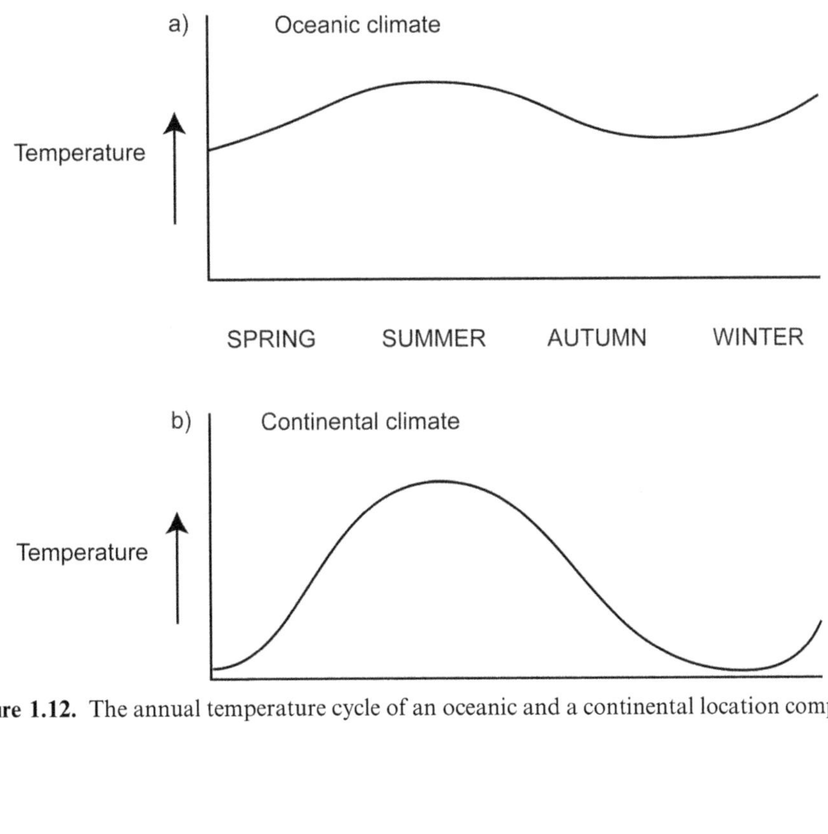

Figure 1.12. The annual temperature cycle of an oceanic and a continental location compared.

1.5.2 Patterns of precipitation

Not only temperature patterns depend on ocean currents and winds. Patterns in the wetness or aridity of large parts of the world's land surface can be understood as a product of circulation.

Why is it, for instance, that the tropics are so moist? Just as with temperature, this is ultimately a result of sun angle. The band of rising air along the equator (the intertropical convergence zone or ITCZ) occurs due to intense solar heating, from the sun being directly overhead. The heating sets up convection in the air, and this rising air sucks in moist ocean winds and water evaporating from the forests. As the air rises it cools, and water droplets condense out as clouds and then fall as rain. This gives the moist tropical rainforest climate down below.

A typical morning in the equatorial tropics begins clear and sunny. As the sun climbs high in the sky, the day becomes hot, but by mid-afternoon clouds begin to build and cover the sky as the heat of the sun sets off convection in the atmosphere. Eventually, by late afternoon the heat of the day is broken by a thunderstorm, leaving the air fresh and mild, and the vegetation moist with rain.

Hundreds of kilometers farther north and south, the air carried aloft in the ITCZ descends back down to earth. It has lost its moisture, which fell as rain as it first rose up from the surface, and now it also warms as it descends (Figure 1.13). The air is already dry, and the warming makes it hold onto its small amount of water vapor

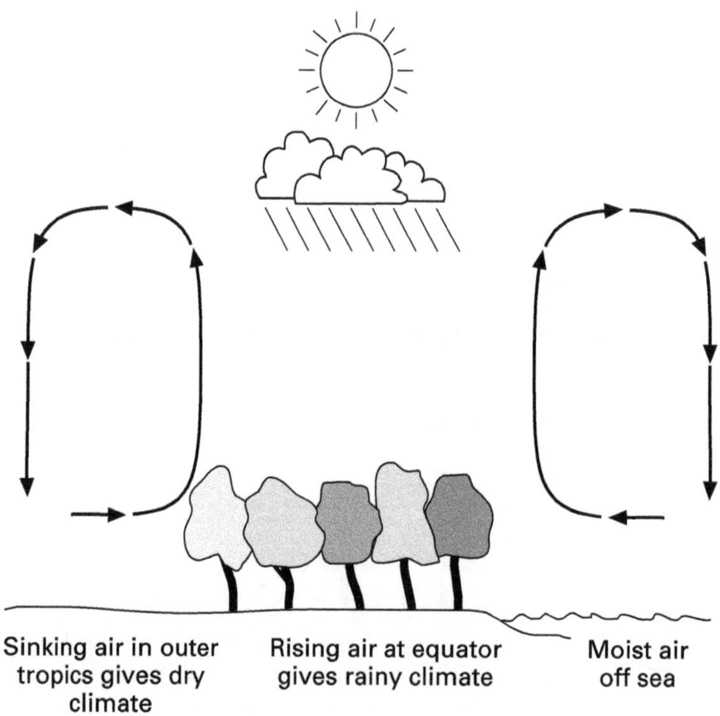

Figure 1.13. How the rain-making machine of the tropics works. The heating of the ITCZ causes water to condense out and fall as rain. When the air descends again, no water vapor can condense out and there is an arid climate.

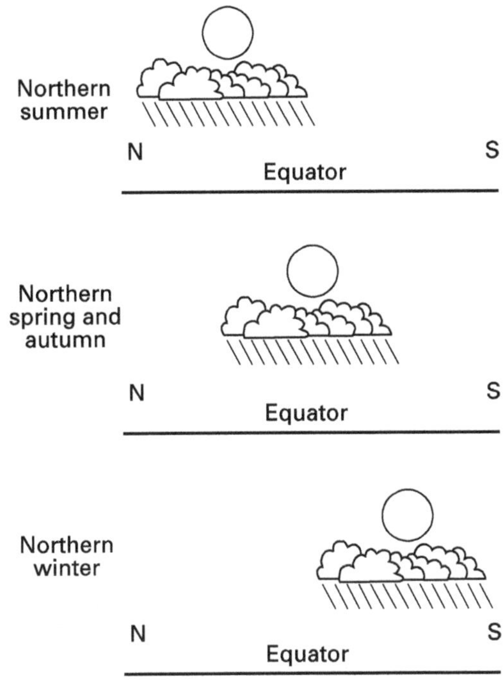

Figure 1.14. How the monsoon rains move north then south of the equator during the year, following the zone where the sun is directly overhead.

even more tightly, so there is no chance of rain falling from it. These bands of descending air, north and south of the Equator, tend to give desert climates with hardly any rainfall. Hence the same mechanism that produced very wet climates along the equator also produces arid climates to the north and south.

The ITCZ does not just stay static. It wavers north and south during the year, because the earth is tilted relative to the sun (this giving winter and summer, as explained earlier). So the highest point of sun in the sky, relative to your point of view on the ground, moves north and south of the equator. Thus, the strongest zone of solar heating is north or south of equator, at different points during the year (Figure 1.14).

The band of rising air near the equator (the ITCZ) follows this zone of greatest heating. In the northern summer, it is slightly north of the equator—although its precise position depends on the layout of land and sea surfaces that can help to drag it either slightly farther south or farther north. In the southern summer, the ITCZ moves to the south. During spring or autumn, it moves between these two extremes, usually crossing the equator itself at these times of the year. Each time the ITCZ passes over the equator, there is an increase in rainfall there—so equatorial rainforest climates have two peaks in rainfall each year (Figure 1.15). However, because they at least get the edge of the ITCZ throughout the year these equatorial locations tend to be quite rainy all the time; the seasonal peak just makes them extra-rainy! Farther away north or south from the equator (Figure 1.16) towards the edges of the tropics,

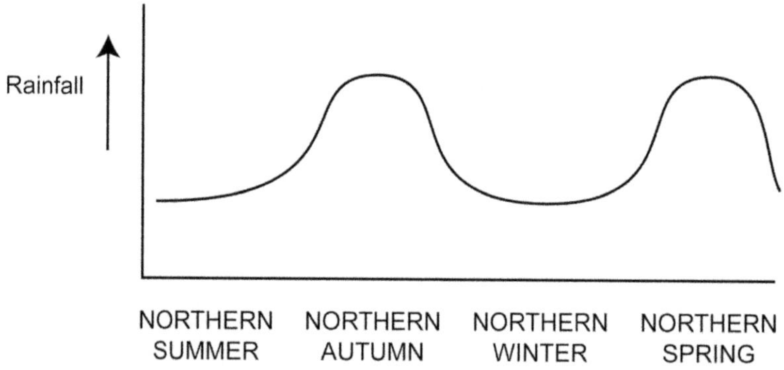

Figure 1.15. Two seasonal rainfall peaks at the equator.

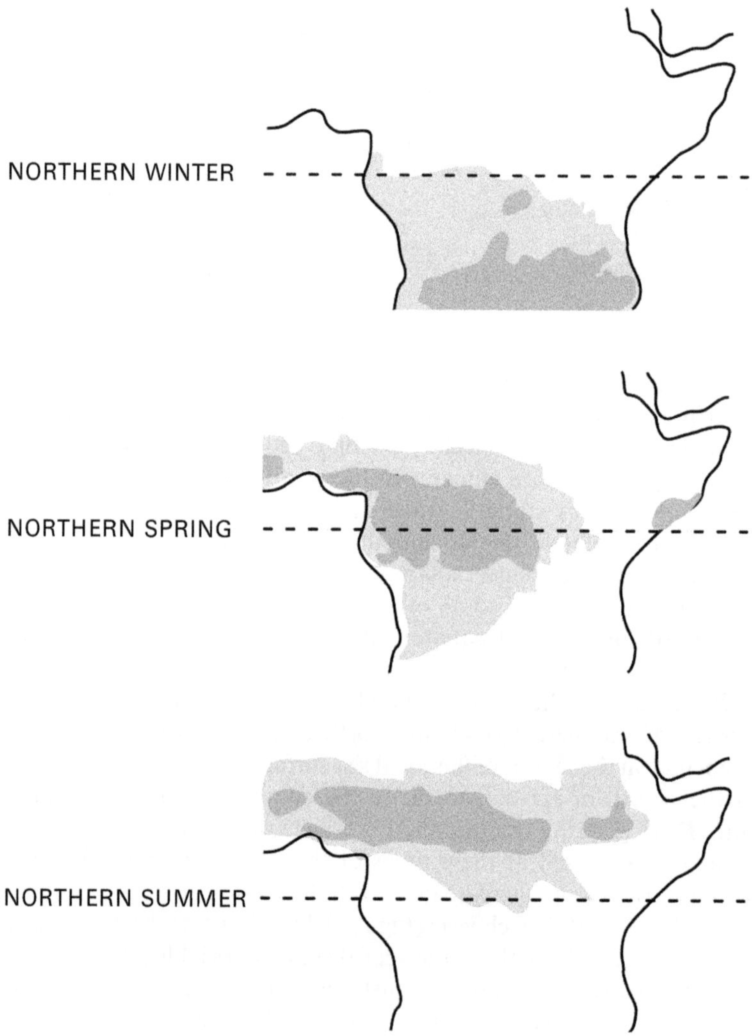

Figure 1.16. The passage of peak rains (shaded areas) from south to north of the equator with the seasons. In autumn, the rains will pass over the equator once again on their way down south, giving the year's second rainy peak on the equator (not shown).

the summer "monsoon" is caused by the arrival of the ITCZ as the sun's summer heating pulls it north (into the northern hemisphere) and then south (into the southern hemisphere). In these places the dry descending air is replaced for a few months by the equatorial climate. In satellite images one can see a "green wave" traveling up through northern Africa in early summer, as the vegetation starts to grow again with the rains.

In India the summer monsoon is especially strong because the mountains of the Himalayan Plateau heat up and feed rising air straight into a belt of upper-level winds known as the jet stream. This pulls up more air to replace itself from lower altitudes, dragging the ITCZ especially far north in this region during the summer, way up into northern India. The pulling effect of the Himalayas on the ITCZ also means that it gives rain to other mid-latitude areas such as Japan and Korea, that would otherwise be much too far north to see an effect from the monsoon.

In winter, when the ITCZ has gone south, there is a "winter monsoon" wind traveling from the north in Asia. In most areas this is dry and cold, but it can carry rain-bearing winds from temperate latitudes if it sucks in some air that has traveled over a moist sea surface.

Winds off the oceans transport water vapor, so areas that get ocean winds tend to be wet. Britain's notoriously damp climate results from it sitting in the path of a belt of eastward-blowing winds off the Atlantic (these are paradoxically known as "westerlies", an old mariner's term, because they blow from the west). Similarly damp climates occur on western edges of land masses that receive other belts of westerly winds which follow the curving ocean gyre currents: for example, New Zealand, southernmost Chile in South America, and the area around Vancouver and Seattle in North America. Such ocean-dominated climates tend to have about the same amount of rain in any month of the year (Figure 1.17a*).

If we travel just slightly closer to the equator, these moist climates begin to lose their summer rain. The moist eastward-blowing ocean winds can only reach these areas during part of the year, the winter (Figure 1.18). This characteristic pattern of wet winters and hot dry summers (Figure 1.17b) is known as the Mediterranean climate: named after the area around the Mediterranean Sea at the southern end of Europe, which provides the archetype of this pattern. Other examples of this same climate occur in southwestern and southeastern Australia, southernmost Africa, south–central Chile and California: all are known for their wine-making and olive cultivation. Mediterranean climates are all on the western side of continents, just beyond the outer edge of the tropical desert belt. The summer dryness is caused by the descending desert-making air of the ITCZ extending a little farther outwards, as the summer sun's heating pulls the whole ITCZ away from the equator (Figure 1.17). This outer belt of descending air pushes away the moist westerly winds of the ocean, resulting in dryness all summer long until it is time for the ITCZ to move back equatorwards again, and the moist westerly winds can get access once again. In effect, the desert climate extends out over these areas for a few months during summer, before it retreats again in winter. But if the ocean is cold,

* See also color section.

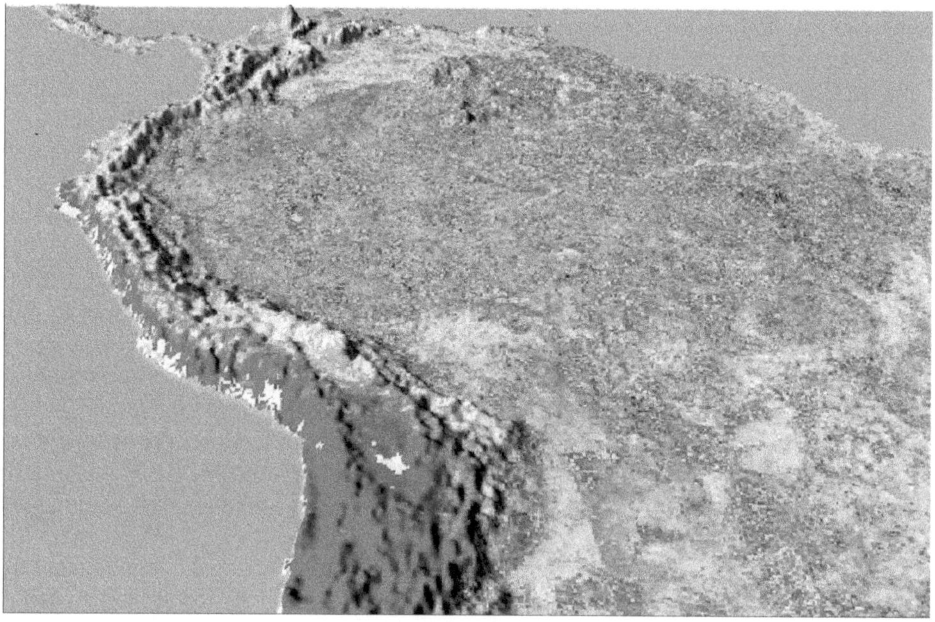

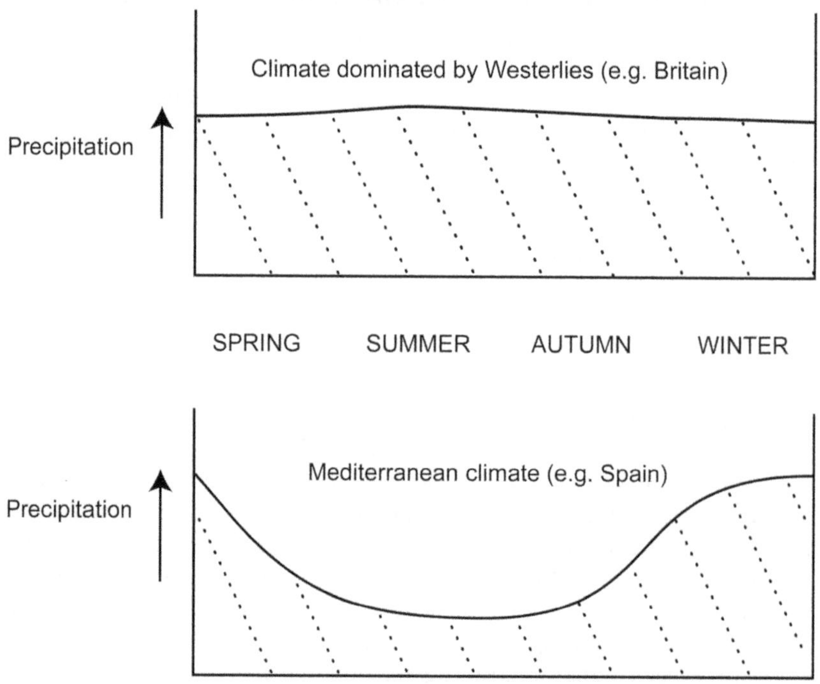

Figure 1.17. A satellite image of the density of vegetation across northernmost South America. By contrast to the green-ness of the Amazon Basin to the east, the coastal strip of Peru is barren of vegetation due to an extremely dry climate. Source: Ned Nikolov, from NOAA NDVI imagery.

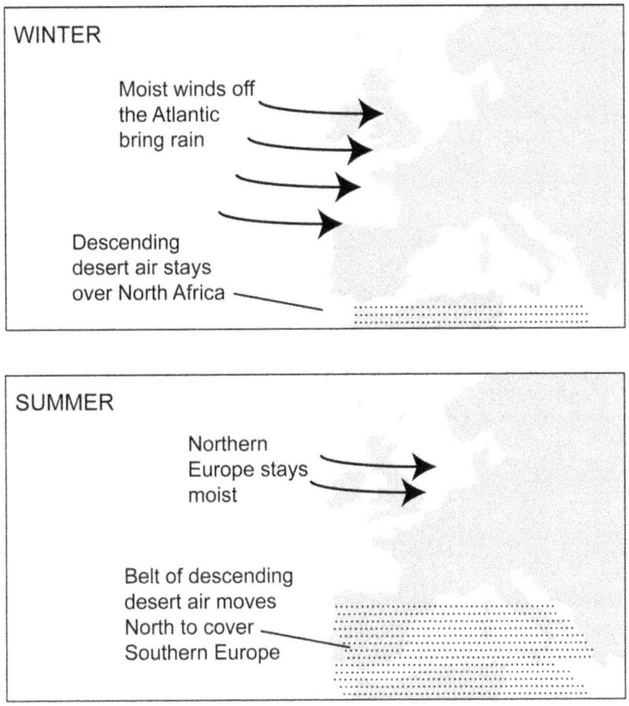

Figure 1.18. The Mediterranean climate. In winter, moist westerly winds cross southern Europe and bring rain. In summer, rain-bearing winds are pushed away by descending air from the "desert belt" of North Africa.

colder than the air, there may be an arid belt along the coast (Figure 1.19). For example, such desert belts occur close to ocean upwelling areas off Peru (Figures 1.17 and 1.20* and Namibia, where winds pulling the surface water away draw up cold deep water to the surface. How does this cause aridity? Because to get rain, there needs to be a cooling effect on already-moist air causing water droplets to condense out to cloud and then raindrops. If the air actually warms as it moves over land, the water vapor is held more tightly in the warmer air and cannot condense out. As an additional influence, over the cold sea surface where water up-wells, the cooling of air above tends to cause sinking within the atmosphere. This too makes rain unlikely, because strong upwards convection is necessary for producing rain.

Another factor which influences which places get rain, and which do not, is topography. If a mountain range is in the path of moist winds coming in off the sea, the stream of air must rise up the windward slope if it is to get over the mountains. This causes the air to cool, and as it cools water droplets condense out to form clouds and then either rain or snow. Thus, the leading edge of a mountain range is usually rainy: a phenomenon known as the orographic effect. By the time the wind reaches over to the far side of the mountain range, it has already lost most of its water vapor, squeezed out as rain on the windward slope. And now descending, it warms and holds onto its remaining water vapor more tightly again—so there is less rain on this leeward side. In fact, this water-squeezing effect can alter the climate for hundreds of kilometers downwind of a mountain range, giving an extensive dry area.

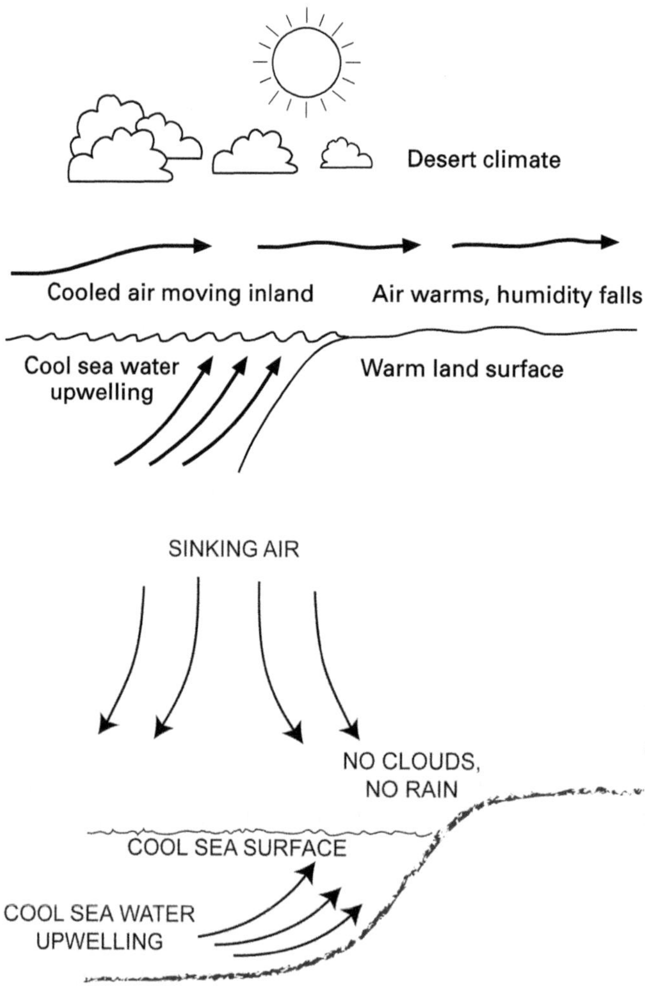

Figure 1.19. (*Top*) Where cold seawater wells up off the coast, air cools and then is warmed as it passes over land. This prevents rainfall, bringing about a coastal desert. In addition to this, the cold sea surface may prevent upwards movement in the atmosphere, likewise supressing rain formation. (*Bottom*) Coastal deserts form in this way.

It is almost as if the mountain is casting a shadow of dryness across the landscape—hence this is termed the "rainshadow effect". Mountain ranges that have a noticeable rainshadow—with a dry zone just inland—include the Cascades in the western USA, and the Great Dividing Range in eastern Australia (Figure 1.21).

The basic climate system explained in this chapter is the blank canvas on which we will now paint a complex picture of the ways in which plants both respond to and actually modify their environment. In Chapter 2 the broad patterns in vegetation produced by this background of climate will be described, and in Chapter 3 the ways in which vegetation can move in response to changes in this background. Chapter 4 will deal with the ways in which plants both respond to and produce their own local climate, the microclimate. Chapters 5 and 6 cross over to how vegetation itself can

Figure 1.20. A view off the coast of Peru. Cool seawater welling up nutrients from the deep supports a very active marine ecosystem, which feeds the abundant sea birds. Desert cliffs on the coast are also influenced by the the cool water suppressing the formation of rain clouds. *Source*: Axel Kleidon.

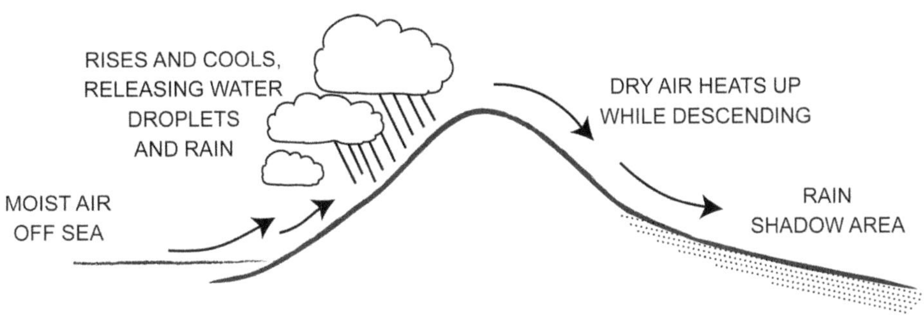

Figure 1.21. The rainshadow effect.

help to make climate on the broader scale, over hundreds and thousands of kilometers. Chapters 7 and 8 will deal with some other important ways in which plants both modify and respond to their natural environment, in terms of the carbon-containing gases in the atmosphere.

2

From climate to vegetation

2.1 BIOMES: THE BROAD VEGETATION TYPES OF THE WORLD

On the broadest scale, certain forms of vegetation occur again and again, scattered between different places around the planet. Depending on how finely you might choose to subdivide them, there are between five and twenty fundamental vegetation types in the world. They include, for example, tropical rainforest and savanna in the tropics, and in the high latitudes temperate forest and steppe. Such broad-scale vegetation types are known as "biomes", and each one of them is distributed between several continents (Figure 2.1a*).

The distribution of biomes is not random—it depends mostly on climate, although underlying rock and soil type and local drainage conditions also determine the precise limits of each biome. Humans have also influenced the vegetation through burning, forestry and agriculture, so that in some places one finds that a biome has been reduced or shifted in area in response to human disturbance during the last few thousand years (Figure 2.1b*).

Exactly how broadly or narrowly a biome is defined can vary between one ecologist and another. For instance, most ecologists would define all the world's tropical evergreen forests (tropical rainforests) as a biome by itself, but others would also tend to lump this and various other types of tropical forest into a single larger "tropical forest" biome. Some would even include all forests—anywhere in the world—as part of a grand "forest" biome.

Each biome is made up of thousands of individual plant species. Although tending to fit in with the general growth pattern and appearance of the biome, each of these species has its idiosyncrasies. A species also has its own distribution pattern, determined by its specific requirements for climate and soil, and also the chance legacies of history.

* See also color section.

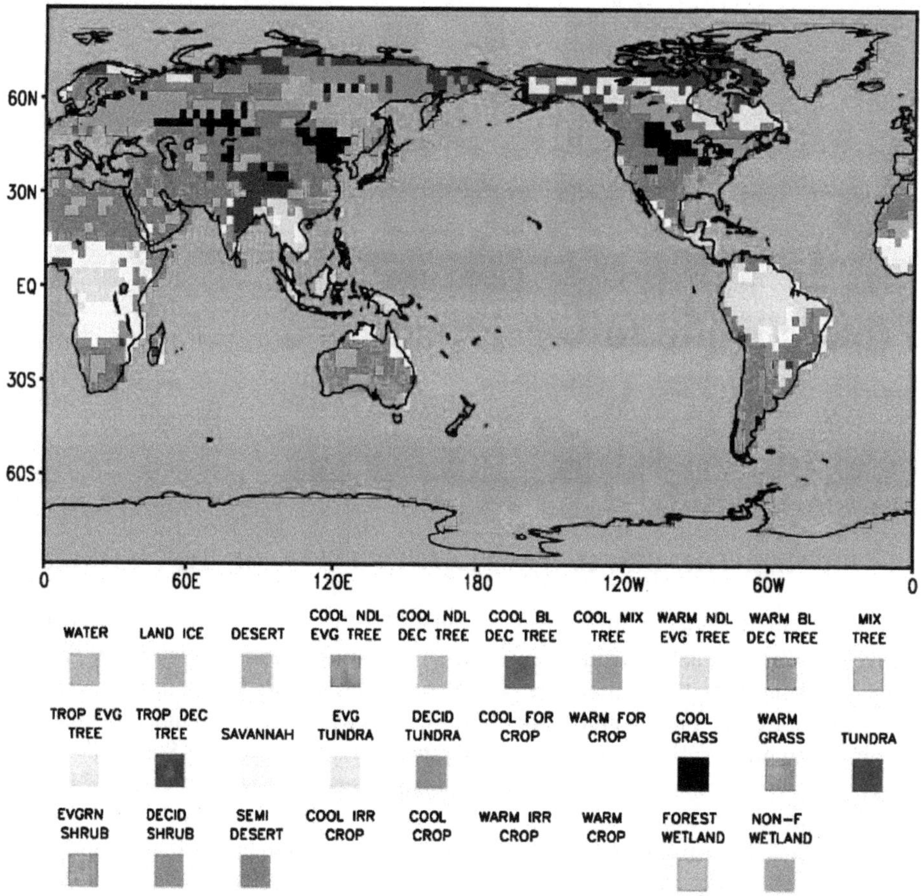

Figure 2.1. (a) Map of major biome distributions. This is for "natural" vegetation as it would be without human disturbance, based on what we know of broad climate–vegetation relationships. The categories vary somewhat between different authors and so show up differently on different maps. *Source*: Chase *et al.* (2000).

In this chapter we will explore the ways in which climate selects and shapes vegetation, both in its general form (as in a biome) and in the detailed appearance and composition of species within it. Then, in the next chapter (Chapter 3) we will take a look at how vegetation can move if the climate changes.

2.2 AN EXAMPLE OF A BIOME OR BROAD-SCALE VEGETATION TYPE: TROPICAL RAINFOREST

In each biome, vegetation looks the way that it does because of selection by the environment. Natural selection has killed plants which had the wrong characteristics,

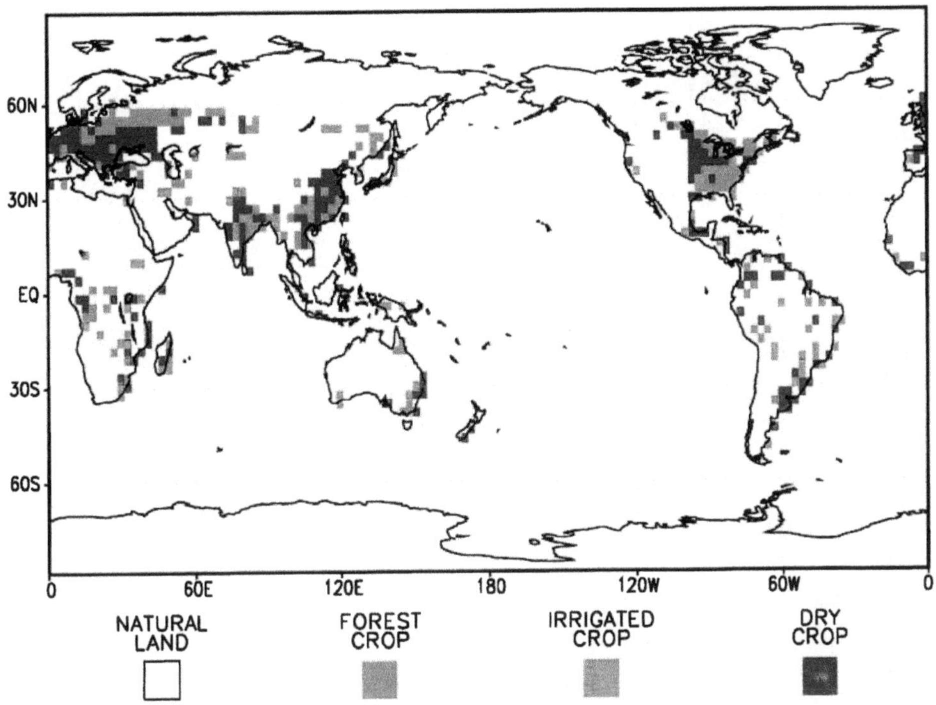

Figure 2.1. (b) Areas of the most intense human alteration of vegetation. Agriculture ("dry" croplands that depend on rainfall, plus irrigated croplands watered by farmers) is extensive. In some areas there is a mixture of croplands and forests regularly harvested on a rotational basis: this is shown here as "forest–crop". *Source*: Chase *et al.* (2000).

and allowed others that had the right features to survive. By this mechanism, plants from many different lineages have evolved to "suit" the climate, often in quite subtle ways. One example that can be used to illustrate the link between form and function in vegetation is the tropical rainforest biome, which is scattered across several landmasses close to the equator. The distribution of tropical rainforest closely follows the equatorial climate zone with year-round rain and warm conditions, occurring in the Amazon Basin, in central Africa, central America and South-East Asia (Figure 2.1a*).

If you were placed in tropical rainforest anywhere in the world, it would look much the same, even though many of the groups of plants are quite different between the regions. This overall resemblance occurs because similarity in climate has selected for various features of the vegetation; evergreen forest with hard glossy leaves, "buttress" roots on the trees that splay out near the ground, an abundance of climbing vines in the forest, and leaves with elongated ends known as drip tips. Another characteristic of tropical rainforest is the presence of epiphytes—plants which grow perched on the branches of large trees. Close similarity in the vegetation between different lands is true within each of the biomes, and it occurs because plants require the same characteristics to exploit the opportunities and survive the challenges presented by the climate.

Figure 2.2. Buttress roots in a tropical rainforest tree. *Source*: Author.

For example, in the tropical moist climate, the soil is often soggy with rain, and the clays that form in tropical soils tend to be particularly slick when they are wet. The intense chemical breakdown of rocks in the warmth and the damp (a process known as chemical weathering; see Chapter 7) tends to give deep soils without any layer of rocks near the surface which the trees could hold onto with their roots. Hence the trees need extra anchorage to avoid falling over in these slippery conditions. The evolutionary response to these unstable soil conditions has been buttress roots—splayed out trunks (Figure 2.2*) that resemble the buttresses on the side of a medieval cathedral. But, instead of propping up the tree, as buttresses on a building would do, these act as points of entry for wide-set roots that anchor the tree more effectively into the ground, much as the ropes on a tent would do.

Long drawn-out tips to the leaves are another very common feature that has a function in the climate of the tropics (Figure 2.3*). Where it rains frequently, the surface of a leaf often accumulates in a layer of water. Since the rainforest leaves are long-lived, often lasting three or four years, if leaves stayed wet over time they would accumulate fungi and lichens that would eventually choke the leaf. The answer to this problem is to drain the water off a leaf each time it rains, and these drip tips (as they are known) help this by concentrating the weight of the surface water down to this central point until its tension breaks and it drips or trickles away.

Figure 2.3. Drip tips on leaves of a rainforest tree shortly after a thunderstorm, with drops of water still draining from them. *Source*: Author.

Epiphytes (Figure 2.4*) occur in the tropical rainforest because of the high humidity and frequent rainstorms. A plant that grows perched high on a tree branch has not got much to root into, usually just some rotted leaf litter and moss. So the supply of water around its roots is very precarious. Only in a very moist environment, where it will be re-supplied with water every couple of days, can such a plant survive.

In other parts of the world outside the equatorial zone, there are different climatic combinations of hot and cold, wet and dry. Similar sets of conditions result in similar types of vegetation. Even though the particular plant species found in the various parts of a biome may be unrelated, they look and behave similarly to one another. This is because natural selection has pressured them in same direction, by selecting plants which had the "right" features, and killing those that did not.

2.3 THE WORLD'S MAJOR VEGETATION TYPES

In the broadest sense the world's vegetation can be divided into several basic structural types, each of which includes several biomes:

Figure 2.4. An epiphyte growing on a tropical rainforest tree. *Source*: Author.

1. There is a set of "forest vegetation" biomes, which have a dense, closed canopy of trees. If you stand and look upwards in a forest, you see few large gaps between the crowns of the trees, which tend to overlap and interlock with one another. Forest biomes include tropical rainforest, temperate evergreen forest, temperate deciduous forest, and cold climate conifer forest (also called "boreal conifer" and "taiga") (Figures 2.5a, 2.6*, 2.7*, 2.8*, 2.9*).
2. "Woodland" biomes are rather like forest but with a more open canopy, with significant gaps between individual trees so that their crowns often do not touch. These include Mediterranean woodland, tropical dry woodland, and boreal woodland. A typical sort of definition of woodland would be that less than 70% of the canopy above is trees, with the rest being open sky (Figure 2.5b).
3. "Shrub" or "scrub" biomes have low woody plants, usually with a rather gnarled appearance and multiple stems instead of a single trunk. They include temperate semi-arid scrub, tropical semi-arid scrub, Mediterranean scrub (garrigue) (Figures 2.5c, 2.10*).
4. "Grasslands" look rather like a lawn or meadow—both of which are human creations—except that these are natural, not cultivated. For example, in the category of grasslands there are the savannas in the tropics, steppe or prairie in the temperate zones, and grassy tundra in very cold climates. Sometimes there may be an open scattering of trees or shrubs (Figures 2.5d, 2.11*, 2.12*).

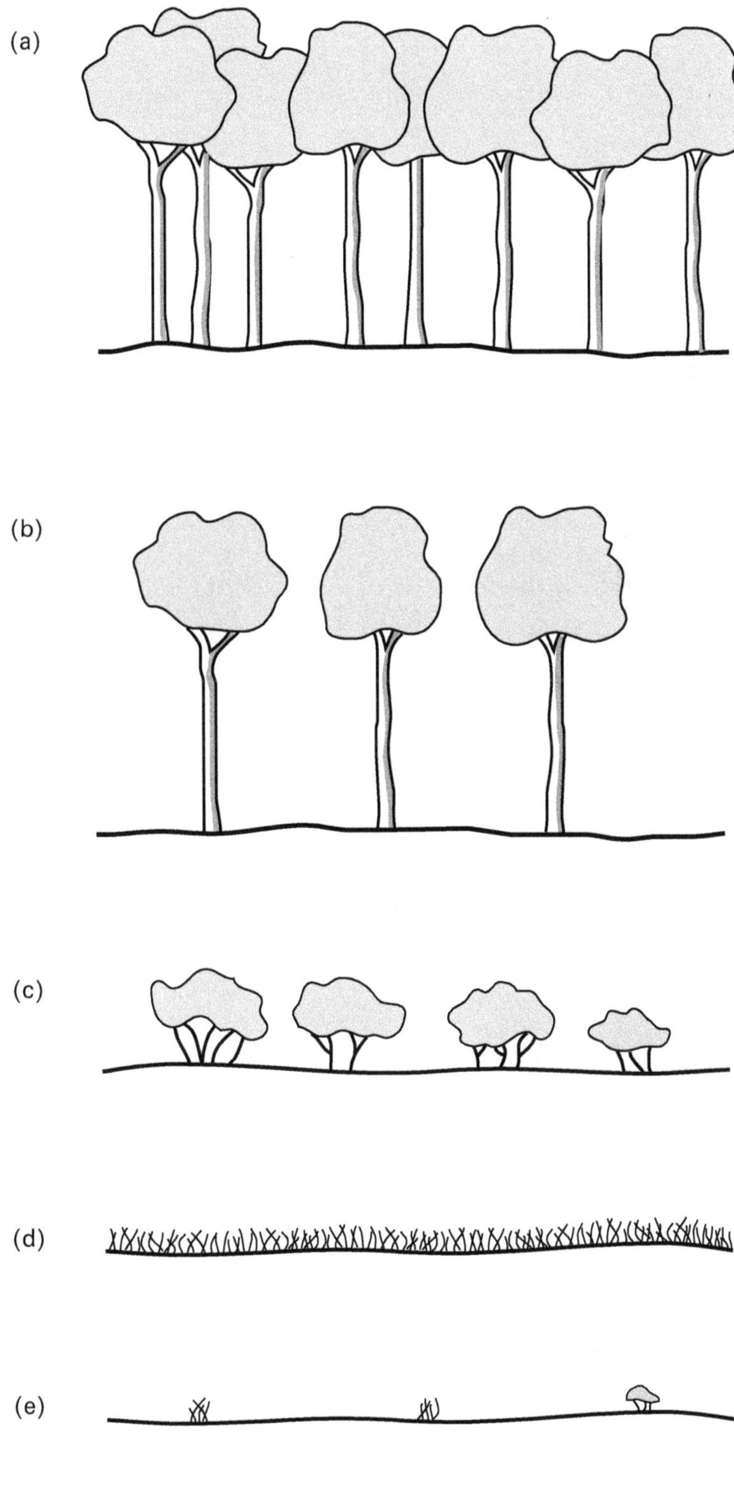

Figure 2.5. General form of vegetation: (a) forest, (b) woodland, (c) scrub, (d) grassland, (e) desert.

Figure 2.6. Tropical rainforest, Malaysia. *Source*: Author.

Figure 2.7. Cold climate conifer forest, mountains of California. *Source*: Author.

Figure 2.8. Semi-deciduous forest during the dry season, with some trees leaf-less, Malaysia, near Thai border. *Source*: Author.

Figure 2.9. Tropical montane forest with a cloud layer blanketing the upper parts of the forest, Sabah, Borneo. *Source*: Author.

Figure 2.10. Deciduous oak scrub, western Iran. *Source*: Kamran Zendehdel.

Figure 2.11. Grassland, California. *Source*: Author.

Figure 2.12. Tundra, above tree line in the Andes, Chile. *Source*: Margie Mayfield.

5. "Desert" biomes are distinguished mainly by lack of vegetation, with differing degrees of openness, or even no vegetation at all. Semi-desert is a sort of transitional open scrub or open grassland, whereas "true" desert has almost no vegetation. People tend to imagine that most deserts are sandy—in fact, more often they are covered by stones or bare rock (Figures 2.5e, 2.13*, 2.14, 2.17*, 2.19).

For more detailed study, these vegetation types are subdivided by ecologists in many different ways, using more specific definitions. It is important to realize that while all of the boundaries are really rather subjective (since vegetation types tend to fade into one another on the broad scale) they are nevertheless useful.

2.4 UNDERSTANDING THE PATTERNS

It is interesting to think about the reasons for the differences and similarities in vegetation structure between different climate zones. These patterns reflect the ecological and evolutionary selection forces on plants, working in consistent ways over large areas, and in different parts of the world.

Figure 2.13. Semi-desert, Mohave Desert, Arizona. *Source*: Claus Holzapfel.

Figure 2.14. Semi-desert, Iran. *Source*: Kamran Zendehdel.

For example, we can start by asking "why is it that some areas of the world naturally support forest, whereas others only support shrubland or grassland?"

2.5 WHAT FAVORS FOREST VEGETATION

Across nearly half of the world's land surface, the original vegetation is forest. In some places it is easy to tell that forest is the natural vegetation because the landscape is still covered by it. For example, the eastern USA and southeastern Canada are at present mostly forested. Even areas in eastern North America that are kept as farmland will go back to forest if left for several decades. Within just a few years an abandoned field will become covered with tree seedlings that have dispersed in from patches of forest nearby. In other parts of the world, it is not always so easy to tell what the natural vegetation would be like. In northwestern Europe, eastern China and Bangladesh, for example, a high human population density and intensive agriculture have removed almost all the natural forest cover in many areas, particularly in the fertile, flat lowlands that are most desirable for farming. In parts of northern Europe and eastern China there is barely a patch of forest in sight, except perhaps a few scattered plantations of poplars all planted in rows. Nevertheless, where the landscape is hilly and difficult to farm, forests are abundant even in these regions, indicating what the lowlands would probably look like if they were not now under cultivation. And we have independent corroboration from the pollen record dug up in the mud of lakes and back-swamps; this shows, for example, that up until a few thousand years ago all the most densely inhabited areas of Europe and China were densely forested.

However, there are many other parts of the world that are presently almost treeless and could not support forest, even before human interference. This is because the climate is simply unsuitable for a dense tree cover. To have forest one needs two things: (1) enough warmth and (2) enough water. Below a certain threshold of temperature or water availability, the only plants that can survive are low shrubs or herbaceous plants.

2.5.1 Why trees need more warmth

So trees need more warmth and moisture than other growth forms of plants such as low shrubs and grasses. In places with a mean summer temperature less than about 7 or 8°C, trees cannot grow in the wild, although they can sometimes survive if they are pampered by humans and protected from competition with other smaller plants. Cool summers prevent trees from growing beyond a limit known as the "tree line" in the Arctic of Siberia or Canada, or in high mountains around the world. Even if a place has quite mild winters but very cool summers below about 8°C it will not have any wild trees, proving the point that it is summer temperature not winter cold that prevents trees from growing. One example is the Faeroe Islands in the north Atlantic, with their cloudy cool oceanic climate.

Why is it that trees do badly in cooler summer conditions, when many small shrubs and herbaceous plants can do just fine? Possibly the problem for trees is that they are in a sense relatively inefficient at growing. They put aside a lot of their energy into woody tissues. They are essentially in for the long haul, to overtop competitors and then reproduce. In many climates, this strategy pays off and trees dominate the landscape. But if conditions are always fairly cool, trees cannot photosynthesize fast enough to sustain basic living processes and also put aside materials into wood. Under such circumstances, their ambitious strategy is rather like trying to pay a mortgage while on a student income! The result of these competing demands is that the trees will simply fail to grow, or they grow so slowly (because they are trying to lay down wood) that the smaller faster-growing shrubs and herbs they are competing with kill them off. This is why tundra (low shrubs and grasses) extends into colder summer temperature zones than the most cold-tolerant boreal (high-latitude) forests. An additional possible reason why trees cannot grow in cold summers is that the actual biochemical process of laying down woody tissue itself requires higher temperatures, although exactly what step in this mechanism might need the extra warmth is not clear. A further reason that trees are replaced by shrubs in cold climates is that shrubs are better at "holding in" the heat from the sun on a cold day. A tree has a loose growth pattern that allows the wind to blow through it and carry heat away; whereas the dense mass of branches of a shrub holds in heat against the wind (more on this in Chapter 4 on microclimates).

The tree growth form drops out sooner along a temperature gradient in certain specialized habitats. In the tropics, in places where mud brought down by rivers accumulates along a shoreline, there are usually mangroves—trees from various evolutionary lineages that are adapted to grow in the salty mud. They need special salt-excreting glands to prevent salt building up in their tissues, and prop roots that prevent the trees from falling over in the soft mud. Yet, whereas mangroves are almost ubiquitous on muddy shores in the tropics (at least when not cleared away by humans), they die out in the subtropics. At mean annual temperatures cooler than about 20°C, there are no mangrove trees and their place is taken by low shrubs, reeds and small herbaceous plants—a vegetation known as salt marsh. Why are there no mangroves in cooler climates, when trees can grow easily on land and further inland in freshwater swamps? Most likely this temperature limit has something to do with the need for a mangrove tree to continuously adjust its anchorage to cope with wave action and the erosion and shifting of mud beneath its roots. In climates that have cool climates, tree growth stops in winter but the wave action and movement of mud does not. Thus, any tree that tried to grow as a mangrove in a climate with a winter would lose its footing and be washed away. In the marine shoreline habitat, the temperature limits on the tree growth form are different because of the peculiar demands of this environment.

2.5.2 Why trees need more water

We can understand then why trees need more warmth, but why do they need more moisture, only existing in the moistest climates? This relates to the fact that trees are

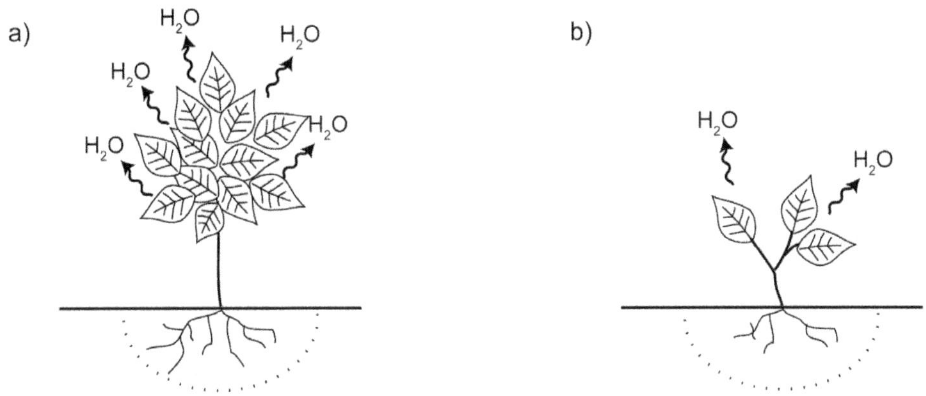

Figure 2.15. A tree's large area of leaves (a) places a high demand in terms of evaporation, which must be supplied from the soil underneath. A shrub (b), with its much smaller leaf area, loses less water by evaporation so it is much better able to survive in a dry climate where there is not much water in the soil.

large—their strategy in life being to overtop and out-compete other plants. They need more water than shrubs and grasses because they have a lot of evaporative leaf area that is essentially placed on top of a single pole, the trunk (Figure 2.15). Plants are limited in how much water their roots can take from the soil underneath them where they are growing, and the bigger the top parts of a plant, the more likely it is to run out of water. So in an arid environment although the tree may start to grow during a rainy spell, it eventually dies when its water supply runs out during a drought period. A low shrubby plant has fewer leaves that evaporate water, relative to the size of its root system and relative to the patch of soil it is rooted into. Thus it makes less water demand on its own area of soil, and it is less likely to die of drought. The result is that along a line of decreasing annual rainfall, forest disappears way before scrub and grasslands do.

So in climate regions beyond the drier and colder limits of forest and woodland biomes, the problem for trees is that they are in a sense "too ambitious" in trying to grow big and overtop everything else, even though this pays off for them in climates which are warm and moist enough.

If fires or grazing are frequent, this can kill off the trees too. Trees cannot easily survive having their expensive top parts bitten or burnt off; they are most susceptible when they are seedlings or saplings—replacing old trees that have died off, or trying to establish in open habitat. Low shrubs can easily re-grow their relatively flimsy inexpensive branches. In contrast, the trees just die as a result of the damage, or become so crippled that they are out-competed by the shrubs and grasses. In a climate that is relatively dry, fires are more frequent and trees also grow more slowly and recover from damage with more difficulty. So climate and other factors can work in parallel against trees.

The advantage usually turns from trees to low shrubs and herbaceous plants quite gradually along a climate gradient. The trees may get sparser and smaller, until there are just isolated individuals, and then none at all. Even where trees are mostly rare in the landscape, often there will be small stands of them here and there where there is a pocket of favorable conditions; for example, a little sunlit cliff in a cold climate, or in a dry climate where a spring emerges, or perhaps where there is a pocket of deep moist soil.

However, sometimes there is a very sharp transition from forested or wooded vegetation to "open country" such as grass or scrub. This may occur at the edge of a frequent grass fire zone, for instance: dead grass standing during the drier part of the year promotes the spread of fires set by lightening. The grass can tolerate fire because it just grows back from underground shoots after a fire, but any young trees establishing amongst the grass are usually killed.The grass fire will burn right up to the edge of the forest, beyond which the lack of fuel and the moist cool conditions prevent the fire from spreading into the forest. So the landscape will either be "all grass"—where any trees are killed by fire—or "all trees" where the grass is shaded out by the dense closed canopy, with no gradation between the two. Such sudden transitions are often seen in savanna zones in South America and Africa, where islands of forest are surrounded by a grassy landscape, and one can step straight from dark moist forest to the dry heat of open grassland in a matter of yards. Forest islands also used to be seen at the edges of the prairie zone in North America, before settlers ploughed up the entire landscape for agriculture.

Another place where the transition from tree cover to grassland can be very rapid is on mountains. Often there is a well-defined "tree line", above which no trees grow. The transition zone from forest, where trees become smaller and sparser until there are none at all, can be as little as 30 metres. Such a sharp boundary is in part possible on a mountain because there is a far more rapid decline in temperature with altitude than there would be when traveling towards higher latitudes (Figure 2.16*). In the high latitudes, the "tree line" is much more gradual with the trees becoming smaller, sparser and more patchy over many tens of kilometers. Adding to the suddenness of the disappearance of trees along a temperature gradient is that their ability to "hold in" the warm air they need collapses as the canopy begins to thin (see Chapter 4). How high up a mountain trees can grow also follows the broader regional warmth of the climate. Where temperature in the lowlands is warmer, there is a higher starting point for the decline in temperature with altitude, so trees will be able to grow correspondingly farther up mountains. In the tropics, tree lines tend to be somewhere up between 3,500 and 5,000 meters above sea level. Beyond about 25°N and 25°S of the equator they begin to move downslope as the whole climate becomes cooler. By the time we reach mountain areas north of 50°N, they are no higher than 1,000 m above sea level. Eventually, in the far north tree lines decline to sea level, the point at which the only vegetation in both lowlands and mountains is treeless tundra. Despite the general trend with latitude, there is a lot of variation in where the tree line occurs between different mountain ranges, even within the same latitudinal belt. One factor which influences it is the "mountain mass effect": small isolated mountains tend to have much lower tree lines than big mountain massifs. Often the difference that this "mass effect" makes to the height of the tree line is more than 1,000 m. This difference is due to larger clumps of mountains trapping a sort of bubble of warm air close to

Figure 2.16. Tree line on a mountain. *Source*: Gianluca Piovesan.

their slopes, better protected from winds and with adjacent mountains radiating infrared towards one another, rather than losing it to the sky. Differences in the aspect and slope angle, and also whether one is on the side of the mountain facing the sun or in shadow most of the day, also affect the position of the tree line more locally on a mountain.

2.6 DECIDUOUS OR EVERGREEN: THE ADAPTIVE CHOICES THAT PLANTS MAKE

In some areas forests keep their leaf cover all year round. In others they drop their leaves part of the year and grow a new set after a few weeks or months. So one finds "temperate deciduous forests" in the northern temperate zone, but "temperate evergreen forests" in eastern Australia, southern China, New Zealand and parts of Chile (Figure 2.1). In some parts of the tropics, mainly near the equator, the forests are evergreen. In other places—mainly the outer tropics—the forests are deciduous. The evolutionary "decision" as so whether the leaves should stay on all year round depends on the energy, nutrient and water economy of the trees. Under some circumstances, there is no benefit to the tree in hanging on to leaves if they are going to be a burden during hard times. It is best to get rid of them and grow a new set when favorable conditions return. In other cases, dropping the leaves would waste an opportunity for photosynthesis, so they are retained all through the year.

In the moistest forests close to the equator, the climate is warm and there is plenty of rainfall all year round. In this environment there is no reason for the trees to drop their leaves at any particular time of year, so the forest stays green year round. The broad laurel-like leaves are held on the trees for several years, before they reach the end of their useful lifespan and are shed.

However, in parts of the tropics where there is a regular dry season (e.g. Thailand), holding onto leaves during the dry months presents a risk of killing the tree by drought. All leaves lose some water, even if the tiny stomatal pores—see Chapter 4—in the leaf are kept shut: the only way to ensure that water loss is eliminated is to shed the leaves. Also if trees keep their leaves during the dry season, they risk losing nutrients unnecessarily through general wear and tear of the leaves, plus herbivores chewing away leaf tissue. This is at a time of year when the stomata must be kept shut so there is no photosynthesis and no benefit to the tree from having the leaves present. Under these conditions the trees will do best by re-absorbing nutrients and dropping leaves for the dry season. They then grow a brand new set of leaves which will photosynthesize rapidly when the wet season returns. So, the reason that forests are deciduous in the monsoonal outer tropics is that this is the best solution to an environmental challenge. To anyone from the northern mid-latitudes, it is a strange experience to walk through the dry deciduous forest with trees bare and leaf litter crunching under foot, looking just like a temperate forest in winter. Yet instead of frost and snow, the air will be roasting hot—often up around 40°C.

Just beyond the reach of the outer dry seasonal tropics, evergreen forests appear again in the warm temperate zone. For example, in southernmost China and the southeastern USA, trees tend to have leathery, long-lived leaves. Evergreen forests also occur in warm Mediterranean climates (such as southern Europe and California) with a relatively dry summer, where the summer drought is not normally long or intense enough to require the trees to shed their leaves.

Temperate evergreen forests can also occur in oceanic climates with quite cool summers—such as in New Zealand and southernmost South America—so long as the winters are mild. Here there is no reason to drop leaves at any particular time of year. There is enough moisture year round, and the winters are mild, so photosynthesis is possible at any time of year. Through most of the mid-latitudes, colder winters mean that there would be a disadvantage in holding on to leaves all through the year. It would be too cold during winter for them to work effectively, and they would just lose water, get tattered and torn, their cells damaged by frost, and chewed by herbivores. With all these damaging influences, they would be thoroughly ineffective by the end of the winter season. Having no strong reason to keep their leaves, and several good reasons not to keep them, the trees shed them as the cold season sets in. An orderly process of dismantling the cell contents of the leaf ensures that nearly all the most valuable substances (such as nitrogen and phosphate-containing molecules, and ions such as potassium and magnesium) are drawn back into the tree. Chlorophyll is broken down for its magnesium ion early on in the process, whereas other less useful pigments in the leaf such as carotenoids and anthocyanins are discarded with the leaf.

Figure 2.17. Autumn leaves in a northern temperate deciduous tree, Norway maple (*Acer platanoides*). *Source*: Author.

Unmasking the colors of these other pigments after the chlorophyll has gone is partly what gives the brilliant colors of autumn leaves in the mid-latitudes of the northern hemisphere (Figure 2.17*). However, adding to the brightness of autumn colours, senescing leaves also increase their levels of red anthocyanin pigments. This is thought to protect the leaf tissue from damaging sunburn while it is still dismantling itself to release nutrients from the cells.

Trees that lose their leaves during part of the year and then re-grow them must take a fairly precise cue from their environment. In the mid and high latitudes, if they put the leaves out too early in the year, these may be damaged by frost and valuable nutrients lost, because the tree cannot easily reclaim nutrients from a frost-bitten leaf. Or in a seasonally dry climate, the tree may die of drought from putting its leaves out too soon. If the leaves are dropped too early, time that could be spent photosynthesizing is wasted. A tree must in effect take a gamble as to the best time to drop its leaves, using the best cues that it has available. Usually in temperate deciduous forests—as temperatures dip close to freezing in autumn—the tree starts to break down the contents of cells in its leaves, and withdraw them back in to be stored in the trunk, branches and roots. A further cue is taken from the declining day length as summer ends. Often trees right next to street lights retain their leaves a few weeks longer because they are "fooled" that there is still more daylight around. If winter-deciduous trees do not receive any cues, they may simply keep their leaves going. In my own experiments, young deciduous white oak trees (*Quercus alba*) grown in a greenhouse in warm temperatures and long artificial day lengths (due to lighting) retained their leaves healthy and green all winter long, and then grew an additional set after several months at roughly the time that corresponded to spring. On the other hand, red maples (*Acer rubrum*) still dropped their leaves just about on cue despite the lack of environmental stimuli.

In the mid-latitude temperate forests of Europe, North America and eastern Asia (extending between around 30 and 50°N, though it depends on the locality), there is an "autumn wave" of leaf shedding that starts earlier in the north and moves progressively southwards as each latitudinal band reaches colder autumn temperatures. Although the timing is tuned by climate, there is evidence that populations of trees of the same species from northern and southern parts of their ranges are genetically programmed to take environmental cues differently. When they are planted farther north, more "southerly" populations tend to keep their leaves longer.

In dry-season deciduous forests in the tropics, it is drought stress that begins the process of leaf drop. If a particular year is unusually wet, the trees retain their leaves longer until the supply of soil moisture is used up, and only then do they drop them.

When good growing conditions return, deciduous trees must also take cues from their environment to regain leaves at the best time. As I mentioned above, for temperate deciduous trees, it is particularly critical not to start producing young spring leaves too early because their soft tissues can easily be damaged by frost. Trees take their cue for the arrival of spring from exposure to a certain number of days of warm temperatures. Increasing day length can also help to act as a trigger for leafing out, and cold temperatures during the winter help to prepare ("vernalize") the buds for breaking with the arrival of spring. Without these requirements, the tree might start leafing out during periods of a few warm days in early winter, only to have all its leaves killed when the true winter cold returns. Just as with the timing of autumn leaf drop, there is evidence that different populations of trees (e.g., elms *Ulmus* in Europe) of the same species can be quite finely adapted in their cooling or day length requirements for leafing out, according to the length of the winter where they come from, to ensure the best balance between leafing out early enough to exploit the arrival of spring temperatures, and leafing out reluctantly enough to avoid being misled by short-lived warm periods during winter. In my greenhouse-grown red maples, populations from warmer climates like Florida leafed out several weeks earlier than ones from colder climates like New England, despite being kept under identical conditions. In the seasonally dry tropics, the most common cue for leafing seems to be the uptake of water by the roots once the rains start. However, some trees that lose their leaves during the dry season may start to produce new leaves just before the rains arrive. It is thought that in this case the cue is an initial drop in temperatures that accompanies the arrival of moist air before the monsoon although in some species, beyond about 10° from the equator, the very slight change in day length that precedes the tropical wet season also seems to be a cue for leafing out.

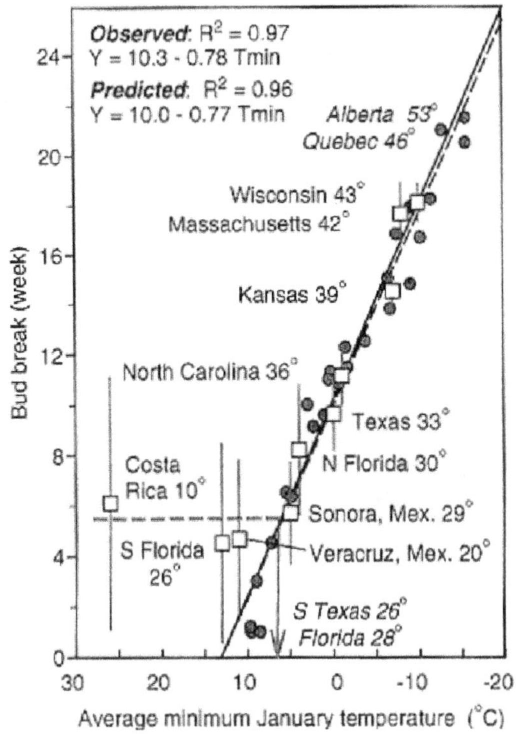

Figure 2.18. The relationship between January temperature and leafing out date in a range of North American trees. The vertical axis is the week of the year, starting from January 1st. From Borchert *et al.* (2005).

In the deciduous forest regions in the mid-latitudes of the northern hemisphere, a green wave of leafing out can be seen sweeping north on satellite images as spring temperatures warm up. The relative timing of this green wave follows the climate so closely that it can be predicted using a simple mathematical formula based on winter temperature (see Figure 2.18). At the northern end of the temperate deciduous biome, leafing out occurs months later than in the south, even in the same species of trees. A similar green wave occurs on the outer edges of the tropics as the monsoon rains move gradually out from the equator.

Leaves must unfurl rapidly to take full advantage of the temperate zone spring, but they must be able to do it without tearing. Most leaves in cold climates have teeth or lobes at their edges (Figure 2.19*), and it has been suggested that this feature may help them to avoid getting torn as they open. Or it may be that the thin leaves of deciduous species need to flex in the wind without tearing when they are fully grown, and the teeth may help them to do this. Another possible explanation for the toothed leaves is that the presence of teeth helps to promote gas exchange for the rapidly photosynthesizing leaves in the spring when CO_2 supply is limiting and evaporation rate is low, due to the relatively cool temperatures. Some types of trees that occur across a wide range of temperatures even show the same temperature-related trend in leaf shape within a species. For example, leaves of red maple populations that come from cold climates in eastern Canada are "toothier" than those from Florida

Figure 2.19. Toothed or lobed leaves are far more prevalent in cooler climate forests. One example is beech (*Fagus grandifolia*) in North America, which has small teeth along the edges of its leaves. *Source*: Author.

(see Figure 2.20), with a gradation in leaf shape with latitude between these two extremes.

Even though the underlying reasons are not well understood, the relationship between average temperature and the prevalence of toothed leaves is very predictable. The cooler the climate, the higher the proportion of trees in the local flora that have teeth on their leaves (Figure 2.21). This relationship is so predictable that geologists use fossil leaf floras as a thermometer for the climates of particular regions in the geological past.

2.7 COLD-CLIMATE EVERGREENNESS

Deciduous forests are a feature of mid-latitude climates with cold winters. Yet at still higher latitudes with even colder winters (as in much of Canada or Russia), evergreen conifers (mostly of the pine family, such as *Pinus*, *Abies*, and *Picea*) are dominant instead. This seems to contradict the explanation for temperate trees losing their leaves—surely here the need to drop leaves in winter is even greater, and yet these are

Figure 2.20. Typical leaf of red maple (*Acer rubrum*) population in Canada (left) compared with one from Florida (right).

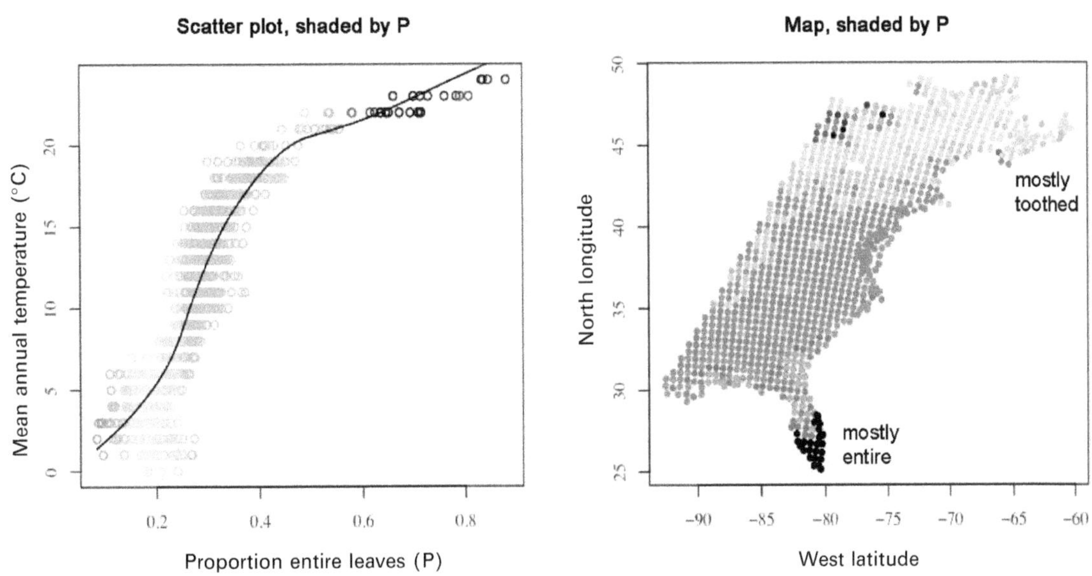

Figure 2.21. The proportion (*P*) of species of trees with "entire" (non-toothed) leaves depends closely on the warmth of the climate. From Adams *et al.*

evergreens. However, another factor has entered the equation, the briefness of the growing season in the high latitudes. The several weeks in spring spent growing new leaves represent valuable time that could be spent photosynthesizing. Similarly, the process of shutting down a leaf ready for it to be shed in the autumn takes several weeks, at a time when temperatures may still be warm enough for photosynthesis. The short summers of the boreal climate may give the edge to plants that can sit tight and hang on to their leaves rather than having to regrow a new set each spring, which is a lengthy process. For leaves to survive the severe winter intact, they must be made tough to stand dehydration and frost; so these conifers have "needle leaves"—thin and hard with a thick waxy coating. Evergreen conifer forest is often also found above the deciduous belt on mountains in the mid-latitudes, where the same conditions of short summers and harsh winters are found.

In fact, although most of the high-latitude forest is evergreen, in the really cold continental parts of east–central Siberia and Canada the winters are so harsh (down to −60°C in some forested parts of Siberia) that even a tough conifer leaf would be damaged by the frost and dehydrated. So in the coldest forest areas on Earth, in north–central Siberia, forests are dominated by the deciduous conifer larch (*Larix*) and small deciduous broadleaved trees such as birch (*Betula*) and aspen (*Populus*). But in these extremely "continental" climates (Chapter 1) the brief summers are quite intensely warm, so the trees can just about do well enough by unfurling new leaves for the summer.

2.8 THE LATITUDINAL BANDS OF EVERGREEN AND DECIDUOUS FOREST

So, moving away from the equator there are alternating bands of deciduous and evergreen forest vegetation (Figure 2.22). This pattern is only found in its most "perfect" form in eastern Asia (Figure 2.23), where the climate is moist enough to support trees all the way along a line from the equator to the high latitudes. However, the pattern is present in a more fragmentary way in many parts of the world.

2.9 NUTRIENTS AND EVERGREENNESS

Evergreenness is not only determined by climate—soils can have a lot to do with it too. Part of the reason why boreal forests tend to have evergreen conifers may be that the soils underneath them are nutrient-deficient. Each time a plant changes its leaves, some nutrients fail to be re-absorbed before the leaves are dropped, and are lost. If nutrients are in short supply, other plants that keep their leaves will grow faster and overtop this plant, and their roots will also grow fast and grab even more nutrients first. Hence there is selection against dropping leaves unnecessarily where nutrients are scarce.

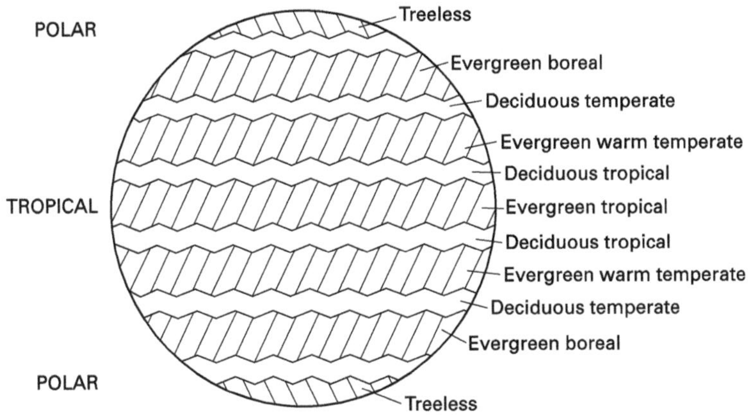

Figure 2.22. Latitudinal bands of alternating evergreen and deciduous forest. Idealized arrangement of evergreen vs deciduous forest types on the earth's surface. In reality, this is complicated by non-forest zones, oceans, and climate and soil differences affecting the evergreen strategy.

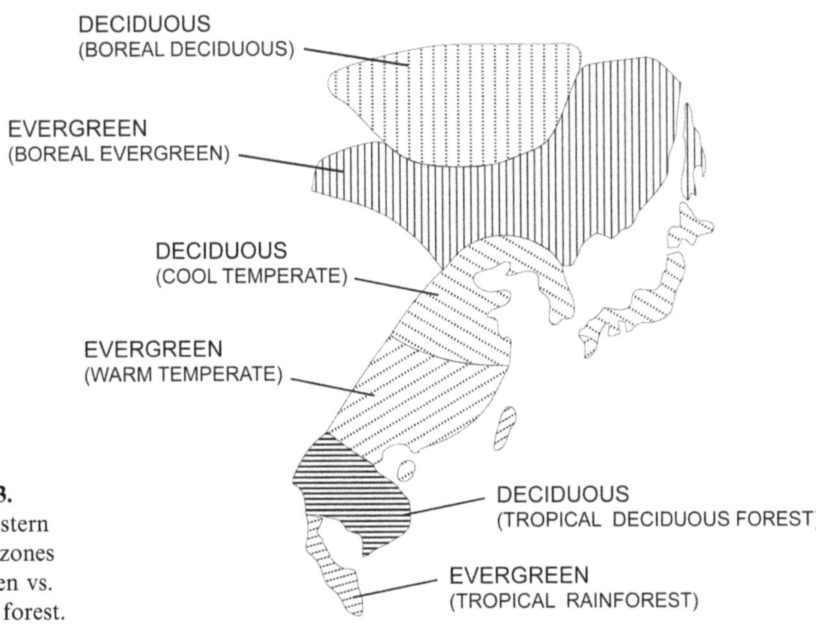

Figure 2.23. Map of eastern Asia with zones of evergreen vs. deciduous forest.

But why are the soils in the boreal zone nutrient-deficient? Partly because there are conifers! The conifers produce a nutrient-poor litter which gives rise to organic acids that cause leaching. To some extent it seems a chicken-and-egg situation, although the fact that there are short summers selecting for evergreenness is probably the underlying cause for conifers being present in the first place.

In other climate zones that have mainly deciduous forest, there can be patches or whole broad swathes of evergreen forest where nutrients are deficient. One example is the local areas of white sand forests in the tropics again, where the trees are holding on and keeping nutrients. In the southeastern USA, conifers (mostly pines, the genus *Pinus*) predominate on the nutrient-poor exhausted soils of the coastal plains. Where the soils are good (e.g. along the Mississippi river floodplain) deciduous trees out-compete the pines. *Eucalyptus*—the Australian "gum tree" genus of some 500 species that predominates across a full range of climates in Australia—is usually evergreen even in areas with a strong dry season. It has adopted the same strategy: "holding on" to its leaves come what may. This probably has something to do with nutrient-poor soils predominating across Australia.

Even in forest biomes, shrubs and herbaceous plants exist as "subordinates". Where trees dominate the vegetation, there is always a contingent of herbs and shrubs to get in quick and reproduce where there is disturbance (before the trees can out-compete them). Other species live as "understory" shrubs or herbaceous ground cover, tolerating low light levels. In effect, they get the scraps of light and nutrients the trees leave behind. But, in defining the biome we pay attention to the trees which are the most noticeable part.

2.10 OTHER TRENDS IN FOREST WITH CLIMATE

There are also trends in the appearance of trees along rainfall gradients. Generally, individual leaf sizes of trees get smaller as you move from a very moist tropical climate (e.g., the central Amazon Basin) to a rather drier hotter climate (e.g., the southern Amazon Basin), even if it is still covered in forest. It has been suggested that this is because drier climates can get hotter (see Chapters 5 and 6 for the reasons for this) and a big leaf cannot loose heat as well as a small one when it is heated under the sun. And being hotter, it looses water by evaporation quicker so the tree is more likely to suffer drought, if water is in short supply. So, perhaps a plant that has small leaves looses heat faster so that it won't loose water so fast? Perhaps this is a good reason to have small leaves, but it begs the converse question of why it is any advantage for a tree in a moist climate to have big leaves. Are they in some sense cheaper to make and maintain, perhaps? So far, no clear answer has emerged on this.

Generally speaking, along a gradient of increasing rainfall in a forest zone there is an increase in the number of leaves per unit area of forest. In the very moistest climates, old growth forest can have on average seven or eight leaves over any particular point on the surface, all soaking up sunlight (the number of leaves stacked above any particular point is known as the leaf area index or LAI). The increase in leaf area is made possible by the abundant water supply. With more moisture around their roots, trees can afford to produce more leaves despite the extra evaporation that this entails. Since more leaves mean more photosynthesis and more seeds and young being produced by the tree, there is a selective advantage in growing as big as possible given the climatic conditions.

2.11 NON-FOREST BIOMES

There are other biomes that only consist of low shrubs and grasses, because the climate is either too cold or too dry for trees to grow.

2.12 SCRUB BIOMES

In climates that are too dry for forest but wetter than desert, one can either have scrub or grassland. Whether scrub or grassland vegetation actually occurs in a particular place depends on a range of factors including soil type, the time of year when rain occurs, fire frequency and the abundance of grazing mammals. It also depends partly on what species of plants happen to have evolved locally; whether they are mostly grasses or mostly bushes. Also, if the soils are thin, infertile and rocky, scrub is more likely than grassland.

Many places around the world that would naturally be forested have been reduced to scrub by human influence, through frequent burning and goat-grazing (see below). Around the Mediterranean, a natural scrub vegetation known as garrigue has expanded greatly in area over the past several thousand years due to these influences. If burning is prevented and goats are kept out, this vegetation often reverts to forest over several decades.

Some scrub areas of the world are strikingly rich in species of plants. For example, there is the very species-rich fynbos vegetation of the Cape region of South Africa, which has some 600 species of heathers (the genus *Erica*) plus thousands of other types of plants packed into an area only a couple of hundred kilometers across.

2.13 GRASSLANDS

Grasslands are as the name suggests dominated by grasses, but usually there are low shrubs and broad-leaved flowers mixed in amongst them too. Grasslands are called "steppe" in temperate latitudes (from a Russian word), while in the tropics they are called "tropical grassland". The term "savanna" tends to be applied to warm climate grasslands that have an open scattering of trees or shrubs. The steppe grasslands fade in to "tundra" in high latitudes and at high altitudes on mountains. Tundra can have a lot of grasses, and mosses, but is often dominated by a rather prickly mass of low shrubs: dwarf willows, dwarf alders and dwarf birches, mixed with lichens. Thus, sometimes tundra can be a grassland, sometimes it is essentially a low scrubland.

Grasses grow from buds right down next to or underneath the ground, and can recover from fire or grazing very easily. They also have long leaves that grow by pushing out from their base, like toothpaste out of tube. This also means that they can regrow very easily if the tops of the leaves are burnt or eaten. In fact, most grasses seem to "need" frequent fires or grazing to keep other plants out. In the absence of either type of disturbance, the grasses are usually out-competed by other plants such as trees or shrubs. Thus, grasses can have a strange indirect arrangement with grazing animals; the grazers kill parts of the grass, but the grass needs the help of the grazers in seeing off the competition.

2.14 DESERTS

Deserts are the ultimate step in loss of vegetation, due to climatic or human conditions that almost prevent the growth of plants. There are different degrees of desert, and the usage of the term has many local variants. What one person in the USA might refer to as "desert" would be considered much too densely vegetated to be called desert in north Africa, where ecologists have the extremely arid Sahara as their standard. Many ecologists from around the world would say that there is no "real" desert anywhere in the USA, because even the driest areas have too much vegetation!

Deserts often have plants which have done away with leaves and instead have swollen stems that store water in large bag-like cells. Or, if they do have leaves, they are swollen and distended, also full of water. This is the "succulent" growth habit, which functions as a reservoir. Water is taken up by the plant's roots when its rains, and this store in its stems and leaves keeps it going for months or years until the next rain storm. Most people call any succulent plant like this a "cactus", but in fact cacti only occur naturally in deserts in the Americas. Several other groups of plants, some of them loosely related to the "true" cacti of the Americas, occur in dry climates in Africa. One example is the curious "living stones" genus *Lithops* in the Namib Desert of southwest Africa, which disguise themselves as pebbles to avoid being eaten by herbivores.

2.15 BIOMES ARE TO SOME EXTENT SUBJECTIVE

It is important to emphasize once again that one biome does not suddenly give way to another over just a few meters, as we might expect from looking at a biome map. Instead of any sudden boundary, biomes tend to fade into one another over hundreds of kilometers. For example, as one moves over a long distance the trees in the forest may become on average more deciduous, or boreal conifers become more common in the vegetation. Patches of grassland mixed in with forest may become more and more frequent, or the trees may become more widely spaced. Biomes are at least in part an abstract human construct, something we use as part of our need to categorize the world around us so that we can work with it. As well as the fact that biomes tend to fade into one another, there is not even a clear and generally agreed definition of what each biome should "look like" in any ideal sense. While there have been many attempts to try to pin the usage of biome categories down more precisely, none has succeeded because ecologists can have differing opinions on where one biome ends and another begins (especially if the ecologists come from separate continents).

Usually, they are loathe to give up using the definition they are familiar with, for one that someone else is trying to impose!

2.16 HUMANS ALTERING THE NATURAL VEGETATION, SHIFTING BIOMES

In some areas the natural vegetation has been almost totally removed—such as where there are now ploughed fields or cityscapes. But, in many other places, the effect of human actions has been more subtle. Often the result of anthropogenic influence seems to be a "downgrading" of the vegetation to something that might be found in a rather drier or colder climate.

For example, a meadow in the English countryside can only exist under human influence; the forest that once covered the land has been removed, and kept from returning by artificially high densities of grazing animals that bite off any tree seedlings. A meadow is in many respects an imitation of a dry Ukrainian or Turkish grassland, which is where many of its characteristic wild flowers (plus the rabbits and sheep that eat the plants) ultimately came from.

In the summer-dry Mediterranean climate zones of the world, the original forest cover has often been completely removed by a combination of agriculture, burning and grazing. Both pollen and historical evidence shows that the vegetation in even the barrenest parts of Greece, Spain and Cypress were once fairly lush forest, usually dominated by deciduous broadleaved trees. After thousands of years of intensive assault from humans and their goats, soil erosion has left thin, droughty soils. The lack of water-holding capacity in the soils favors tough, prickly vegetation known as garrigue or maquis that would once have been more typical of drier climates in North Africa and the Near East. As we shall see in Chapter 6, the lack of trees may also be due to a regional drying of climate, that was itself partly caused by the loss of forest.

2.17 "PREDICTING" WHERE VEGETATION TYPES WILL OCCUR

Knowing that biomes are in a general way related to climate, ecologists have wondered if it is possible to predict which biome will occur in a particular place, using some simple set of rules based on climatic conditions. As well as providing a satisfying explanation of the present-day world, these predictive schemes are useful in enabling ecologists to look both forwards and backwards in time. They can be used (1) to predict how biomes will shift in the future in response to human disruption of climates (e.g., under global warming due to the "greenhouse effect"; see Chapter 3), and (2) to reconstruct past climates from fossil "biome indicators", or conversely to reconstruct past biome distributions from certain climate indicators.

Perhaps the earliest serious attempt to express how climates relate to vegetation was by the German climatologist Vladimir Koeppen (1846–1940), who presented his global scheme in 1918. Koeppen noted that particular types of vegetation (biomes, essentially, though he did not use this term) are associated with particular climates, such that a map of vegetation can more or less be predicted from a map of climates.

The sort of feature that Koeppen used to demarcate a climate zone was the mean rainfall, and the extremes of monthly temperatures. The tropical zone, for example, included areas with every month of the year on average warmer than 18°C. Polar climates, by contrast, had a mean temperature for the *warmest* month of less than 10°C. Using formal rules like these, Koeppen marked out several very broad ecological zones which had different combinations on the scale of warmth and dryness. So for example he distinguished zones of "wet tropics" and "dry tropics". He also recognized that the distribution of rainfall during the year was very important. For example, one of his major categories is for areas with a Mediterranean climate—a marked dip in rainfall during the summer and plenty of rain during the cool winter. Mediterranean climates occur in several parts of the world and tend to have similar-looking vegetation and even closely related genera of plants between the different places.

Although, in many respects, Koeppen's scheme does broadly predict types of vegetation that will occur in different parts of the world, ecologists were aware of its imperfections. In many areas, what Koeppen's scheme would predict does not quite match what is seen on the ground. These mismatches prompted others to try to come up with schemes for linking climate and vegetation, which used slightly different features of climate chosen from a consideration of what would really matter in the ecology of plants.

In 1967 an American ecologist, L.R. Holdridge, put forward a rather different scheme that incorporated the balance between precipitation and evaporation. He wanted to emphasize that in a warm climate a certain amount of rainfall goes much less far in terms of keeping plants alive, because evaporation is so much stronger in the heat. The net "water balance" is surely what really determines whether a plant experiences drought, and the sorts of plants that will be able to survive in a place. As an example, lowland England—which has a notoriously damp climate—has an annual rainfall of around 700 mm. This is enough to sustain closed forest vegetation and to keep lawns green year round. Yet an area in equatorial Africa which has this amount of rainfall will be a dusty, dry place most of the year, with only an open scrub vegetation. In the much warmer climate in Africa, water evaporates faster and so more rainfall is needed to keep things moist. The important difference is not the amount of rainfall, but how rainfall compares with the temperature, and Holdridge's scheme recognized this.

Holdridge also emphasized that, in terms of judging the favorability of the climate to plant growth, only temperatures over a certain threshold should really matter. Holdridge drew the line at 0°C; he suggested that we should not bias temperature averages during the year by counting anything lower. Below that threshold level, plants are essentially dormant, so we can ignore those parts of the year—no matter how cold—because it makes no difference. So, for example, one might have a climate that is −30°C for six months of the year and 30°C for the other six months of the year. This would have a mean temperature of 0°C, implying that just about no plant could grow there, yet as a matter of common sense we know there would be forest vegetation able to thrive in the warm temperatures during half of the year. Taking a simple yearly average would obviously be a misleading way of classifying the world in terms of vegetation and it would have much less predictive value. In Holdridge's scheme, months below 0°C on average default to 0°C, and the average temperature derives only from the "important" temperatures, which are those above freezing. With these sorts of refinements, Holdridge's scheme did rather better than Koeppen's scheme at "predicting" vegetation based on climatic rules.

Holdridge also came up with his now-famous "triangle" (Figure 2.24) with three axes of classification in terms of climate. The world's vegetation types were arranged like cells in a honeycomb within this triangle, each with its particular range of temperature and water balance. While visually appealing and easy to read off, it would be surprising if the world's vegetation types neatly fitted in this way on the diagram in a perfect geometric pattern! Although it does better than Koerner's scheme, the Holdridge model has not been found to be very practical at predicting in many parts of the world. It seems that vegetation does not always follow the rules, perhaps because other climatic factors are really more important, and also because different types of soil, exposure, relief, and many other geological and geographical factors strongly influence vegetation. Partly this must be because Holdridge's scheme does not recognize patterns in the seasonality of rainfall or temperature, which can be all-important (e.g., does the rainfall all come in glut in part of the year, leaving the rest of the year dry?). In this sense it is more limited than Koeppen's old scheme. Nevertheless, the Holdridge scheme is still quite widely used because of its familiarity, and one often sees maps of "Holdridge vegetation-climate zones" presented for particular parts of the world.

Many ecologists have tried to build from the legacy of Koeppen, Holdridge and others to come up with schemes that are better at predicting vegetation from climate. These schemes are particularly useful when it comes to predicting how the ecology of the world may look in the future under the global warming of an increasing greenhouse effect. Basically, one generates a climate for the high CO_2 world on a computer, and then slots in the biome categories using the vegetation–climate scheme. The result is a vegetation map for the future-changed climate, that one can compare with the present-day vegetation.

Examples of vegetation schemes that are used by modelers include the aptly-named BIOME3 scheme. In their major aspects, such schemes tend to resemble Koeppen or Holdridge's schemes, but they have many minor refinements in terms of where the boundaries are drawn. In many cases, what has been altered in these latest schemes are the lowest temperatures that occur in local climate records, which appear to predict the limits of certain growth forms of plants. For instance, in the BIOME3 scheme, the "tropical rainforest" biome is said to be limited to areas where the mean temperature of the coldest month is above 15.5°C (rather than 18°C as Koeppen suggested). The explanation put forward for this is that 15.5°C is closely correlated with the real factor that limits where tropical rainforest can occur: the occurrence of occasional frosts on the time scale of decades. These frosts cannot be tolerated by many typical rainforest tree species, so their distribution limit (and the drip tips and buttress roots that go along with them) ends there.

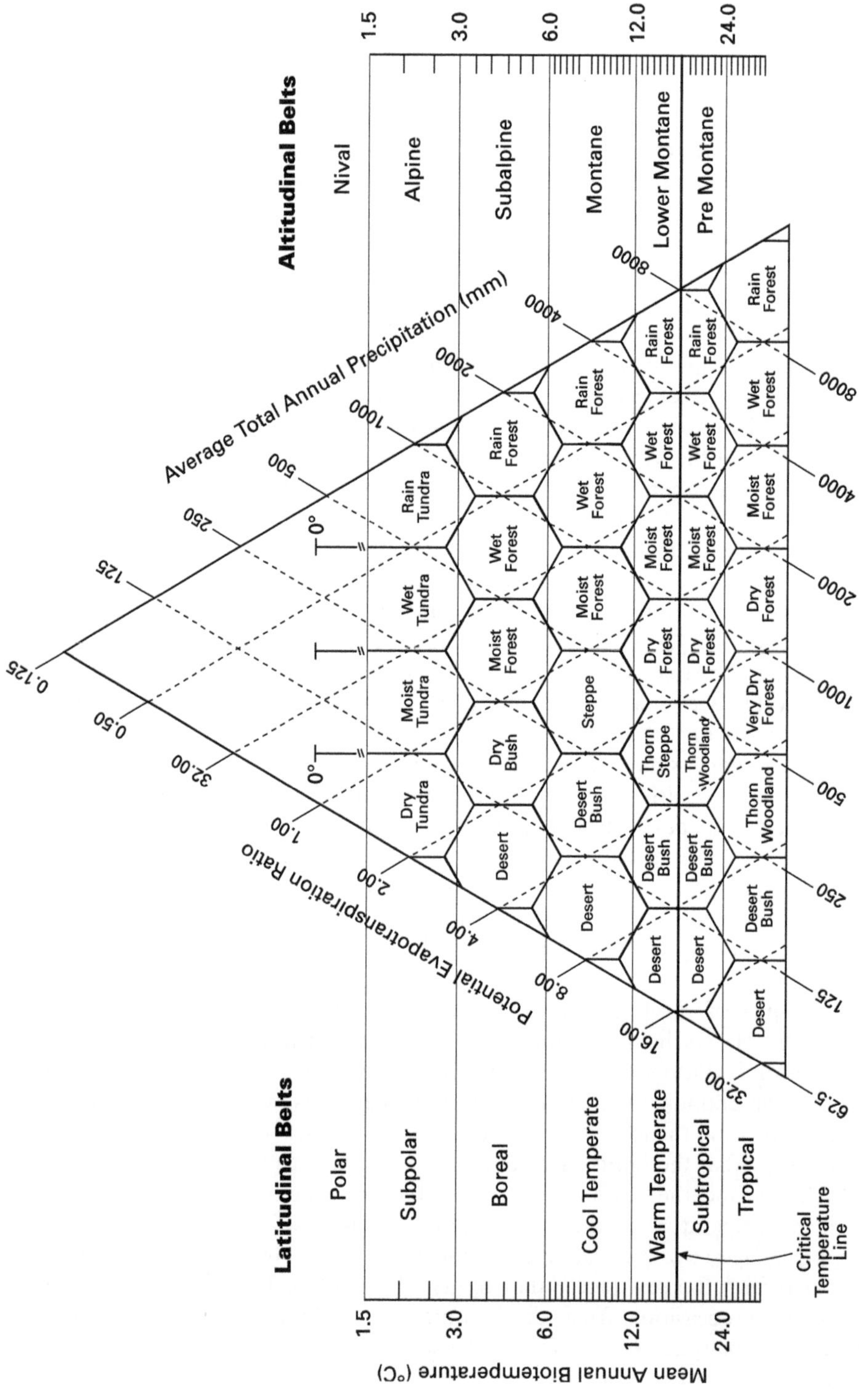

Figure 2.24. Holdridge's predictive scheme for relating biomes to climate.

Though they are useful, generalized bioclimatic schemes such as BIOME3 can never get it completely right. The vegetation–environment relationship is just too complicated to be completely predictable. Also, it is important to remember that such schemes are ultimately based on people just looking at biome maps drawn by ecologists, and choosing something in the climate that seems to correlate well with these limits. Although that choice may have a reasonable plant physiological basis, it is ultimately only chosen because it corresponds to what is on the map. In the case of tropical rainforest, the limit is drawn at 15.5°C, because that is the temperature at the point where, in vegetation maps of southwestern China, tropical rainforest reaches its most northerly point in the world, at 26°N. Although the vegetation boundaries on a map might look as if they are beyond dispute, the real world tends to be much more complicated. In many areas of the tropics (including southern China) the equatorial forest grades almost imperceptibly over hundreds of kilometers into the vegetation of cooler and drier climates—with drip tips and buttress roots becoming progressively less common—so that there is no single point where one can truly objectively say that equatorial forest ends and another vegetation type begins. For example, many ecologists would disagree with the idea that the southwest Chinese evergreen forest is really tropical rainforest at all. In order to make sense of the world, it is necessary to chop it up into neat categories such as biomes. But we should also bear in mind that even the maps that bioclimatic schemes are based on are somewhat subjective.

2.18 SPECIES DISTRIBUTIONS AND CLIMATE

Each biome is made up of many species of plants, and each one of these species has its own particular "distribution range", the area in which it grows naturally. Sometimes, towards the edge of a biome a lot of plant species seem to reach their limit at just about the same point. For example, in eastern North America, quite a few species of deciduous trees die out at the southern edge of the boreal forest in Canada. This is not surprising, because there must be a point along a temperature gradient at which the strategy of being deciduous is no longer quite so viable. Thus a lot of trees that use this strategy will tend to reach their limit at about the same place.

However, most of the species of plants present in each biome have their own idiosyncratic distribution ranges that do not show much relationship to the boundaries of the biome. In many cases the species range boundaries do seem related to aspects of climate, though not necessarily the same factors that define the edges of the biome.

Often, if one plots the distribution limits of a particular species it turns out to correspond quite closely to a climatic parameter such as the mean temperature of the warmest summer month, the annual rainfall or the yearly minimum winter temperature. It is perhaps too easy to keep trying different climate parameters until one "fits", but often the correlation between a parameter and the species range limit is so striking that it is hard to believe that it could be just coincidence. It seems that beyond a certain extreme of temperature or rainfall conditions, each species of plant is physiologically unable to survive (e.g., it cannot survive the frosts, or the summer drought, etc.). But the tolerance limits vary greatly between different species of plants, according to their own anatomical and physiological peculiarities.

In parts of the temperate latitudes, many plant species have what is called an "oceanic" distribution pattern that roughly follows coastlines, even though they may extend inland several hundred kilometers away from sea shores. The oceanic distribution tends to occur because these are plants that do best under cool summers and/or mild winters, in the climates which result from the moderating influence of the ocean. One example of an oceanic species is the ivy, *Hedera helix*, which is concentrated along the western fringe of Europe. Another is the cross-leaved heath (*Erica tetralix*) (Figures 2.25a*, b) which occurs along the western edge of Europe from Norway down to Portugal.

Moving farther inland, the "oceanic" species drop out and are replaced by certain other species which seem to thrive under the hotter summers, colder winters and lower rainfall conditions. These are known as "continental" species, because they are associated with the more extreme continental climates. In England, continental species of wildflowers—whose ranges tend to extend to the steppe environments of Ukraine—are found in meadows in southeastern England, especially on warmer drier south-facing slopes and on sandy soils which tend to imitate the warm droughty conditions of the steppe grasslands. In the wetter, cooler west of England, these species are absent. An example of a continental species in Europe is the stemless thistle, *Cirsium acaule*, which towards the more oceanic northwestern limits of its range is confined to the warmer, drier more "continental" south-facing sides of hills.

The individual shapes of ranges cannot always be put down to climate. They also seem to be affected by soils, and in some cases chance aspects of history such as where that particular species managed to survive during glacial times and how far it managed to disperse out of these refuges before reaching topographic barriers (see Chapter 3). An example of this historical effect from Europe is the purple-flowered rhododendron, *Rhododendron ponticum* (Figure 2.26*). It thrives when introduced to Britain and Ireland, and has escaped to fill many woodlands there, yet its natural range was confined to the mountains of Spain, the Balkans and Turkey. The same plant turns up as fossils from an earlier warm period in Britain, so we know that it once grew there too. It seems that *Rhododendron* was pushed back by ice age cold and aridity and then never managed to regain its former range, largely due to bad luck by being hemmed in by nearby areas of unsuitable climate. It was only when humans helped it out by importing it as a garden shrub that *Rhododendron ponticum* managed to make the leap to favorable climates in northwest Europe.

2.18.1 Patterns in species richness

When the ranges of individual species are superimposed on one another and counted up, striking patterns in the total numbers of species become clear. Species richness, as it is called, tends to be greater at the warmer end of each biome in the mid and high latitudes, and in the wetter parts in the tropics. In general, there is a strong trend towards more species of trees in forests at lower latitudes. This trend is most obvious in eastern Asia where the climate is uniformly moist from north to south and the only

Figure 2.25a. The cross-leaved heath (*Erica tetralix*), a plant with an oceanic distribution. *Source*: Christian Fischer.

Figure 2.25b. The cross-leaved heath (*Erica tetralix*) shows a typical "oceanic" range, along the western side of Europe next to the Atlantic Ocean.

Figure 2.26. *Rhododendron ponticum.*

major trend in climate is in terms of temperature (Figure 2.27). Some areas of the world show trends related to both temperature and rainfall: for example, the species richness of the deciduous forest in eastern North America which increases towards the south but also decreases into the dry interior of the US (Figure 2.28).

No-one is quite sure why species richness tends to be higher in warmer and moister environments. A range of hypotheses have been put forward during the last 150 years, but each of them starts to look paradoxical when examined in detail.

One popular idea amongst ecologists notes that the latitudinal difference in tree species richness correlates strongly with net primary productivity, the growth rate of vegetation. According to this idea, if there is a bigger "cake" of resources enabling and resulting from faster growth, there is more chance for species each to take their own "slice" (or niche). However, when we look in detail there is not really much evidence that species are on average more specialized in species-rich environments than in species-poor environments.

Another idea suggests that, because the world was nearly all warm and moist around 50 to 60 million years ago when the flowering plants were busy diversifying,

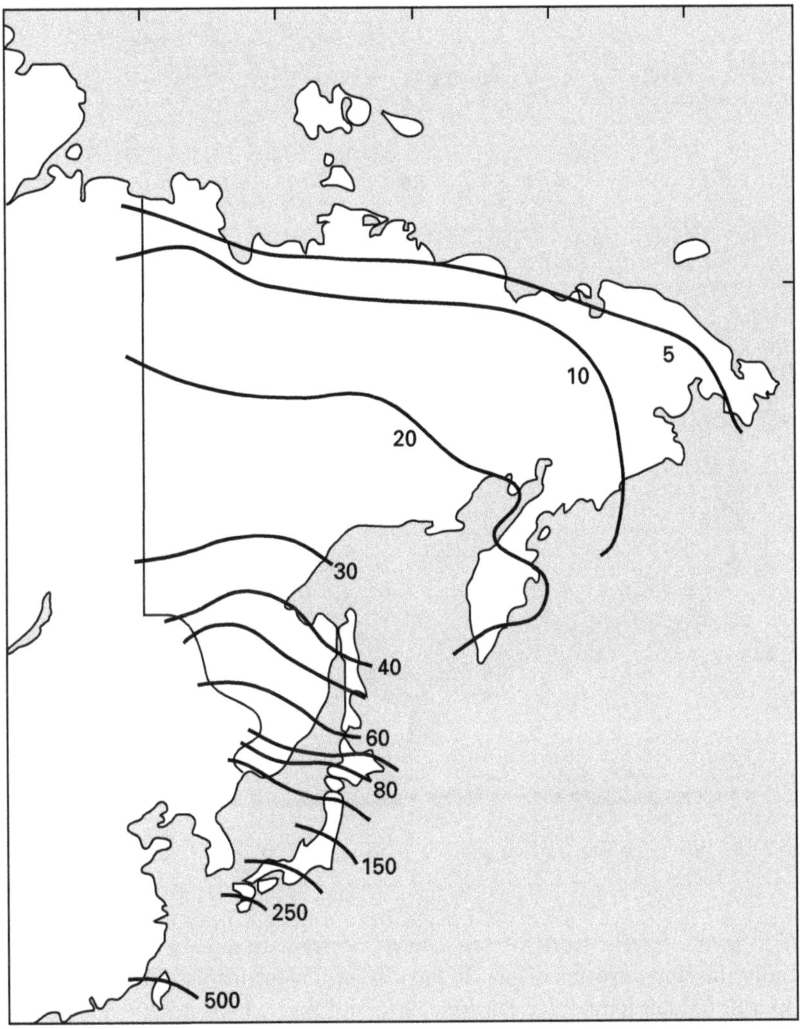

Figure 2.27. Tree species richness map of parts of eastern Asia (eastern Russia, Japan, Taiwan). These are the numbers of wild tree species occurring per cell in a geographical sampling grid, based on published tree species range data. There is a very strong latitudinal gradient. *Source*: Redrawn from Author.

most lineages became fundamentally adapted to living in the tropics. Over more recent time, the cold and dry environments that have become much more widespread have presented a new challenge that few lineages of plants have been able to adapt to. If this is the case, surely we would expect to see the levels of botanical richness of the high latitudes increasing in the fossil record, as more groups of plants overcame these barriers. Also, the earliest groups of plants that made it out into colder and drier environments should have been busy diversifying into more and more forms over

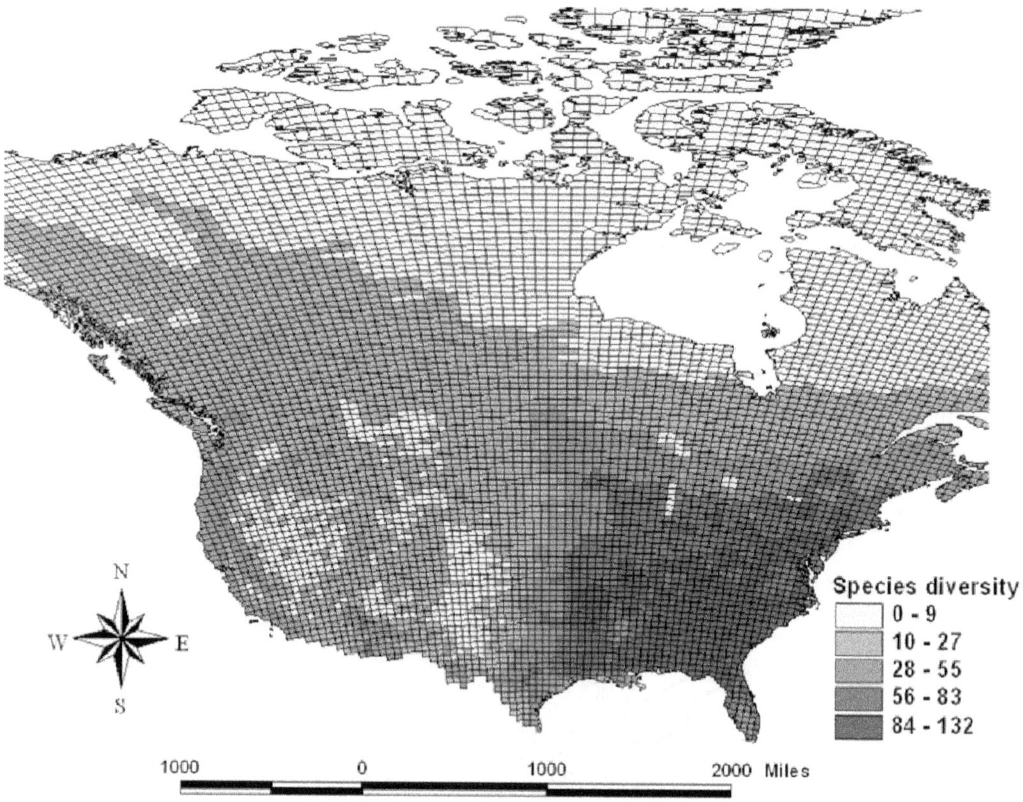

Figure 2.28. Wild tree species richness for North America. From Adams (2009).

time. Yet. in the fossil record we see almost no signs of such a build-up in diversity. Essentially the same groups of plants have been important for the past 30–40 million years in the colder temperate forests, with nothing much added. Contrary to the expectations of this hypothesis, diversity in the temperate forests has if anything declined somewhat over the past few million years (see below). Essentially, then, the causes of these grand geographical gradients in richness remain a mystery to ecologists. For a lot more discussion on the topic, see my other book *Species Richness* (Adams, 2009).

Even if we cannot really explain latitudinal gradients, certain other broad-scale patterns in species richness can be explained more convincingly in terms of past events which destroyed most species of plants in some places but left many more to survive in others. For example, the tree flora of temperate eastern Asia is a lot richer in species than climatically similar parts of Europe and North America, even though all three areas show a strong underlying trend in species richness which parallels the average temperature. The reason for this difference between the regions may be the fact that during glacial phases over the past 2 million years, the climate in parts of east Asia stayed a lot moister than anywhere in Europe or the eastern USA. Pollen evidence from ancient lakes shows signs that moist, closed forest persisted in eastern Asia. In North America, and Europe especially, the climate was much drier with open woodland, scrub or grassland. Many drought and cold-sensitive types of trees that existed in all three regions before about 3 million years ago would have been able to survive in Asia, whereas they died out in Europe and North America.

In this chapter we have considered how vegetation is shaped by climate in a relatively static sense. Even when including the extremes of its changes, we have merely touched upon the ways in which plants might have moved from one place to another when climate shifted. Chapter 3 is devoted to these transformations in vegetation in response to climate.

3

Plants on the move

3.1 VEGETATION CAN MOVE AS THE CLIMATE SHIFTS

Biomes are fundamentally determined by climate, as are the ranges of most individual plant species. Whenever the global climate changed in the past, so did the form and species composition of vegetation in each part of the world. In the past couple of decades, geologists have become increasingly aware just how much the earth's climate can change, and often on far shorter timescales than would have been thought possible. Each of these changes must also have had profound effects on vegetation: in this chapter we will deal with these effects and consider what they might mean for the present world, which seems to be warming rapidly.

3.2 THE QUATERNARY: THE LAST 2.4 MILLION YEARS

Even before we humans began our grand experiment with greenhouse gases, we had always lived in a time of dramatically unstable climate. Such variability was unusual even against the standards of the changeable history of the earth. The impression of ever-lasting stability one might get from seeing the world over a few decades is an illusion: on the timescale of a several thousand years, the climate in any place in the mid-latitudes can plunge to near-Arctic temperatures, and then after a few hundred or a few thousand more years shoot back up again to be warmer than today. On the same timescale, smaller fluctuations in temperature can also chill the tropics, but more importantly for the ecology of these regions there are large fluctuations in rainfall. There have been several times in the past few tens of thousands of years when the tropics were far drier than today, and there were also times when they were much moister. All these changes must have had dramatic effects on the distribution of biomes, and the individual species within them.

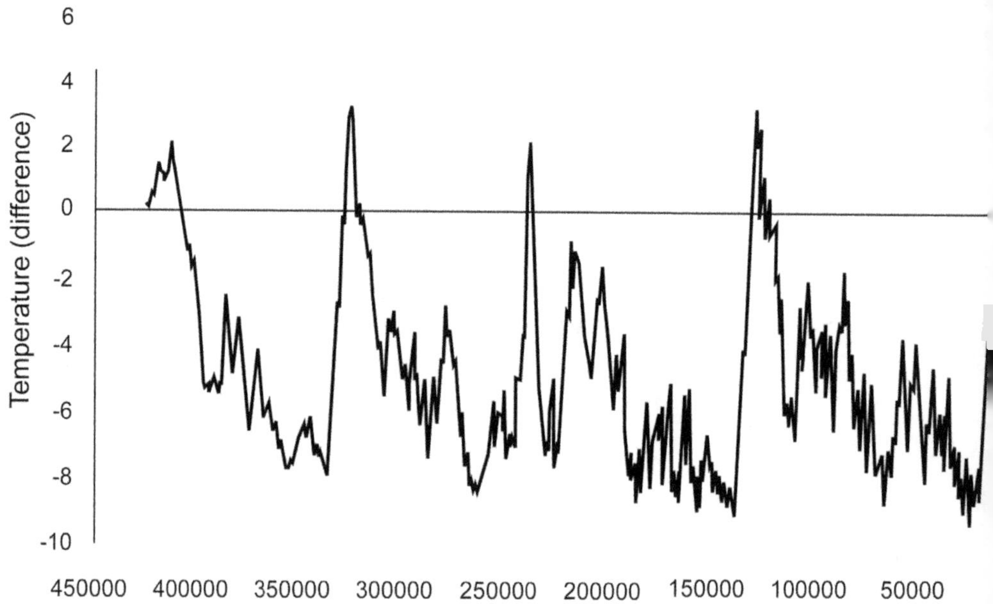

Figure 3.1. The global temperate history of the last 450,000 years showing a sawtooth patt which appeared by 700,000 years ago. Temperature in °C compared with 1960–1990 baseline. *Sou* CDIAC.

Instability in the earth's climate started in earnest about two and a half million years ago, and since then it has intensified into broader, longer term temperature fluctuations. Over the last 700,000 years, these big swings in climate have tended to occur on roughly a 100,000 year rhythm (Figure 3.1).

Each pendulum swing in climate begins with a warm phase much as we are in now, lasting maybe 15,000 years. These warm phases are known as "interglacials", and the latest one that has so far lasted for about the last 10,000 years or so is called the Holocene. In each of the last several interglacials for which we have a good climate record, global temperatures reached an early peak and then slowly declined afterwards. Eventually, the rate of decline became faster and often ended in a sudden, dramatic plunge in temperatures around the world. After a few thousand more years there might be a partial recovery of temperatures, but this would be a very temporary respite. Within a few hundred or a few thousand more years it would plunge again, often further than before. The decline would continue as a halting, reversing trend—two steps forward, one step back—until eventually after tens of thousands of years the earth's temperature arrived at a low point, known as a glacial maximum. As well as being much colder than the present, these glacial maxima tended to be much drier on a global scale. All the time as the earth's temperature was declining, great ice sheets would build over North America and Europe until at the glacial maxima they covered the northern half of both these continents, reaching several kilometers in thickness.

After maybe 10,000 years in this glacial maximum state, the earth would then be seized by a sudden warming. It is thought that these warming phases often occurred over just a few decades, raising annual temperatures by 5, 10 even 20°C—depending on the geographical region—to begin the next interglacial. The ice would begin a dramatic meltback, but taking several thousand years to disappear completely, because of its sheer bulk.

This overall pattern in climate change is known as the "sawtooth" cycle, so-called because it begins with a sudden rise in temperature to a peak, followed by a much slower decline (Figure 3.1). It is thought that the sawtooth cycle, plus the many sudden jumps in temperature that occur within it, results from a complex series of amplifying factors within the earth's climate which magnify small triggering changes into something much bigger. Such amplifying factors are known as "positive feedbacks" (see Chapter 5). The underlying control on the timing of the 100,000 year cycle seems to be caused by a series of wobbles in the earth's position relative to the sun. These affect the relative proportion of sunlight that hits the earth's northern hemisphere during summer rather than winter, and this is ultimately thought to control temperature change through a complex assortment of amplifiers, some involving melting of snow and ice, others involving capture of heat by vegetation (see Chapters 5 and 6).

As recently as 16,000 years ago, during the most recent glacial maximum, our planet was a very different place. Seen from space, the outlines of continents would have been recognizable and yet oddly unfamiliar, because sea level was lower and land extended out for many kilometers. What are now separate land masses were joined together by plains that are now drowned below the sea. For instance, Alaska was joined to Siberia by low-lying land across the Bering Straits, and most of the islands of southeast Asia formed a single land area. Another striking difference about that time would have been the huge white ice sheets—ice as thick as a mountain range—covering Canada and northern Europe. And the land surfaces themselves, in areas that are now dark with dense vegetation when seen from above, would then have tended to be much lighter with the yellows, reds and browns of bare soil. The global climate at that time was colder and more arid, and for the most part regions that would naturally now be forest were covered by drier vegetation such as open woodland, scrub, grassland or desert (Figures 3.2, 3.3). The great forest belts of Canada, Europe, Siberia and eastern Asia were almost absent at that time, because the climate was too dry or too cold for any dense tree cover. In the tropics also, the large block of rainforest that covers central Africa seems to have been largely absent, and replaced by savanna. The great Amazon rainforest was fragmented and shrunken down to a smaller core area surrounded by savanna or scrub.

Although the world was generally a lot drier during the last glacial, a few areas were instead wetter. For example, the southwestern USA was much moister than nowadays, with dense scrub vegetation and deep lakes in areas that now have only sparse semi-desert and dry salt pans. The salt flats at Salt Lake City, Utah were part of a huge lake—Lake Bonneville—which stretched hundreds of kilometers through

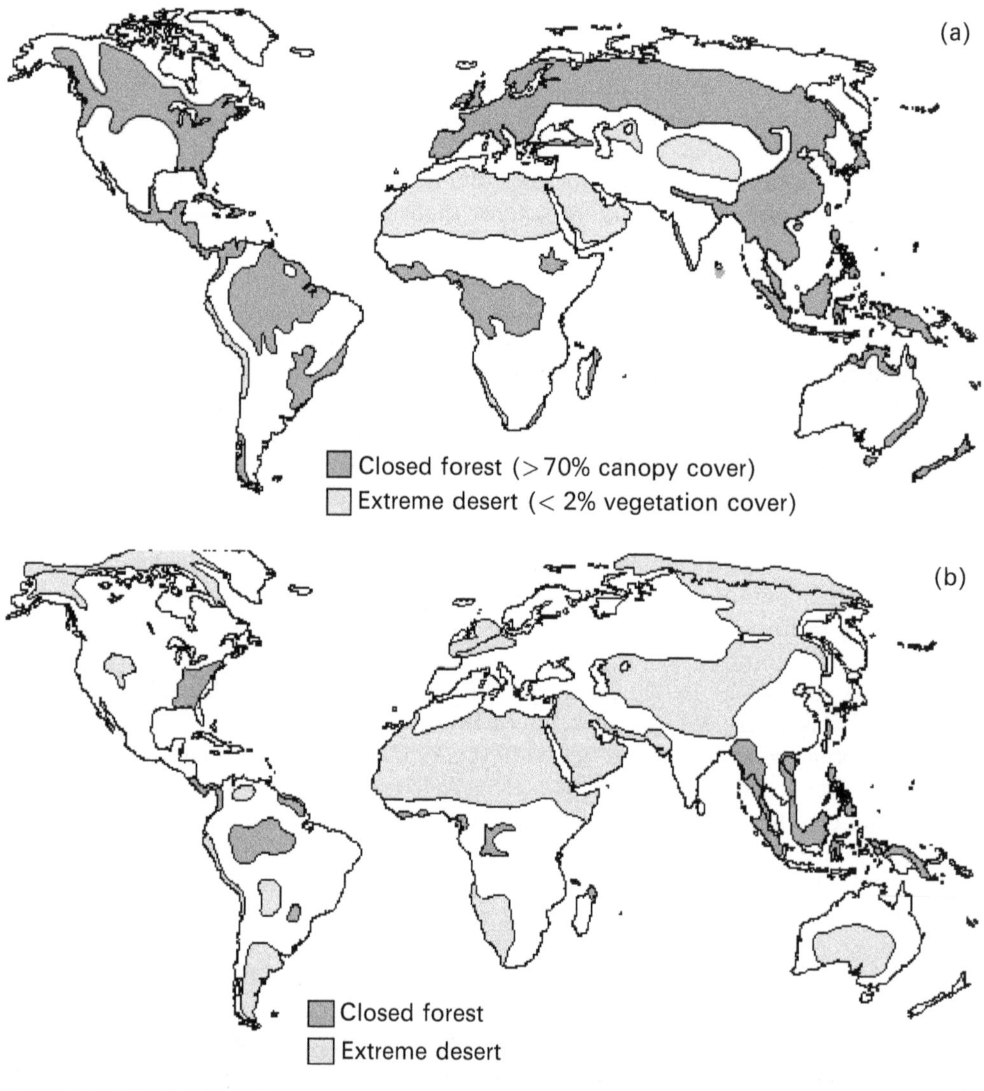

Figure 3.2. Distribution of forest vs desert, (a) present day and (b) last glacial maximum (18,000 ^{14}C years) compared. *Source*: Author.

the mountain valleys. The reason for the wetter climate in the American southwest at that time was that it was receiving the belt of rain-bearing winds (linked to the northerly part of the ocean gyre) off the Pacific that nowadays hits Seattle and Vancouver; this wind belt had been diverted more than 1,000 km farther south by the presence of the vast ice sheet that covered Canada and northern Washington State. Although weakened and giving much less rain than it does now, it was able to make a real difference to the ecology of the region.

In the mid-latitudes of Europe, above about 45°N, plants that nowadays grow above the Arctic Circle were common where there is now temperate forest. For example, the pollen of the Arctic rose, *Dryas*, turns up commonly in the muds of ancient European lakes from that time. Temperatures in the southeastern USA—in Tennessee, for example—were comparable with the climates we presently see at the border with Canada (Figure 3.4*). The cooling during glacial phases affected not just the high and mid-latitudes, but the tropics as well. There are numerous indications, from preserved pollen and ancient glacier limits, that the tropical lowlands were perhaps 5 to 6°C cooler on average than nowadays. On mountains in the tropics, vegetation zones were moved downslope by about 1,000 m because of colder temperatures (although possibly complicated by effects of decreased CO_2 on plants; see Chapter 8).

During the glacials, the difference was not simply that biomes changed places. Some biomes that are widespread today possibly did not exist at all during the last ice age. One example may be the lowland tropical rainforests: even just 15,000 years ago they apparently did not occur anywhere in a truly modern form. Pollen records extracted from lakes show that tropical mountain species of trees lived mixed in together with the lowland rainforest species of today. For example, in South America, South-East Asia, and Africa, the coniferous tree *Podocarpus* that is now usually only found high on mountain slopes was present down close to sea level. Does that make the vegetation different enough to put these forests in a different biome from tropical rainforest? Some ecologists would say it is, others would say it is not, because biome categorizations are always to some extent subjective.

Conversely, some biomes which were widespread during the last ice age do not exist today, or at most they only barely exist. An example of such a "vanished" biome is the steppe–tundra. This open and rather arid vegetation type covered most of northern Eurasia and parts of North America during the last ice age. It combined tundra and steppe plants that do not normally grow together nowadays, plus others that are nowadays more typical of sea shores. Ecologists have wondered what caused this strange combination to prevail over such vast areas. Was it due to peculiar climates at that time—types of climate which no longer exist? Or perhaps it had something to do with the effects of low CO_2, bending the ecological requirements of plants so that species that now live in quite separate environments could grow side by side (see Chapter 7)? A type of vegetation rather resembling the steppe–tundra does actually still occur as isolated patches on south-facing slopes in the mountains of northeastern Siberia; at least, it combines a more limited subset of steppe and tundra plants, with some extra species that apparently did not occur in the glacial steppe–tundra. What makes for this vegetation is a combination of short but warm and dry summers, with extremely harsh winters. This might have been the sort of climate that was much more widespread during the last glacial. But, once again reflecting the slippery nature of biome categories, many ecologists who study the ice ages do not accept that this eastern Siberian steppe–tundra is the same biome as the steppe–tundra that once covered northern Eurasia.

* See also color section.

(a)

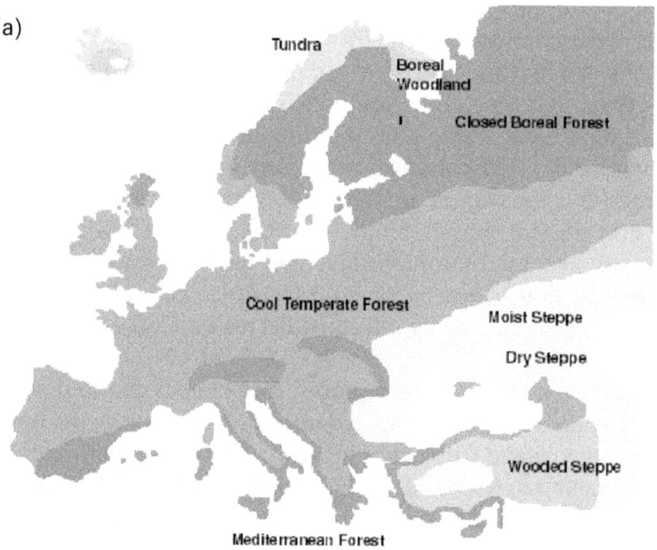

(b)

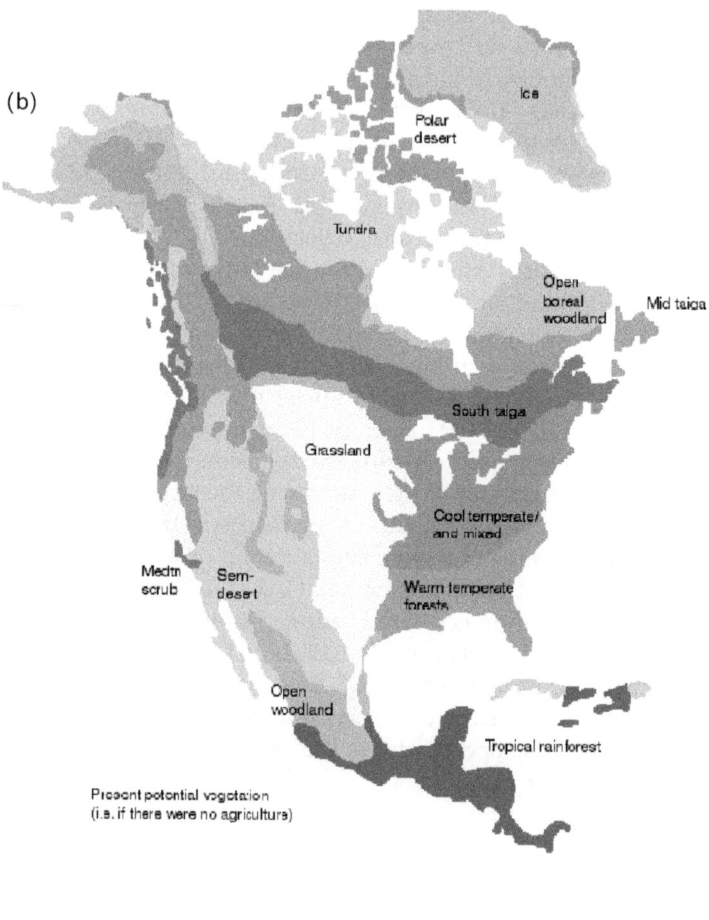

Figure 3.3. Biome distributions of Europe, North America at the present day (a, b) and last glacial maximum (22,000 14,000 ^{14}C years ago) (c, Part (d) is a map of the result of a collaboration between the author J. Adams, A. Beaudoin, O. Davis, P. and H. Delcour and P. Richard *Source*: Author.

(c)

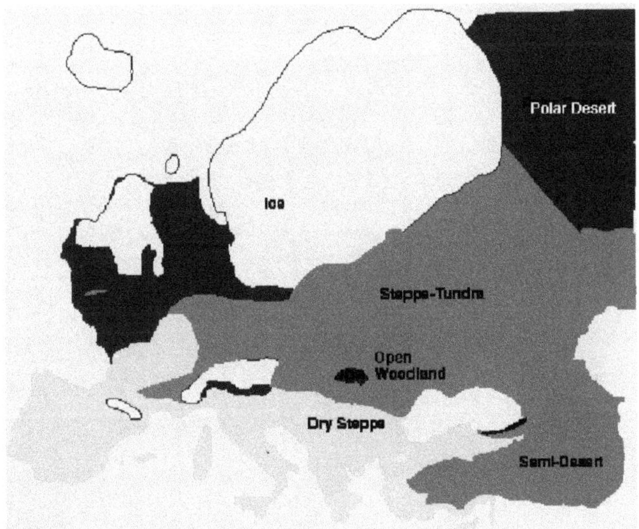
Polar Desert
Ice
Steppe-Tundra
Open
Woodland
Dry Steppe
Semi-Desert

(d)

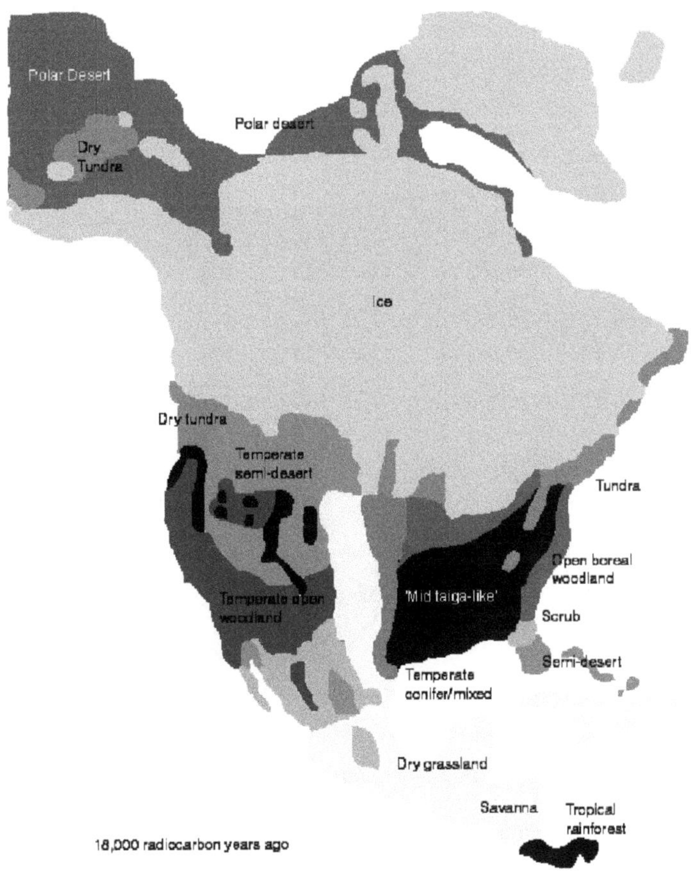
Polar Desert
Polar desert
Dry
Tundra
Ice
Dry tundra
Temperate
semi-desert
Tundra
Open boreal
woodland
Temperate open
woodland
'Mid taiga-like'
Scrub
Semi-desert
Temperate
conifer/mixed
Dry grassland
Savanna
Tropical
rainforest
18,000 radiocarbon years ago

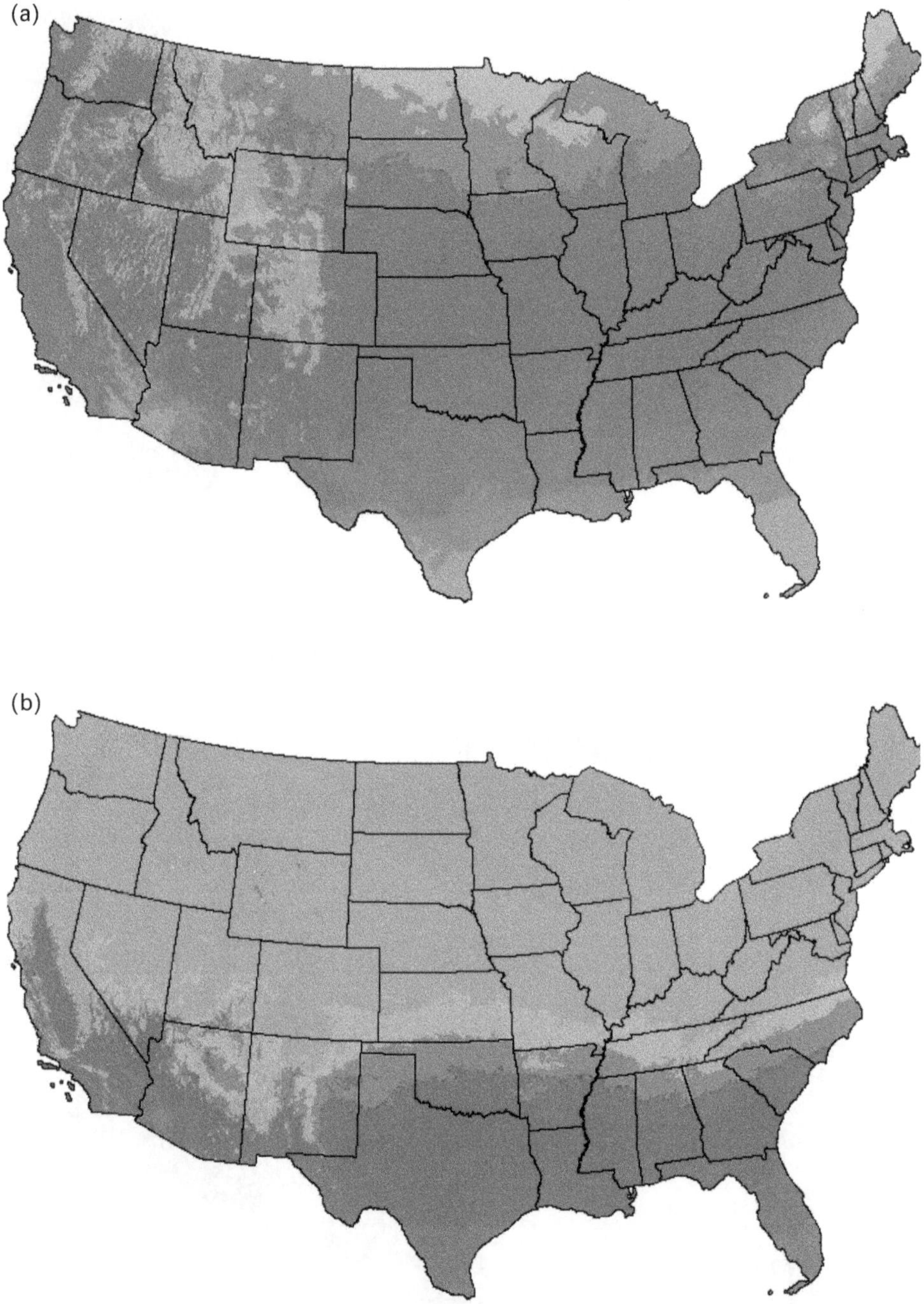

Figure 3.4. (a, b) Temperature zones in the USA for the last glacial maximum 20,000 years ago and the present day compared. Climates now associated with the border region with Canada (lighter grays) came down south as far as Tennessee and North Carolina at that time. *Source*: Author, with William Hargrove.

3.3 BIOMES IN THE DISTANT PAST

In the distant past, millions of years ago, both biome distributions and the types of biomes which existed were far more different from now. For example, 55 million years ago the world seems to have been much warmer and much moister than it is now. There was forest almost everywhere on land—even at the north and south poles—and no desert or dry grassland existed anywhere, apparently (at least no-one has found evidence of them). The warmth of this time was so dramatic that subtropical palms and alligators occurred as far north as Spitsbergen Island, well within the Arctic Circle. If tundra occurred anywhere then, it must have been confined to the tops of very high mountains.

Still further back in time, before the rise of the flowering plants about 120 million years ago, there could not have been quite the same "tropical rainforest" that we see today. In its place, in the wet climates close to the equator, various sorts of conifers as well as other gymnosperms such as ginkgo trees seem to have made up the tropical forest of the time. There were drier climates then, but no grasslands, because grasses had not yet evolved. Their place was apparently taken by ferns that may have grown in extensive savanna-like meadows.

3.3.1 Sudden changes in climate, and how vegetation responds

It used to be thought that all climate change in the geological past was very slow, taking thousands or even millions of years. With more detailed understanding of climate indicators in the geological record, we now know that many past climate changes occurred extremely rapidly. During the glacial–interglacial cycles of the past couple of million years, it seems that climates in the mid-latitudes often took just a few decades to switch from near-Arctic to temperate conditions, or back again. For instance, in the mid and high northern latitudes, the sudden global warming that occurred at the end of a cold phase known as the Younger Dryas 11,500 years ago seems to have been largely completed in under a century, with most of the change occurring in less than 50 years. Some geologists who work on this timeframe suggest that most of that change actually occurred in under 5 years. Similarly abrupt changes in temperature and rainfall may also have occurred in the past in the low latitudes: for example, in the Saharan and Arabian deserts according to some interpretations of data from sediments.

If and when climates changed so suddenly in the past, how long did the biomes take to move? We can get clues to the speed of change in vegetation from the pollen record preserved in lake and sea floor sediments. Each particular species or genus of plants tends to have its own distinctive-looking pollen grains. Thus, when a particular type of tree or herbaceous plant moves into an area it is possible to pinpoint its arrival from looking at the preserved pollen in sediments. The best evidence is heavily biased towards the mid-latitudes of the northern hemisphere, where there are a high concentration of botanists, a relative abundance of research funding and conditions favorable to preservation. Data from hundreds of sediment cores in North America, Europe and eastern Asia gives a general picture of how the temperate and boreal

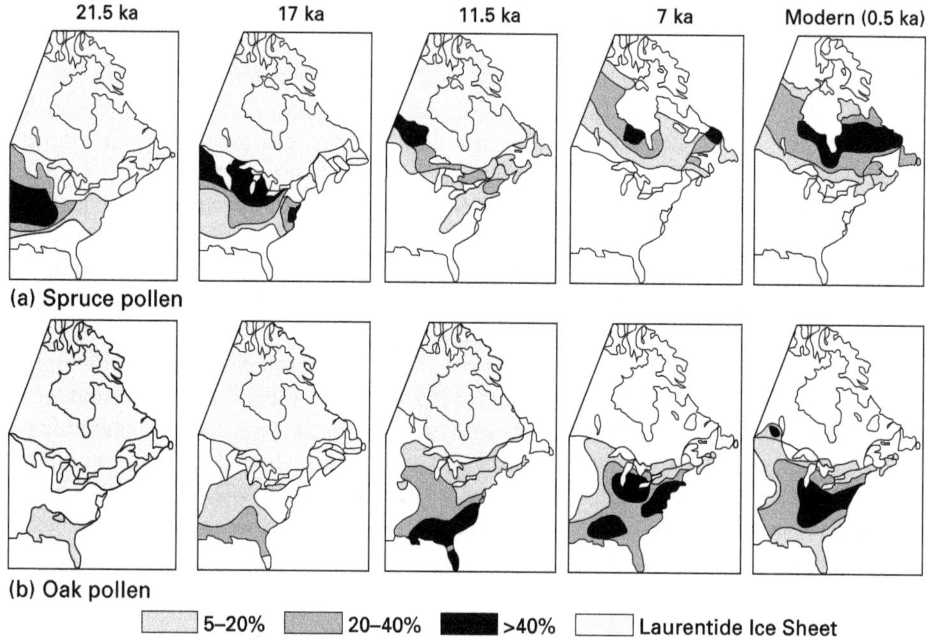

Figure 3.5. Maps of migration rate of eastern North American trees (a) spruce (*Abies*) and (b) oak (*Quercus*) in the pollen record, starting from the last glacial maximum (ka = thousands of years ago). From Davis *et al.* (1988).

forest biomes migrated and changed after the last glacial ended. It is possible that some of the general lessons learned from these regions also apply to vegetation change in the tropics after the last glacial (Figure 3.5).

Looking at the pollen evidence from the mid-latitudes, one thing that is clear is that the new biome distributions suited to a changed climate do not snap into place overnight. In many areas the sudden warming phases seem to have left vegetation way behind, so that it took hundreds or even thousands of years to catch up. For example, just before a sudden warming event 14,500 years ago (known as the Late Glacial Interstadial), the climate of England had been much colder than it is now. The summers were too cool and dry for trees to grow, so there was a tree-less tundra vegetation (except perhaps a few shrubby clumps of birch—*Betula*—in moist valley bottoms). Suddenly, the warming event hit (Figure 3.6), and we can see the change by an influx of warm climate beetle species and plankton, as well as more direct chemical indicators of temperature which turn up in the lake sediments. It seems that in almost no time at all, insects that can live in open grassy vegetation spread northwards from where they had survived in southern Europe.

Warmer-climate snails, which we might think of as being slow-moving, turned up only a few decades later than the insects. But the trees that could have thrived in this climate remained absent for several hundred years. Forest had still failed to arrive when eventually another sudden cold snap occurred around 12,500 years ago

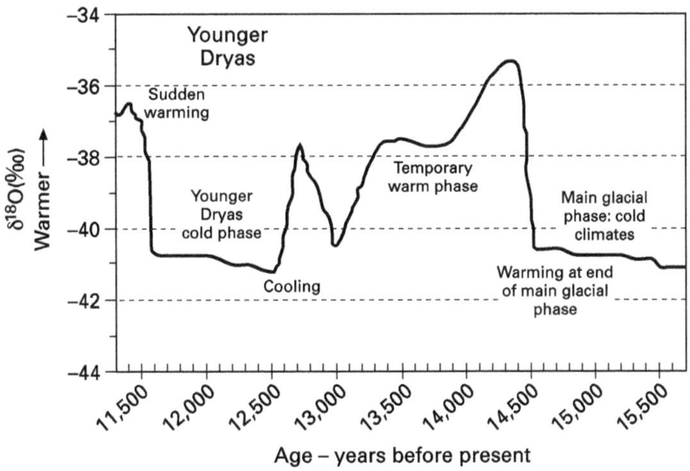

Figure 3.6. Temperature history of the late glacial. There were two main sudden warming phases, separated by a cold phase known as the Younger Dryas.

(Figure 3.6), making the climate too cold for forest once again. What had existed in the meantime before this cooling was a meadow-like grassy vegetation, consisting of just a few tundra and weedy grassland plants that could also grow well under the warmer summers. This was apparently a climate perfectly suited for trees, yet it was a landscape almost without any trees!

The reason for this delay in the arrival of forests may have been that before the warming, the only populations were more than a thousand kilometers away in southern Europe, mostly as scattered woods clinging to rainy mountainsides and moist gulleys. The rarity with which they show up in the pollen record of the cold phases shows just how restricted the populations of many common European tree species must have been at the time.

It simply took a long time for these tree populations to begin to disperse outwards from their glacial-age refuges, and establish extensive populations that could then send seeds further on their way. Something that slows trees down particularly is that it takes quite a few years for them to mature to the stage where they produce seed, so each little "hop" northwards tends to take decades.

One might expect that tree species with light, wind-dispersed seeds would have spread north fastest in Europe and North America. In fact, there is no sign of this difference in seed characteristics showing up in the migration speeds recorded in the pollen record. Some light-seeded trees spread relatively fast, others much more slowly. Exactly what caused the differences in rates of spread is not certain. Some of the heaviest-seeded trees that rely on animals to help spread them were also amongst the fastest to spread (e.g., the hazel tree, see below). It is uncertain whether trees sometimes made huge leaps of tens of kilometers in a single generation with a single "lucky" dispersal event, such as a bird carrying an acorn a very long way before accidentally dropping it.

It has been suggested that prehistoric humans played an unintentional role in dispersing many of the European trees that have edible seeds, such as beech, hazel and some oaks. People could have gathered nuts and then migrated many kilometers in search of game, accidentally dropping the seeds along the way and allowing such sudden jumps in ranges of trees. Some studies by Oxford University anthropologist Laura Ravel and colleagues of Amazonian forest Indians in the present-day world suggest that they often plant seeds of useful trees from the rainforest along trails or out into forest patches in the savanna, to ensure that the products of the trees are always on hand. If this sort of deliberate planting occurred in the past, it could have aided the spread of trees out from their refuges after the last ice age, in both temperate and tropical environments. An example of a tree that could have been spread this way is the hazel tree in Europe, which perhaps explains why it turned up so early in eastern England.

The influence of humans on the world's vegetation might thus be even more pervasive than we would expect, extending back even to the broad-scale migrations of biomes after an ice age. Whole continents could actually turn out to be gardens, at least in a very loose sense.

Some geologists who have studied the post-glacial movement of trees suggest that the delays in migration in northern Europe were actually dictated by the climates of the time. Although warmer, the climate just after the end of a glacial may have been quite hot and dry in summer—too arid for trees to establish. However, this does not seem to tally with the evidence for fossil beetles and other invertebrates which indicate a moist, warm and mild climate even while trees were absent.

An additional factor that may have slowed down the migration of trees northwards in Europe was a lack of the symbiotic mycorrhizal fungi that trees need to help them take in nutrients, especially in rocky, newly colonized soils. Often when they are planted on mine tailings, trees do much less well if they lack these fungi. In the landscape that followed on from these warming events, there were actually a few shrubby birch (*Betula*) trees that had been confined to moist valley bottoms during cold phases. Even during several hundred years in the new warmer, moister climate of the interglacial, they steadfastly refused to spread out across the landscape, and lack of mycorrhizal fungi seems one possible explanation for this.

Something that is obvious from the pollen record after these past warming events is that each tree species migrates in its own individual way, not depending on other types of trees being around it. In both Europe and North America, different species of trees spread out from their glacial-age refuges at very different rates, the slowest arriving thousands of years after the fastest. They also took quite different routes that probably reflected chance dispersal events, and also the locations of the source areas where each had survived the last glacial. Some ecologists have also suggested that the different routes trees took were individual responses of species to changing combinations of climatic parameters, each species adjusting its range to the climate that suited it at the time.

After the most recent big warming event 11,500 years ago marking the end of a cold phase known as the Younger Dryas (Figure 3.6), trees gradually spread northwards from where they had been surviving in southern Europe. They eventually arrived in northern Europe where they blanketed the landscape in forest. But, it was a matter of chance which tree species got there first, and the earliest arrivals often

Figure 3.7. The hazel tree (*Corylus avelana*), shown here with its distinctive flowers known as "catkins", blanketed much of eastern England after the end of the last ice age. This seems to have been due to sheer luck in arriving before other trees.

initially multiplied to form great forests of only one type of tree. For example, after this warming event the hazel tree *Corylus avelana* (a small nut-bearing tree; Figure 3.7*) initially formed great uninterrupted forests in eastern England. The most reasonable explanation for this is that hazel was simply the first tree to arrive from the south following the change in climate. After dominating for hundreds of years, it began to lose out as other tree species arrived and began to compete with it. Nowadays, hazel is still a common tree in northwest European forests, but always mixed in with and overtopped by other species of trees.

Although there was an element of unpredictability in terms of where and when each tree species turned up as the ice age ended, there were some kinds of trees which seem to have been destined to spread in early and play an important part in the forests. For example, the pollen records show that around 10,000 years ago Scots pine (*Pinus sylvestris*) and silver birch (*Betula pendula*) spread quickly across northern Europe, and in most areas they became the major trees of the first forests of our present interglacial. It is thought that in part these two species did so well at first because they were good at getting to the right place at the right time. Both were able to disperse their seeds on the wind. However, many other wind-dispersed species were actually much slower in migrating, so it is uncertain how much this helped them. Perhaps more important is that pine and birch had already been surviving quite far north in Europe as isolated groves of trees during the glacial time, because they are both very cold-tolerant—so they did not have too far to move when the climate warmed. Another reason for their early success, though, may be that pines and birches were good at surviving on the relatively shallow, poorly developed soils of lands that had previously been ice-covered or barren steppe–tundra during the ice age. In most areas, within a few thousand years other tree species arrived or became more abundant, steadily pushing out the earlier colonists. In England, Scots pine even seems to have gone completely extinct before it was reintroduced as trees planted by humans. In other areas of colder climate (northern Russia and Scandinavea, and high mountains around Europe) both pine and birch are still an important part of the forests, because the other tree species that might outcompete them cannot live in such cold. The fossil pollen records of previous interglacials show that there was always a similar pattern of early abundance for birch and pine in northern and central Europe, followed by a decline after a few thousand more years. This seems to make the point that this broad-scale pattern of succession was not just chance: it was the inevitable result of certain underlying ecological factors.

It is a moot question as to whether some trees are still migrating out from their glacial refuges to fill their potential ranges. Certainly, once forest has established across a broad area, this is likely to slow down the migration rate of any new species arriving, because large competitor trees dominate the space where seedlings of the new arrival would otherwise be able to establish. After the last major warming event 11,500 years ago, the trees that migrated certainly seem to have spread north a lot faster through western Europe—which had previously been a barren, tree-less landscape—than through eastern North America which retained a covering of trees throughout the last glacial. However, since most trees in both regions stopped moving north several thousand years ago (or have even retreated slightly south since then), it is thought that they have probably reached the boundaries of their own climatic limits.

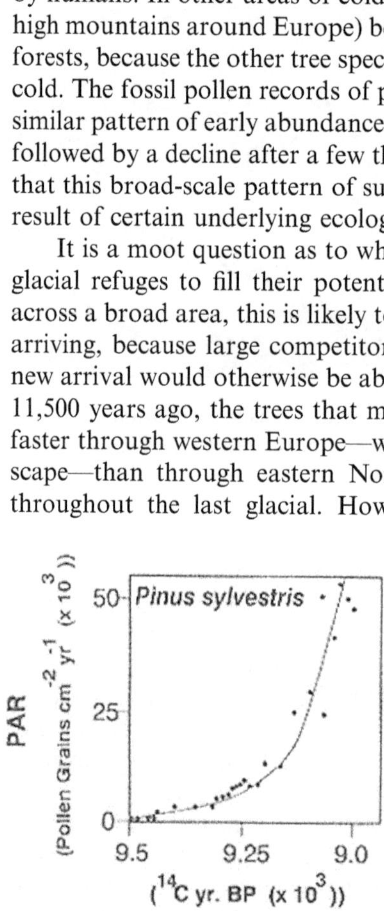

Figure 3.8. Increase in pollen abundance of Scots pine (*Pinus sylestris*) in eastern England after the end of the ice age, indicating an expanding population of the tree after it colonized the area.

Even when a species of tree arrived in a particular area, it only became an important part of the forest after its population had had a chance to multiply up. In eastern England after the last big warming the pollen record from lakes suggests that tree populations doubled every 31 to 158 years, depending on the species (Figure 3.8). It took several doubling times (varying from several centuries to a couple of thousand years) before each reached a roughly stable "plateau" level of abundance. Interestingly, a rare piece of information from tropical tree communities—the Queensland rainforests of northern Australia—indicates very similar doubling times for tree populations following the global moistening of climate that occurred in the early part of the present Holocene interglacial.

3.4 THE INCREASING GREENHOUSE EFFECT, AND FUTURE VEGETATION CHANGE

The greenhouse effect keeps the earth warm enough for life, and lack of it makes mountains cold (Chapter 1). But now the warming from the greenhouse effect is intensifying, as humans push more and more of the so-called greenhouse gases into the atmosphere (see Box Section 1.1). Leading amongst these gases is CO_2, released by fuel-burning and deforestation. Its concentration is already 40% higher than it was 200 years ago, and it looks set to double by 2050 (Chapter 7). Methane is another important greenhouse gas, produced largely by flooded rice fields and by cattle. Its concentration has more than doubled since the year 1800, as human population and agriculture have expanded. Together with several other more minor greenhouse gases resulting from human activity, these additions should already have produced a warming of something like 1°C since 1800.

If the increase in greenhouse gas concentrations continues as expected, by around 2100 there could be somewhere between a 3 and 5°C increase in global temperature. This forecast is based on sophisticated climate models, known as general circulation models (or GCMs), that divide the land surface, oceans and atmosphere into their basic climatic components (more about these models in Chapter 5).

The amount by which greenhouse gases have increased over the past 200 years should already have been enough to produce a noticeable warming of the global climate. The GCMs suggest that the warming already will have been around 1°C. Temperature records from around the world, and a variety of other indicators of climate, seem to confirm that there *has* been a warming during that timespan, enough to have had at least some noticeable effects on vegetation. The warming seems particularly rapid and steady during the last 25 years, suggesting that the effect of greenhouse gases is now starting to dominate against the background of natural climatic variability (Figure 3.10).

3.5 RESPONSE OF VEGETATION TO THE PRESENT WARMING OF CLIMATE

There are of course many aspects of plant ecology that seem tightly controlled by temperature: the broad-scale distribution of biomes across the continents is one example (Chapter 2). Temperature also determines exactly how high up a mountain trees can grow, and the precise time of year that trees start to leaf out, or when spring flowers appear. It also determines how fast a tree can grow, with variations in climate showing up in the width of the annual rings.

Because of the amplifying factors that operate near the poles (Chapter 5), GCMs predict that climate should be changing most dramatically close to the poles; and indeed climate station data show that these areas are warming particularly rapidly.

Land and ocean ice data seem to corroborate the view that this warming is both intense and sustained; sea ice around the Arctic has decreased rapidly, glaciers everywhere are melting back fast, and permafrost is thawing in areas where it has been stable for centuries or millennia. Already there are signs that the rapid warming that has occurred over the past few decades has had some effects on biological processes, in at least some parts of the world.

In the mid and high latitudes of the northern hemisphere, most of the temperature-sensitive aspects of plant behavior are showing at least some signs of shifting in response to the recent warming trend. However, the trend cannot be found everywhere, partly because the warming itself is somewhat patchy, and perhaps also because other environmental factors can intrude and complicate the picture.

As one would expect from the temperature trends, vegetation around the Arctic has begun to change. On the broadest scale, satellite data show a greening of the Arctic since the 1980s, especially in the northernmost parts of Canada and Alaska, and northwestern Siberia and Scandinavia (Figure 3.9*). The general pattern of the warming appears to correspond to a natural climate fluctuation that has always

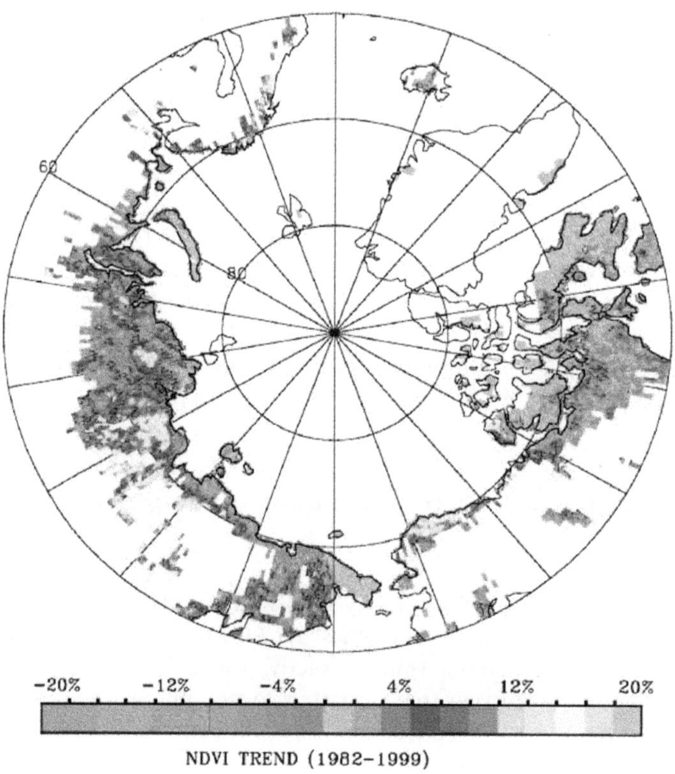

Figure 3.9. The greening trend around the Arctic from satellite data. *Source*: data from Stowe *et al.* (2004), figure by Zhou and Myeni (2004). (Note: NDVI is a measure of the "green-ness" of the image. The higher the NDVI the more vegetation.)

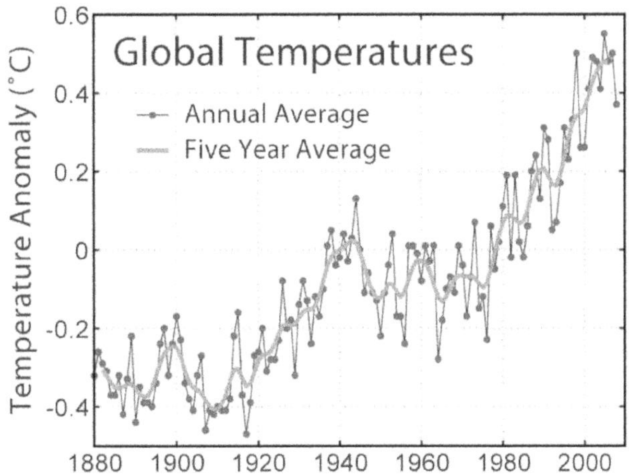

Figure 3.10. Instrumental temperature record of the last 120 years (from CDIAC).

occurred across these regions: the Arctic Oscillation. What is unusual now is how intense and sustained this phase is. From all the climate indicators that we have available, there has been no other period in the past thousand years where the Arctic experienced such warm temperatures for so long. This suggests that something beyond the natural background of climate fluctuation may be at work. On the ground, this warming translates into noticeable changes in the structure and species composition of tundra vegetation in northern Alaska and Canada. Many of the small ponds that dot the landscape have drained, as a result of the layer of icy soil (permafrost) that held them in place melting away, so terrestrial vegetation is taking over from the aquatic communities that lived there before. Shrubby vegetation of dwarf willows and alders is pushing into the grassy tundra on Alaska's north slope. On the far northern islands of Canada, where climate has always been too cold for a continuous covering of tundra, comparison of aerial photographs taken in the 1940s and today shows that there has been an expansion of shrubby vegetation out from the most sheltered spots, which were the only places it was able to grow before (Figure 3.11a, b). It seems then that the landscape in the far north is changing, because of the warming that has occurred during that period.

At the other end of the world, on the rocky edges of the Antarctic Peninsula, a noticeable warming has occurred over recent decades. On the west side of the Peninsula, temperatures have gone up by 2.6°C since the 1940s. This warming has resulted in a veritable population explosion of the only two types of vascular plants native to Antarctica: a grass (*Deschampsia antarctica*) and a tiny member of the cabbage family (*Colobanthus quitensis*). At sites where these two species have been monitored over more than 30 years, they have expanded from scattered plants and clumps to form the first "lawns" on Antarctica.

Mountain tops around the world also seem to be experiencing warmer temperatures. Mountain glaciers are melting back almost everywhere, a strong sign that there is warming going on. Change in vegetation on mountains is harder to find and

Figure 3.11. (a, b) Arctic shrub cover change in northern Canada. The numbered areas in the foreground show the change most clearly. *Source*: Stow *et al.* (2004).

interpret than melting of glaciers, but it is certainly widespread. Some of the most striking changes are in the Ural Mountains of western Russia, where the treeline over a very broad area has migrated 60–80 m upslope. Similar upwards migration of the treeline has occurred in the mountains of Scandinavia, in the western USA, the Alps and the mountains of Tasmania. However, in some areas of the world, mountain vegetation has not responded, even where meltback of glaciers is occurring nearby. Partly this reluctance of the vegetation to change may relate to the extreme conditions at tops of mountains: soils are thin and poorly developed, and plants cannot establish themselves and grow very easily even if the climate is now warm enough. So, a response to warming will often take time, perhaps the time taken for microbes to get to work breaking down minerals that seedlings will eventually be able to use. Also, plants grow and develop slowly in the cold climate of high mountains, even if some warming has occurred. It can take them a long time to respond to the warmth by becoming larger or setting more seed. Such is the slowness of ecosystems in the high mountains that some ecologists believe that treelines are now rising in response to a warming event that occurred 150 years ago, not the current burst of warming!

Even where the treeline has not moved noticeably, a shift in composition of the existing forest further downslope may reveal the effects of increasing temperatures. For example, in northeastern China, Changbai Mountain has not shown much change in the treeline, but my co-worker Yangjian Zhang has shown that the ancient forest on its upper slopes has thickened and the species composition has shifted towards more warmth-demanding species—certainly the trend that would be expected for a climate warming.

Near the top of mountains in Tasmania (the big island off southeastern Australia), the high alpine tundra zone—which has several beautiful types of cushion plant—is disappearing as trees are able to seed themselves higher and higher up the mountains, due to less severe winters and warmer summers. The mountains in Tasmania are only just tall enough to have a tundra zone at the top. At the rate things are going, in a few more decades these mountains will probably be forested right up to the top; there will be no tundra zone left in Tasmania and these alpine plants will only survive in cultivation.

In ecology, there always seem to be some exceptions to a trend. In parts of the timberline in the lowlands of northern Siberia, trees are retreating south as the tundra expands. The change in tree cover seems to have gone totally in the opposite direction to what would be expected from the warming trend observed across the region. However, the retreat is apparently due to a reduction in rain and snowfall seen in the climate records, rather than any trend towards coldness.

Although the most striking shifts are evident at the coldest limits of the world's vegetation, changes may also be occurring almost unnoticed in other parts of the world that are not being so closely watched, or where there is not such a striking boundary in vegetation structures as on the edge of a biome. In many parts of the world, trees now seem to be growing faster on average than at any time in the last few centuries, perhaps as a result of the warming. In some areas, the trees in native forests are growing faster and dying younger than they used to several decades ago. This increased "turnover", as it is called, may be a result of warmer temperatures allowing each tree to grow and complete its life cycle more rapidly. Such speeding-up of forest growth has been detected across much of the western USA in a recent study by Mark Harmon of the University of Oregon. A similar trend has been found by Oliver Phillips and colleagues across the Amazon Basin, and also in a widespread study in the rainforests of central Africa. However, the pattern is not present everywhere: for example, in at least some tropical forest localities in Malaysia and Central America, in a certain percentage of the forest plots within South America and Africa, and in parts of the eastern USA, forest growth and turnover seem to be slowing down. Exactly what is at work behind the general trend towards faster forest growth and turnover is not clear, although a change in climate is widely suspected. Other factors might be important too: for example, the direct CO_2 fertilization effect (see Chapter 8). In Europe, faster forest growth might be partly due to the beneficial effects of nitrogen and sulfur pollution on the trees, making up for nutrient-deficient soils.

3.6 SEASONS AS WELL AS VEGETATION DISTRIBUTION ARE CHANGING

For a long time, naturalists and gardeners have recorded dates of flowering and leafing of the plants around them. These records happen to provide another interesting measure of responses to climate change.

In Europe, it is quite evident that the seasonal patterns in vegetation have been shifting in response to warmer temperatures. In Britain, for example, a long tradition of amateur natural history has ensured an abundance of information on the detailed distribution and behavior of plants, stretching back many decades. Temperatures in England are on average 1°C warmer than they were 50 years ago: in response, certain wild flowers are now blooming about a week and a half earlier, and autumn leaf fall has been getting later.

For instance, a study of the long-term records of the respected botanist R.S.R. Fitter in England since the 1950s has shown that many plants (about 16% of a sample of 385 species) have been flowering several days later during the last decade, compared with their long-term average over the previous four decades. His study would probably have shown even more striking trends if the earlier data had been divided up into individual decades so the 1950s could be compared with the period since about 1995 when warming has been most dramatic. Not all plants in Fitter's study showed the same trend: about 3% of the sample actually flowered significantly later during the last decade, indicating the complexity and the diversity of plant responses to warming.

Another study across Europe by Annette Menzel and colleagues looked at the records of 542 plant species between 1971 and 2000. They found that 70% of these species had followed a trend towards earlier leafing out, flowering or fruiting during this period, and only 3% had followed the opposite trend in becoming later. On average, spring seemed to be getting earlier by about 2.5 days per decade. Autumn leaf fall was also starting later in many species (see below), also likely as a response to warming, and adding to the longer growing season.

There are similar indications of a trend towards earlier flowering in North America—for example, in the flowering times of cultivated trees and shrubs. One study compared flowering times of hundreds of plants growing at the Arnold Arboretum in Boston between 1980 and 2002, with old records of flowering times of the exact same individuals between 1900 and 1920. In the last two decades, the plants flowered 8 days earlier on average than they did back then. This seems to be a response to the warming of about 1.5°C in average temperature that shows up in the climate record for Boston. A similar trend has been noted amongst wild trees in the same general region: the time of maple sugar sap flow in northern New England has moved forward in the year by at least a week and a half since the 1960s.

Satellite studies that compare the timing of a "green wave" of emerging leaves in spring are finding evidence that the change in seasons is more dramatic than the more localized ground-based observations suggest. Franz Badeck and colleagues of Potsdam in Germany conclude that the satellites suggest a consistent trend all across the northern hemisphere's mid and high latitudes, with spring getting between 4 and 8 days earlier with each decade that has passed since the 1970s.

Across Europe, records kept by naturalists show that the leaf color change into autumn has become 0.3–1.6 days later each decade since the 1950s. In some areas the length of the total growing season between leafing out and leaf fall has increased by up to 18 days over the past 50 years. Satellite images of the timing of autumn leaf color change seem to confirm that autumn has been getting later over the past two decades.

Because climate can be fairly fickle from one year to the next, a colder-than-average spring nowadays can still be later than a warm spring in the 1940s or 1950s. But it is the long-term average trend that we should pay attention to, and in terms of averages there is no doubt that the length of the growing season has increased during that time.

There are various other examples of ongoing change in vegetation seasonality that seem to be a result of warming in the mid and high latitudes. It isn't occurring everywhere, and where it is occurring it often fluctuates and even temporarily reverses for a few years. But the predominant pattern in vegetation looks like a response to increasingly warm climates.

One must also bear in mind the possibility that *part* of this change in seasonal patterns in vegetation might be due to the increasing direct CO_2 fertilization effect (see Chapter 8). Conceivably, it might in some ways tend to mimic the effects of temperature increase, with earlier leafing out and later leaf fall. On the other hand, one experiment on the seasonal growth patterns of grasses under increased CO_2 showed that when they were CO_2-fertilized some species flowered later in the season, not earlier! If this is more generally true of plants, climate change might be pushing them in one direction (earlier flowering) while increasing CO_2 is pushing them in another (later flowering).

Another important thing to consider is that part of the trend we see might be due to the fact that cities have got bigger over time. Bigger cities have in themselves

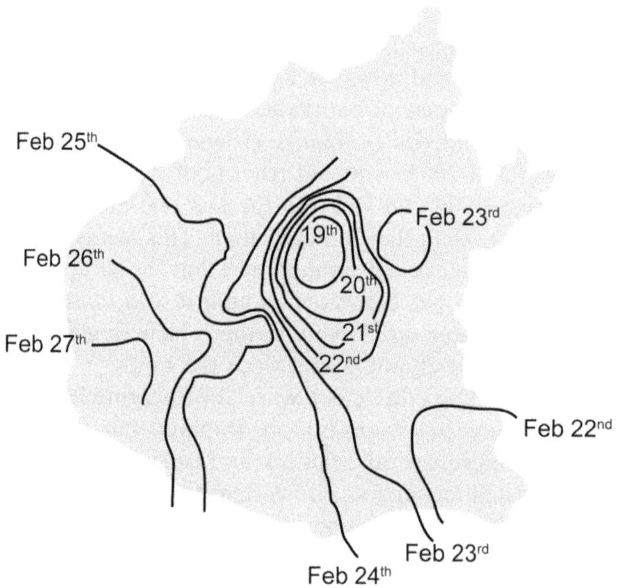

Figure 3.12. The heat-trapping effect of large cities alters the time of arrival of spring. In Tokyo, Japan, cherry trees flower earlier in the warmer central districts than on the outskirts. Contours mark differing average flowering times in spring.

tended to pour out and trap more heat, producing their own local climates (this is known as the "urban heat island effect"), and plants that grow within urban areas will be affected by this. In some big cities, such as Tokyo (Figure 3.12), flowering times of trees are more than a week earlier in the center than in the outskirts. Surely, if cities have grown then the warming effect on local climate will have grown too, producing a change in the seasonal rhythms. However, even far out in the countryside, away from any growing city, the trend in vegetation over the last few decades seems to be much the same. This implies that the changes in plant seasonality are not solely due to this "heat island" effect.

Trends in seasonal timing observed in wild vegetation and in cultivated plants are in a sense reassuring. They show that plants are flexible enough in their biology to respond to climate change fairly rapidly, at least up to some point. In some areas of the world, observations of changes also indirectly bolster the evidence that global warming is really occurring, because few long-term climate station data exist there. The plants are in effect acting as weather stations, helping to show up what appears to be a global warming trend!

3.7 WHAT WILL HAPPEN AS THE WARMING CONTINUES?

Eventually, global warming may get to the point where the present distribution ranges of species of plants and animals are left far behind the areas that they could potentially live in. For example, the potential range of sugar maple (*Acer saccharum*) might shift hundreds of kilometers to the north, way up into Canada towards the

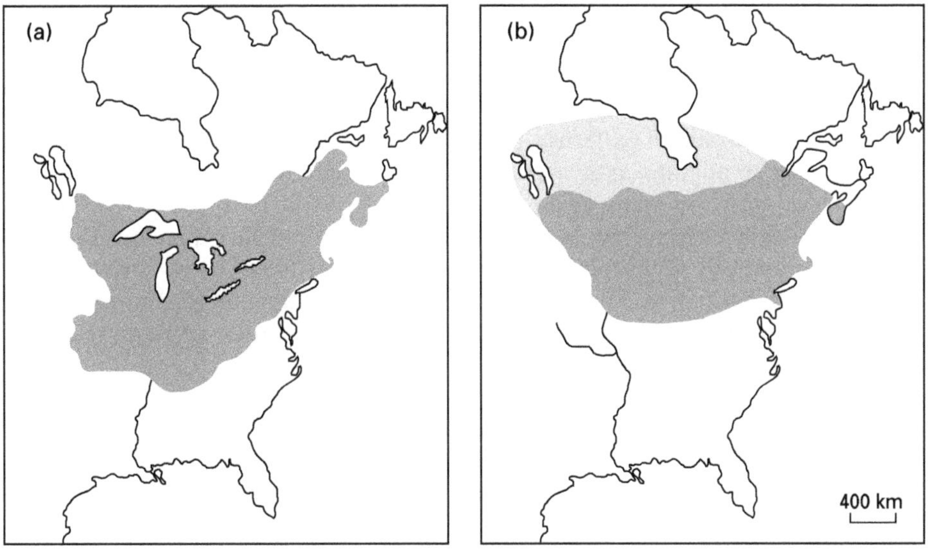

Figure 3.13. Sugar maple extends from southeastern Canada to the south–central USA (a). By 2090, sugar maple may be able to grow way up to the southern Hudson Bay (lighter gray area on map). Meanwhile its natural range in the southern USA will tend to be lost as temperatures there become too warm for it (b). *Source*: Redrawn from IPCC.

edges of the Hudson Bay (Figure 3.13). When this situation of a much warmer climate arises, the new vegetation zones and communities won't just snap into place overnight. Seeds of plants will have to physically spread over the landscape, across hundreds of kilometers.

From what we can see of present-day plant distributions, many species will be left growing in temperatures that are too warm for them, at least in the lower-latitude parts of their distribution ranges. Sugar maple, for example, extends down to Tennessee and if the warming causes climates to shift north it may be left out-of-sorts in the south of its range. If the southern range limits of northern temperate species contract faster than they can expand at their northern limits, they may ultimately go extinct. It is unlikely that a widely distributed species such as sugar maple, which occurs across a very broad range of temperatures, will ever be completely driven extinct by global warming. However, there are many other species of plants that have a much more restricted geographical range, and presumably a much narrower climate tolerance range. These are the ones most likely to lose their original ranges when the climate changes. If they cannot migrate quickly enough, they will also be unable to reach any new potential range created by the warming—and that will be the end of them. One example is the southern Torreya (*Torreya taxifolia*), a coniferous tree which occurs in the wild in just a few small areas of northern Florida.

If we assume a very glum scenario in which nearly all the world's plant species are unable to move their ranges over the coming decades (for whatever reason), then extinctions are likely to begin. For example, Thuiller and colleagues (2005) took the current distributions of 1,350 wild plant species in Europe, and assumed each has a range that is limited by the combination of temperature and rainfall conditions in which it occurs. Then as a "what if" scenario, they used various different climate model predictions for Europe between around 2050 and 2080, with altered temperature and rainfall patterns. In all of these scenarios they found that if species don't manage to migrate, many of them will be left with hardly anywhere to live in a few decades' time. As many as a fifth of plant species in Europe were predicted to become critically endangered, having lost 80% or more of their previous range. Up to 2% (around 30) of Europe's plant species seemed likely to go extinct, having nowhere at all left to live. This is of course not a mass extinction for Europe's plants, but it does seem to set the scene for later disaster if temperature keeps on rising after 2080.

If species do manage to migrate in response to climate, there will be fewer extinctions—and the faster they migrate, the better they will be able to save themselves. But in European landscapes broken up by farming and towns, it is especially uncertain how fast these wild plants will be able to shift themselves around as the climate changes.

How long will it take for the plant communities of the greenhouse world to take shape? Will it be decades, centuries, millennia; or will species just never manage to shift themselves this far? And what will happen to the communities and vegetation types that exist in each place now, when climate warms. For example, will the trees in our forests just die, before others can spread north and replace them?

Questions such as these are very hard to answer, but there are some clues from the past, when warming events—of comparable speed and magnitude to that which we are anticipating over the next century—actually occurred (see above). Such events were associated with the immensely unstable climates of the last 2.4 million years. There were repeated sudden warming and cooling stages, apparently taking only decades in many cases. For instance, around 11,500 years ago at the end of the cold phase known as the Younger Dryas, a very sudden warming event around the North Atlantic was largely completed in 75 years. This sudden jump is comparable in size and speed with the projected "greenhouse effect" warming over the next 100 years.

From the evidence of responses to past sudden climate changes, it looks like vegetation will remain out of equilibrium with climate for hundreds and indeed thousands of years following the onset of greenhouse effect warming. For example, in Britain, after a similar sudden warming event 14,500 years ago, vegetation remained out of balance with climate for hundreds of years (above). Trees seem to have been unable to spread north into Britain fast enough to exploit a warm climate which would have suited them, and the landscape remained covered in a sort of meadow vegetation. We might find similarly strange situations arising in a future greenhouse world over the next several centuries: vegetation types that no longer match the climate, without warm-climate plants having spread in from the south to take their place. When warmer-climate plants do eventually start to arrive, particular species may start to dominate out of all proportion, just as the hazel tree did in England after the warming event 11,500 years ago (above).

We can estimate how fast trees spread after past warming events, from the time delay between them turning up in the pollen record at each lake and then the next ones slightly farther north. For Europe, when trees spread north after sudden warming events, they moved at peak rates of between 0.02 and 2 km a year, depending on the species. In North America the rates of movement were rather slower, between 0.08 and 0.4 km per year. Going simply by these past figures for migration rates, it seems that the geographical ranges of a few species in Europe might almost be able to keep track within a moderate greenhouse warming scenario, where climate moves north at about 5 km a year. They would perhaps show a migration lag of a few decades. The future rate of warming may be expected to vary with latitude, according to climate model predictions, so the rate at which tree species' ranges will need to shift to keep step with climate warming will be greatest at the more northerly latitudes. If the reported migration rates from the past are representative of the northern coniferous and temperate zones of the world, it appears that at all latitudes most tree species would be left far behind, but might catch up on a timescale of centuries or millennia if the warming stabilized.

However, just relying on reported migration rates of trees in the pollen record is a very simplistic way of trying to forecast their future responses to climate change. For one thing, we can never quite be sure that each species arrived in northern Europe exactly when it shows up in the fossil pollen record. Some species might have arrived hundreds of years earlier as very isolated individual trees, and then spent a long time building up their numbers to the point where their pollen wqs abundant enough to get noticed in lake sediments. If these trees really did turn up in the north so much earlier than we find them, it would mean that their capacity to migrate fast in response to climate is even greater than we had thought. On the other hand, part of the reason so many trees turn up in northern Europe so fast after the end of the ice age may be that they actually had small, scattered ice age populations that survived quite far north into central Europe all the way through the cold of the ice ages (Stewart and Lister, 2007), instead of being confined to southern Europe. If trees were really already living so far north, it is no surprise they turn up quickly after the end of ice age because they did not have so far to move. So, the estimates on how fast trees can potentially migrate with global warming depend on having a rough idea when each species made it north and where it was coming from, and at present there is some uncertainty about this. Because of these, we should regard the potential future migration speeds of trees based on pollen evidence with some caution, and not make too many firm pronouncements about what this means for forests under greenhouse effect warming. It is difficult to figure out all the other factors that could have affected the rate of movement of trees. In some areas, such as Europe, there were no pre-existing forests in place before the warming, and this probably allowed trees to move faster than they would through the now-forested landscapes. However, nowadays humans often harvest trees from the woods, even clear-cutting whole swathes of forest. Such open areas may provide an ideal opportunity for migrating trees to establish themselves. On the other hand, the open spaces of fields or suburban land between forest patches could present a new barrier to migration.

In North America, there was already forest covering the eastern USA at the time of the sudden warming 11,500 years ago, but its species composition altered in response to the change in climate. The pollen record from lakes shows that many different tree species spread north, but it generally took between several hundred and several thousand years for them to reach their final limits under the new warmer climate (above). It is likely that, if left to themselves, forests in the mid-latitudes will take a similar period of time to adjust to greenhouse effect warming.

What about the cooler-climate species of trees that already made up these forests before the warming event? Did those trees already in place die in response to the change in climate? Reassuringly, there is relatively little evidence that the rapid warming event at the end of the last ice age in North America was associated with any sudden death of the forests. It seems that the trees already present at the time were tolerant enough of warmer temperatures to survive. Where they disappeared from the forests it seems to have been a gradual process over hundreds of years brought about by competition from other trees that moved in from the south, allowing them the chances to spread their ranges northwards. This makes it likely at least that the forests we see in the present world will not all die when the climate suddenly warms by several degrees. There is an exception to this in southern New England, where a study by Dorothy Peteet and colleagues shows that various cold climate conifers suddenly vanish from the pollen record at the end of the Younger Dryas. Their disappearance seems to take less than 30 years suggesting that these tree populations just died with the warming. This is an ominous sign for the fate of at least some of the world's forests under global warming.

We do not really know just how much warming will occur, especially if some of the feedbacks mentioned in the later chapters in this book start to kick in. Eventually, the temperature rise might start to exceed what is survivable by some species. The temperature increases are likely to be particularly drastic in the high latitudes, where various positive feedbacks (see Chapters 5 and 6) will tend to amplify the greenhouse warming. At least one major tree species in Siberia—the Siberian larch (*Larix siberica*)—seems unable to cope with mild winters regardless of competition from other species and will simply die in place. If it is grown in the mild climates of western Europe, Siberian larch thrives for about 25 years and then suddenly dies, apparently unable to defend itself against attack by fungi in its environment. Whether winters in Siberia will ever become as mild as they now are in western Europe is a moot point, but it does show that there are limits to what cold climate plants can tolerate, beyond which they will simply die. For all we know, other plants from the north might turn out to be even less tolerant of warmth than Siberian larch is.

3.7.1 Movement of biomes under greenhouse effect warming

The climate predictions of GCMs coupled to vegetation schemes provide some clues to what the final distribution of vegetation types in the greenhouse world might look like. Warming of several degrees C is enough to push the ranges of northern temperate trees hundreds of kilometers polewards beyond their present limits. At the same time, range limits in the south are likely to contract as well (although the picture from the last glacial suggests that this may be a slow process, dependent on other warmer-climate competitors moving northwards to out-compete them).

Movement of many different temperature-limited biomes outwards from the equator seems likely. Some areas that now have temperate climates with frosts are predicted to become tropical. For example, in a moderate warming scenario, by 2100 tropical rainforest is predicted to be the "right" vegetation type for southern Louisiana. However, even if they are expanding at the edges, the core areas of tropical rainforest that exist at present might start to suffer under global warming. A model study by Peter Cox and colleagues suggested that, as sea surface temperatures warm due to the greenhouse effect, the Amazon rainforest will experience severe droughts added to by a collapse of the rainwater recycling mechanism within the Amazon Basin (see Chapter 6). This collapse of water recycling is intensified by decreased transpiration from the leaves of forest trees under the higher CO_2 levels of the future (Chapter 8). There is also a widespread fear amongst ecologists that tropical forest trees might not be able to stand temperatures that are much warmer than at present, such that even a fairly slight warming in climate will cause them to die. Already in years of warmer-than-average temperatures, the world's tropical forests seem to grow less and take in less carbon: a sustained warming could be enough to push them over the edge. Generally, though, we do not know much about the upper heat tolerance limits of tropical rainforest, so it is hard to say exactly what will happen. As if to prove this point, a year after Cox *et al.*'s study was published, the Amazon region suffered an unprecedented drought associated with a sudden warming in the equatorial Atlantic.

In the mid-latitudes of the USA, one study using various climate model scenarios by Bachelet and colleagues suggested that with a certain moderate amount of warming there will be a net increase in forest, spreading out over desert and grassland areas as a result of increased rainfall. But they also suggested that if the temperature keeps on increasing above a certain limit, the climate will get drier overall and forest will retreat. Something that complicates many of these modeled future scenarios is that they also include a direct effect of increased CO_2 on the physiology of plants. As we will explore in Chapter 8 of this book, the influence of higher CO_2 on the growth and water balance of plants is a big uncertainty that adds to the difficulty of forecasting effects from climate change alone.

The greatest shifts in vegetation are predicted to be seen in the high latitudes where warming is predicted to be strongest, and where the most dramatic warming is in fact already under way.

Changes in the amount of rainfall and snow, and in the precipitation/evaporation balance, are seen as being more difficult to predict than temperature. Different GCMs come up with very different conclusions for the amount and distribution of change in rainfall based on only slight differences in their assumptions. Overall, it looks like the changes in moisture balance will not be dramatic over the next century as the global climate warms, with perhaps more rainfall giving slightly less arid vegetation overall across the greenhouse world.

The way in which biome-based models divide up the world tends to give the impression that the only changes which occur during warming are at the boundaries between biomes. However, there are major differences in species composition and physical form of vegetation *within* each biome, and it is important to remember that we can expect changes in these just as much as at the boundaries.

All the biome-predicting vegetation schemes we have considered here so far are "static": they simply state what vegetation types will be in balance with a changed climate. They do not tackle the problem of how long it will take for the new vegetation to arrive in a new place and then grow to maturity. We know from the history of past change that vegetation can remain out of balance with climate for hundreds or even thousands of years. To get a better idea of the time course of changes in vegetation as the earth warms over the next century, ecologists have come up with dynamic vegetation schemes, which gradually "grow" new vegetation suited to the changed climate produced by a GCM. Dynamic schemes do not just assume that forest can spring up in grassland or desert areas fully grown overnight; they recognize that it will take decades to mature from seedlings. Examples of such schemes are the MAPSS scheme, and the DOLY scheme.

Although they are likely to give a more realistic simulation of the time course of events as climate warms, these dynamic schemes do not simulate the complex processes of migration of species which will be necessary in order to alter biome distributions. They simply assume that the vegetation of the future is already in place as seedlings, waiting to grow when the climate changes. Yet, as we know from the aftermath of sudden warming events in the earth's recent history, the time taken for migration can cause a major delay in the adjustment of vegetation to climate.

In the modern world, the process of migration could take even longer than it did after ice ages. The distributions of many plant species are broken up by agricultural landscapes, making it hard for them to move across sterile fields that are regularly ploughed and sprayed with herbicides. For instance, in western Europe many types of plants that normally only live in forest would somehow have to hop between isolated woods that may be kilometers apart from one another. The problems of migration may be particularly great for species of European and North American wild flowers known as "ancient woodland species", because they only seem to be found in very old, established fragments of forest, seeming unable to colonize young forest. One example is the bluebell, *Hyacinthoides non-scripta* (Figure 3.14*), which forms a beautiful blue carpet in English and Welsh woodlands in the spring. Others include the various types of *Trillium* of the forests of eastern North America. It is not clear whether such species would ever be able to migrate in response to climate change under present circumstances, given the extra handicap that they suffer due to their restrictive requirements.

However, it is possible that humans can come to the rescue, helping many wild plants to overcome what is in the first place a human-made problem. In the northern mid-latitudes, which are so intensively farmed, deliberate planting of species north of their previous range could allow them to exploit the warmer climate, and make up for loss of range at their southern boundaries. It may require a concerted mass movement of volunteers to plant young trees and flowers farther north in their new potential climate range. However, it is also important to remember that many species of trees and shrubs are already planted well outside their natural ranges in parks, gardens and

Figure 3.14. A bluebell (*Hyacinthoides non-scripta*) woodland in spring, southwestern England. Being a flower confined to ancient forests, the bluebell is unlikely to migrate easily across human-influenced landscapes as the climate changes. Credit: Keith Hulbert.

forest plantations. Beyond their ranges they may exist as poorly-performing and poorly-reproducing individuals, unable to compete with the wild species around them in the current climate. Yet, as climate warms they may come into their own and form a natural part of the vegetation. In effect, part of the flora of the future greenhouse world may already be in place, waiting for the warming to happen.

4

Microclimates and vegetation

Climate on the broad scale, across hundreds of kilometers, brings about the broad-scale distribution of vegetation types (Chapters 1 and 2). However, even looking at the world much more locally, we see that there are also very substantial differences in the average climate. For example, a south-facing slope has a different climate from a north-facing one. The year-round temperature and rainfall conditions under a tree will be different from those just a few meters away in the open. The temperature right at the soil surface is different from the temperature a few centimeters under the surface.

Such local differences make up what are known as "microclimates". These are little climates that exist to some extent everywhere and vary on a scale of a few tens of meters, a few centimeters or even a few millimeters. Such differences are all-important to plants, and also the animals that live amongst them.

Microclimates help to explain part of the patchiness in vegetation that occurs on smaller scales; they determine which plants can grow where. They are also important in understanding how so many different species of plants manage to coexist, without them all being out-competed by one strong species. And microclimates can explain certain features of growth form, leaf shape and physiology of plants.

Furthermore, microclimates are the building blocks of climate. The broad-scale climate is in part the product of these countless little climates, added up and averaged out. If we really want to understand how climate on the global scale is made, including how plants themselves help to form it (Chapters 3 and 4), we have to understand microclimates.

4.1 WHAT CAUSES MICROCLIMATES?

Microclimates are caused by local differences in the amount of heat or water received or trapped near the surface. A microclimate may differ from its surroundings by receiving more energy, so it is a little warmer than its surroundings. On the other hand, if it is shaded it may be cooler on average, because it does not get the direct heating of the sun. Its humidity may differ; water may have accumulated there making things damper, or there may be less water so that it is drier. Also the wind speed may be different, affecting the temperature and humidity because wind tends to remove heat and water vapor. All these influences go into "making" the microclimate.

4.1.1 At the soil surface and below

Soil exposed to the sun heats up during the day and cools during the night. Within a few centimeters of the surface, the temperatures during the day can be extreme: 50°C or more in a dry desert climate when there is no water to evaporate and cool the soil. Even high on mountains, exposed dark soil surfaces heated directly by the sun can reach 80°C—hot enough to kill almost any lifeform.

At night a bare soil surface cools off rapidly and by morning it may end up more than 20°C cooler than during the day. Yet, only 15 cm down the fluctuation between night and day is only about 5°C, because the day's heat is slow to travel through soil. Thus, the soil at depth has its own quite separate climate: a microclimate distinct from that at the surface. Down at 30 cm there is essentially no difference between temperature of night and day because the soil is so well insulated from the surface; it stays at about the average temperature of all the days and nights combined over the last few weeks. At about 1 meter depth, there is no difference between temperatures in winter and summer—the soil remains right at the yearly average without fluctuation.

These differences are all-important to plant roots and the small animals and microbes that live within the soil. At depth, the extremes of heat or cold are much less and survival is often easier. But in high latitudes where the average annual temperature is too low, below −3°C, the soil at depth always remains frozen, for it is never reached by the heat of the summer. Water that once trickled down into the soil forms a deep layer of ice, known as permafrost, that may stay in place for many thousands of years. Where there is permafrost, roots cannot penetrate and plants must make do with rooting into the surface layer above which at least thaws during the summer. In the far north, patches of trees in the tundra seem to promote the formation of permafrost in the soil underneath themselves. The freezing of the soil eventually kills the roots, causing the trees to die and give way to tundra again. Permafrost forms under these tree patches because, in the shade cast by the leaves and branches, there is no direct heating of the ground by sunshine in the peak of summer. The frozen sub-surface of the soil never thaws out, and that equals permafrost. This is despite the fact that the covering of trees absorbs sunlight and heats up the air above the ground in the warmer months, and warms the local and regional climate overall (see Chapter 5). This extra warming does not reach into the ground, however; at least not strongly enough to compensate for the lack of the intense direct heating of the sun that would be found on open tundra soil in summer.

4.1.2 Above the surface: the boundary layer and wind speed

If we now go upwards from the soil surface into the air above, there is another succession of microclimates. When wind blows across bare soil or vegetation, there is always some friction with the surface that slows the wind down. This slowing down causes the air just above the soil to form a relatively still layer known as the *boundary layer*. Within a few millimeters of the soil surface, the friction is severe enough that the air is almost static (Figure 4.1). Air molecules are jammed against the surface, and the molecules above them are jammed against the air molecules below, and so on. Moving up a few centimeters or tens of centimeters above the surface, the dragging influence of friction progressively lessens as the "traffic jam" of air molecules gets less severe, and there is a noticeable increase in average wind speed because of this. In fact, what with the decreasing friction from plants, trees, buildings, etc. the average wind speed keeps on increasing with higher altitudes, until it really tears past a mountain top. It is no coincidence that the strongest wind gust ever recorded was at the top of a mountain (372 km/hr at the summit of Mount Washington, USA). Even so, mountains are not always windy. Some days in the mountains will have hardly any breeze, when the weather favors calm conditions.

In a sense there is a succession of boundary layers, each on top of one another and with the air higher up moving faster. "Boundary layer" is really a relative term: it is a layer of slower moving air caused by being closer to a rough surface, below a faster moving one above that is less affected by the surface. The term "boundary layer" is used at many different scales in the field of climatology, and can be very confusing because different sub-disciplines each use the term in their own way at the scale they are most interested in.

The boundary layer fundamentally affects the heat balance at the surface and in the air above, up to the height of a few centimeters or a few meters. If sunlight is hitting the surface, being absorbed and heating the surface up, heat is being

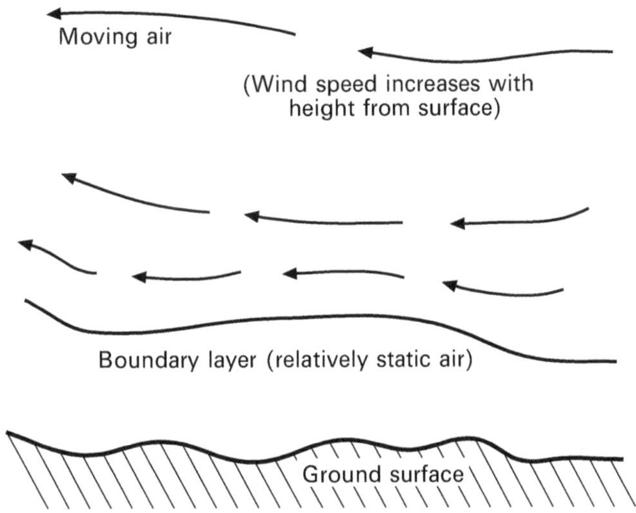

Figure 4.1. The boundary layer over a surface. *Source*: Author.

conducted gradually to the air above it. The relatively static air in the boundary layer will be able to heat up as it is close to the surface, and because it stays still and accumulates heat it will be quite a bit warmer than the mixed air in the wind above. As this boundary layer air is not being continually whisked away, the surface will not lose heat as fast either. In effect, the warmed boundary layer air acts like a blanket over the surface. The thicker the blanket, the warmer the surface can become. If the surface below the boundary layer air consists not of soil but of living leaves (as it does above a forest canopy, for instance), this extra warmth can be very important for their growth and survival. In a cold climate, there may be selection on the plants to maximize the thickness and the stillness of the boundary layer. In a hot climate, on the other hand, the plants may be selected to disperse the boundary layer, to prevent the leaves from overheating.

So, in a layer of still air the temperature can be several degrees higher than the mixed-in air just above it. This can make a lot of difference to the suitability of the local environment for particular plants and animals. For instance, in a tundra or high mountain environment, at the very edge of existence for plants, this small amount of shelter can determine whether plants can survive or not. On the upper parts of mountains, with strong winds and short grassy vegetation, a local boundary layer can make a big difference to the temperature the plants experience. If a spot is sheltered—for instance, between rocks or in a little hollow—the wind speed is also lower; there is a small space of static air with almost no wind movement. On a mountain slope in the mid or low latitudes, the intense sunlight can deliver a lot of energy directly to the surface. If the shelter of a hollow prevents this heat from escaping to the cold air above, it can become much warmer and types of plants that require more warmth are able to survive.

By making their own boundary layer climate, plants can turn it to their advantage. The upper limit to where trees can grow on a mountain—the treeline—occurs below a critical temperature where the advantage shifts from trees towards shrubs or grasses. Trees themselves standing packed together create a layer of relatively still air amongst them that can trap heat, but there comes a limit up on a high mountain slope at which this heat-trapping effect is no longer quite enough for trees to form a dense canopy. In a looser canopy, much of the heat-trapping effect collapses and suddenly beyond this point the trees are left out in the cold. This effect helps to produce the sudden transition in vegetation that is often seen at a certain altitude up on many mountains.

Often, right above the treeline on a mountain, dense woody shrubs take over. It is thought that shrubs can thrive at mountain temperatures too cold for trees because they can create a strong boundary layer against the wind among their tightly packed branches. Wind cannot blow between the branches, so the sun's direct heat is not carried away as fast, and their leaves can thrive in the warmer temperatures of the trapped air (Figure 4.2). Trees, by contrast, have a much looser growth form; so, if they are standing out on their own the wind can blow straight through their branches and carry away the sun's heat. Shrubs—with their heat-trapping growth form—can keep their leaves as much as 19°C warmer than the trees, making all the difference between success and failure in the high mountains.

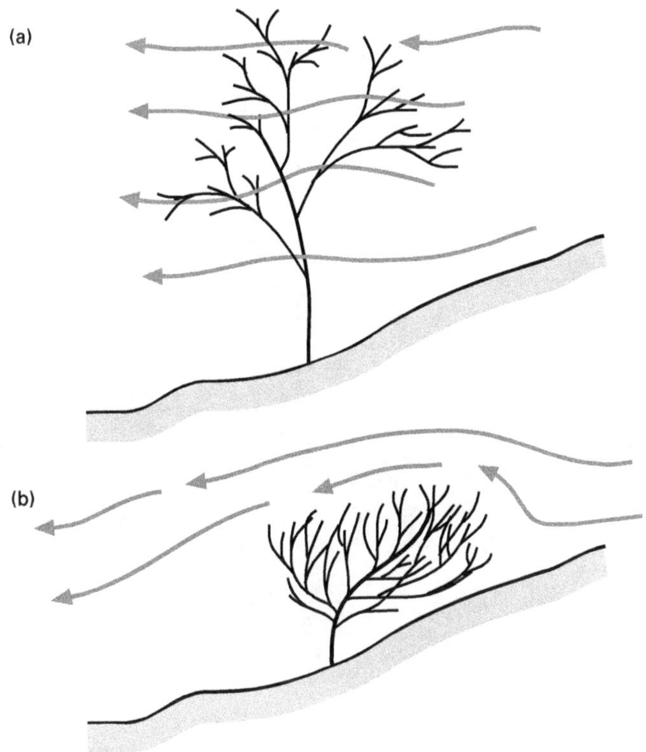

Figure 4.2. Shrubs trap more heat amongst their branches than trees do, because the wind cannot blow between the tightly packed branches of a shrub. *Source*: Author.

Higher even than shrubs can grow on a mountain is the "alpine" zone of cushion plants (Figure 4.3*). These exquisite little plants, from many different plant families in mountains around the world, form a little dense tussock of short stems and tiny leaves. Many of them look at first sight like cushions of moss, but they are flowering plants—often producing a flush of pretty flowers on their surface in the summer. The cushion plant growth form seems to be adapted to a version of the same trick that mountain shrubs use. A cushion plant, which needs all the heat it can get, creates a miniature zone of static air in the small gaps down between its tightly packed leaves. Leaves within the tussock are heated directly by the sun, and because the wind cannot blow between them everything within the tussock stays warmer. The plant is able to photosynthesize, grow and reproduce in an extreme environment by creating its own miniature boundary layer and microclimate amongst the leaves. Measurements show that on sunny days in the mountains, the leaf temperature of these cushion plants is often 10 to 20°C higher than the air immediately above. One reason why such alpine cushion plants are difficult to grow in sunny, warm lowland climates is that they are so good at trapping heat. They essentially fry themselves when ambient temperatures are already warm, raising their own leaf temperatures to levels that would also kill any lowland plant.

* See also color section.

Figure 4.3. An alpine cushion plant, *Silene exscapa*. The growth form of cushion plants maximizes trapping of heat in the cold high mountain environment. *Source*: Christian Koerner.

Many cushion plants use an additional trick to trap heat: above the dense cushion of leaves is a layer of hairs—transparent, and matted. These act like a little greenhouse, letting in sunlight and trapping warmed air underneath because it is not carried away by convection or by the breeze. This miniature greenhouse significantly increases the temperature of the leaves underneath, presumably resulting in more photosynthesis and better growth.

4.1.3 Roughness and turbulence

Although an uneven surface creates a boundary layer by slowing the air down, it can actually help set the air just above the boundary layer in motion by breaking up the smooth flow of the wind. The surface of a forest canopy, with lumpy tree crowns and gaps between them (Figure 4.4*), can send rolling eddies high up into the air above. This turbulent zone created by the canopy often reaches up to several times the height of the trees themselves. A more miniature turbulent layer will also be created above scrub vegetation when the wind blows across open ground between the bushes and then jams against their leaves and branches. Generally, whatever the height of the biggest plants in the ecosystem, the rolling turbulence that they create will extend for at least twice their own height into the atmosphere above.

Figure 4.4. The lumpy, uneven tree crowns of tropical forest create turbulence in the air that flows over them, Perak, Malaysia. *Source*: Author.

The turbulent microclimate created by air blowing over uneven vegetation surfaces also helps to propel heat and moisture higher up into the atmosphere, altering the temperature on the ground and feeding broader scale climate processes. In Chapters 5 and 6 we will see various case studies where changes in vegetation roughness seem to affect climate quite noticeably.

4.1.4 Microclimates of a forest canopy

The canopy and understory of a forest are like two different worlds, one hot and illuminated by blinding sunlight, the other dark, moist and cool. Parts of a large forest tree can extend all the way between these two worlds, and trees will often spend their early years in the deep shade before pushing up into the light above. Both the canopy and the understory microclimates present their own distinct challenges, and the plants need adaptations to meet these.

It is remarkable how hot the surface of a temperate or tropical forest canopy can become on a sunny summer's day, with leaf temperatures exceeding 45°C. In tropical rainforests, although it is cloudy and humid much of the time, a few sunny hours are enough to dry out the air at the top of the canopy and really bake the leaves.

It is critical that a leaf exposed to strong sunlight keeps itself cool enough to avoid being killed by heat. A leaf can lose heat very effectively by evaporating water brought up by the tree from its roots; the heat is taken up into the *latent heat of evaporation*, vanishing into water vapor in the surrounding air—it is the same principle by which sweating cools the human body. Evaporation from the leaves occurs mostly through tiny pores known as stomata, which are also used to let CO_2 into the leaf for photosynthesis (see Chapter 8). When the evaporation occurs through these stomata, ecologists call it "transpiration". As we shall see in the later chapters of this book, both the heat uptake and the supply of water to the atmosphere by transpiration are also important in shaping the regional and global climate.

Slowing down heat loss by transpiration presents a dilemma for the plant. On one hand, if its stomata are open and it is transpiring, a leaf can keep cool. However, keeping cool in this way gets through a lot of water. If the leaves "spend" too much water, there is a risk that eventually the whole tree will die of drought because its roots cannot keep up with the rate of loss. Even if there is plenty of water around the tree's roots, the afternoon sun can evaporate it from leaves faster than the tree can supply it through its network of vessels. If water is indeed limiting, the leaves will shut their stomata to conserve it. Tropical forest leaves in sun-lit microclimates also have a thick waxy layer, to help cut down on evaporation when water is in short supply.

If leaves close their stomatal pores and swelter, they risk being damaged by heat. It is thought that certain chemicals which are naturally present in leaves, such as isoprene, may help to protect their cells against heat damage in situations where they cannot evaporate enough water to keep cool. A breeze over the forest canopy will always help the leaves to lose heat even without any transpiration going on, and the faster the wind blows the better the leaves will be able to cool. The size and shape of leaves can also be important in avoiding heat damage. A big leaf is at all the more risk of overheating than a small leaf, because it creates a wider, thicker boundary layer that resists the cooling effect of the breeze. These sorts of problems are thought to limit the size that leaves of canopy trees can reach without suffering too much water loss or heat damage. The only exceptions are big-leaved tropical "weed trees" such as *Macaranga*, that can have leaves 50 cm across. They seem to keep themselves cool by sucking up and transpiring water at a high rate. Perhaps because of the risks of overheating, in temperate trees the "sun leaves" (see below) exposed at the top of the canopy tend to be smaller than the "shade leaves" hidden down below, even on the same tree.

The most intense aridity in the forest is likely to be felt by smaller plants that grow perched on the branches of the big trees: the epiphytes. In tropical and temperate forests where there is high rainfall and high humidity year-round, these plants are able to establish themselves and grow even without any soil to provide a regular water supply. But, because they are isolated from the ground below, and only rooting into a small pocket of debris accumulated on the branches, epiphytes are at the mercy of minor interruptions in the supply of water from above. When it has not rained for a while, epiphytes up in the canopy can only sit tight, either tolerating dehydration of their leaves or holding in water by preventing evaporation from their

Figure 4.5. The eery gloom of tropical montane forest shrouded by cloud. Cloud water condensing on the leaves constantly drips down from the canopy, watering the trees. Photo: Genting Highlands, Malaysia. *Source*: Yang Ren Kit.

waxy leaves. Some epiphytes live rather like cacti within the rainforest, having thick fleshy leaves that store water for times of drought. One very important group of epiphytes in the American tropics, the bromeliads, tends to accumulate a pool of rainwater in the center of a rosette of leaves. They are thought to be able to draw upon this water reserve to keep themselves alive when it has not rained for a while. Other bromeliads are able to tolerate drying out and then revive and photosynthesize each time it rains. One well-known example is Spanish moss (*Tillandsia*) which festoons trees in the Deep South of the USA.

Sometimes trees can in effect water themselves. High up on many tropical mountains, around 2,000 m above sea level, are "cloud forests" which thrive in the layer where clouds tend to hit the mountain slopes (Figure 4.5*). The cloud droplets condense on leaves in the forest canopy and drip to the ground. Walking under the trees when clouds shroud the mountain, cold water condensed from the fog continuously drips onto the back of one's neck. Often this contributes 30% or more of the water that reaches the trees' roots. Similarly, in northern California where coastal fogs constantly roll in off the sea, the water captured from fog droplets plays an important part in the survival of the giant redwoods (*Sequoia sempervirens*).

4.1.5 Under the canopy

In the cooler forest understory, out of the direct sun, overheating is not a problem and leaves can grow bigger than at the top of the canopy. Many of the types of plants that grow down near the floor of the forest have large plate-like leaves 30 cm or more across; undivided leaves this size are hardly ever seen up in the forest canopy.

On the forest floor, the overwhelming impression is of stillness and quiet. The calls of birds up in the canopy are muffled by the leaves. There may be barely any breeze even as branches of the trees far above wave about in the wind. Friction with the leaves and branches of the tree crowns slows down the wind, so only the uppermost parts of the canopy get the full force of it. The wind speed tends to be at its least in the lower part of the canopy where the high density of leaves blocks movement of air. Down below on the more open forest floor, a light breeze may sometimes blow through between the trunks of the trees.

While overheating is not a problem on the forest floor, and dehydration is much reduced, the plants that grow there have their own problems to cope with. In a really dense forest—such as primary tropical rainforest or under a dark boreal conifer forest—more than 97% of the daylight may be filtered out by the canopy. The light levels are so low that it can be difficult to get a good photo without using a flash. In this twilight, photosynthesis can only be carried out slowly, and there is just a sparse layer of plants on the forest floor, many of them barely making a living.

The spectrum as well as the amount of the light is very different at the forest floor compared with the canopy. There is almost no UV, and blue and red light have been filtered out by chlorophyll so what is left is mostly green. The ratio of red to far red light is also shifted by the sunlight passing through leaves above, with chlorophyll in the photosynthetic cells absorbing most of the red.

Since there is not much photosynthesis going on under the canopy, CO_2 is not used up quickly. Yet there is plenty of decay of fallen leaves and branches, pushing CO_2 into the air. So, CO_2 levels near the forest floor will often be several times higher than they are in the earth's atmosphere in general. In contrast, up in the rapidly photosynthesizing canopy, CO_2 levels can be much lower than the "average" of the broader atmosphere. Effectively, within the forest there is a "carbon pump", taking CO_2 by photosynthesis and pulling it down (as dead leaves and other material) to decay on the forest floor.

Out of the drying influence of the sun, under a dense canopy of leaves the relative humidity can be much higher than above the canopy. Rain that has fallen and dampened the ground also adds greatly to the humidity. Where a stream passes through the forest, the evaporation of water from it tends to give even higher humidity and cooler temperatures to the nearby areas of forest floor.

Perhaps 50 meters above, the intense heating of the upper canopy by the sun tends to form a stable layer of air—less dense because it is warmer—floating within the canopy during the day. This stable cap of warmer air helps to seal off the forest floor from the world outside. CO_2 gas released from respiration tends to build up during the day, until the inversion layer disperses in the evening.

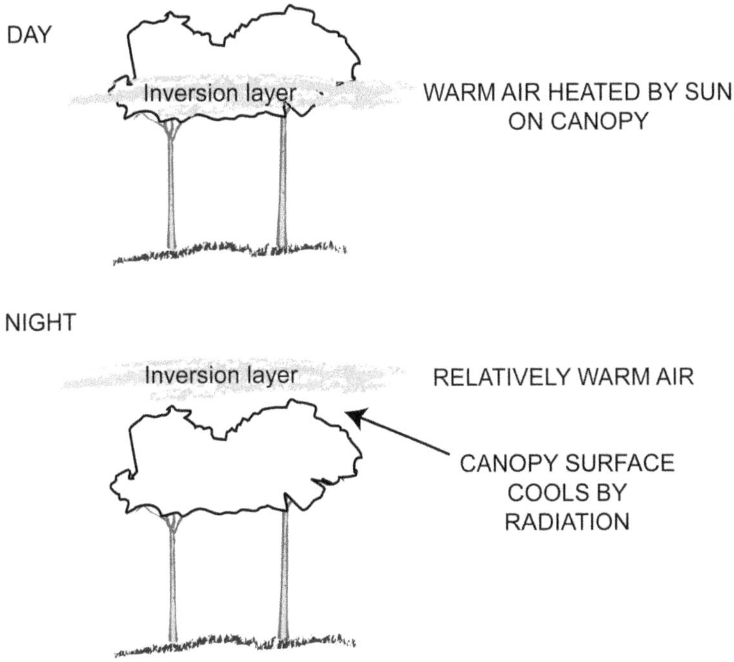

Figure 4.6. During the day, heating of the canopy leaves by the sun gives an "inversion layer" of warm air floating within it (*top*). After the sun goes down, the canopy loses heat to the night sky, and the air floating just above it is left relatively warm, as it fails to cool down as fast.

However, the forest's interior is not totally insulated from the air above. To some extent, turbulence created by wind blowing over the upper surface of the canopy drags moist air up from within the forest, and spins dry hot air down inside. Convection rising from the hot leaves of the canopy also has a similar effect by sucking air up from below. This amount of air exchange with the surface tends to limit how much the forest can "make" its own interior climate by shade and by evaporation of water.

After sunset, air movement above the canopy tends to settle down. As the surface of the canopy cools off—radiating to the night sky—another "inversion layer" may now form above it as the daytime air stays relatively warm (Figure 4.6). The evening chorus of monkeys and other creatures in tropical rainforests seems to take advantage of the boundary of this inversion layer to bounce sound sideways across the canopy, allowing them to send their signals much farther than they would be able to during the day. It is called an "inversion layer" because it turns upside-down the normal pattern of temperature with altitude, which is for things to get cooler higher up.

If part of the forest has been cut, air blowing to the forest interior from the open ground at its edge is likely to have a very different temperature and humidity. The air entering the forest understory from recently cleared land has been heated by the full force of the sun, and there is not the dense mass of leaves to evaporate water and keep the air cool and moist. This "edge effect" of dry hot air blowing in can alter the ecology of the forest floor and the lower parts of the canopy, with its influence extending some tens of meters into the forest. The presence of edges in both tropical and mid-latitude forests has been found to have noticeable effects on the types of understory plants that will grow there. Close to the forest edge, the plants that require low light levels and high humidity (see below) are replaced by tougher species that can cope with intense sunlight and dehydration.

Even with the edge effect diluting the influence of the forest, the contrast in temperature and humidity is immediately apparent to a casual observer stepping from underneath trees to an open area in direct sunlight. Studies of the microclimates of small grassy clearings around 10 or 20 metres wide have shown that they are around 2–4°C hotter during the day than the understory of undisturbed tropical forest, even without the direct heating effect of sun on the leaf surfaces which adds much more to the heat loading on plants. Extensive clearings of several hectares or more can get warmer still; generally speaking, the bigger the clearing the hotter it gets. The increased air temperature is due to the sparser leaf cover of the clearing: fewer leaves mean less evaporation of water to cool the air. But at night the situation is reversed. Temperatures stay slightly warmer under the closed forest than out in the clearing, because the dense canopy of leaves above blocks the loss of heat to the night sky. In the clearing there is no such blanket of leaves overhead, so infra-red is radiated out to the sky more easily.

4.1.6 Big plants "make" the microclimates of smaller plants

The plants that live on the forest floor—at low light levels, milder temperatures and higher humidity—are specialized to a microclimate made for them by the canopy trees that absorb most of the sunlight. Their photosynthetic chemistry is specialized to low light levels and they cannot cope with direct sunlight. These forest floor plants tend to have soft leaves, because leaves underneath the canopy have no need to be "tough"—they are not blown about by the wind, nor are they dehydrated in direct sunlight. An example of one of these forest floor plants is the African violet (*Saintpaulia*), a common house plant which requires shade. As many houseplant owners know all too well, it dies quickly when exposed to direct sunshine.

Some forest floor plants have peculiar adaptations to help them gather as much as possible of the light that falls upon them. Certain herbaceous plants—such as the South-East Asian vine spike moss (*Selaginella willdenowii*) and some species of *Begonia* (Figure 4.7*)—have a bluish sheen (known as iridescence) to their leaves. This is caused by little silica beads within the epidermis of the leaf. Experiments have suggested that these beads help the leaf to focus in light from a range of directions, sending it straight into the photosynthetic cells below. In *Selaginella* each cell underneath a silica bead has a single large chloroplast which seems to be precisely located to receive this focused beam of light.

The leaves at the top of a tree also make the microclimate for the leaves below them. Even on the same tree, leaves that are out in full sunlight develop slightly differently from those in the shaded branches down below. The "sun leaves" are thicker with more layers of photosynthetic cells packed in, to take advantage of the abundant light. The lacquer-like cuticle on the upper surface of a sun leaf also tends

Figure 4.7. This species of *Begonia* lives in the understory of mountain rainforests in South-East Asia. The bluish metallic "sheen" of many species of rainforest understory plants is thought to come from the refractive effect of silica beads which help to gather in light for the leaves. *Source*: Author.

to be thicker, to help reduce unnecessary evaporation. On a sun leaf there are more stomata—the pores which open to let CO_2 in—so that the leaf can take advantage of high light levels to bring in more CO_2 for photosynthesis when it has enough water. As soon as evaporation through the stomata becomes too intense and the leaf is in danger of dehydrating, the stomata are clamped shut and the leaf relies on its cuticle to prevent further water loss.

The chemistry and color of sun leaves also tends to be different from shade leaves. Shade leaves tend to be a darker green because they are richer in a particular dark green form of chlorophyll (chlorophyll *b*) that is good at harvesting light at low intensities and at the wavelengths that get past the filtering of leaves above. Sun leaves have more of the chlorophyll *a* form which exploits high light intensities more effectively. The upper epidermis of sun leaves is also packed with natural sunscreen compounds such as flavenoids which absorb most UV light and prevent it from damaging the sensitive photosynthetic cells below. Just putting a shade-grown tree seedling out into direct sunlight shows how important this protection is: in a few days the shade-grown leaves are bleached and useless.

Tree seedlings often survive for years in the shaded forest floor environment. Depending on how much light they are getting, they may either stay more or less the same size, or slowly grow up into the canopy. These young trees from the forest floor can go through different phases in their life, with physiological adaptations to different light levels. For example, many of Australia's *Eucalyptus* trees have an "early" phase with an entirely different leaf form, suited to growing at low light levels within the darker forest interior. Typically, the juvenile leaves of such eucalypts form disk-like pairs with the stem in the center, while the adult leaves are long and strap-shaped. It is thought that the disk-like leaves—arranged along the stem like a kebab—are good for harvesting light coming down from a small gap in the canopy above; it helps to keep the photosynthetic area "all lined up" within a shaft of sunlight as the seedling grows its stem up to follow the light.

Often a young tree on the forest floor will only really be able to start growing fast when a bigger tree—or a large branch—falls from the canopy to give a patch of sunlight that illuminates its leaves. This is the turning point that gives the young tree enough energy to fix enough carbon to lay down wood and grow tall, rather than merely surviving.

The subtle range of opportunities provided by microclimates is thought to help maintain the species diversity of forests and other plant communities. A tropical rainforest can have more than two hundred species of trees packed into a hectare, and in ecological terms it is difficult to explain how they can all manage to exist side-by-side. Simple ecological theory suggests that eventually just one species that can compete more effectively should increase its numbers and push the rest out, so that it dominates the forest. Yet, obviously this does not happen. Ecologists suspect that part of the reason such exclusion of species does not happen is that small differences in light level, as well as soil texture and nutrient levels, determine which tree species gets established in any particular spot on the forest floor. It is thought that for trees in particular a critical stage which determines whether a species grows in any one spot is its early growth as a seedling and small sapling. Each species might be adapted when it is a seedling to a narrow range of light intensities, or light of particular wavelengths or angles. If it finds itself in its *forte*, it will out-compete seedlings of other species. Once that critical seedling stage is passed and the tree has established itself, it is essentially guaranteed a place in the forest. Although this is quite a compelling theory, there is still only limited evidence that this sort of specialization on the forest floor is important in the competition and survival of forest trees.

4.1.7 The importance of sun angle

Just as sun angle makes the difference overall between temperatures at different latitudes of the earth, it makes a significant difference on a local scale too. If a slope is angled towards the sun when the sun is low in the sky, it gets more of a full beam and so the surface temperature of soil or leaves (and the air just above) will be warmer. On a slope that is in the "wrong" direction relative to the sun, much of the day is spent in shadow or being sunlit at an angle, so it will be colder than if it had been on the flat.

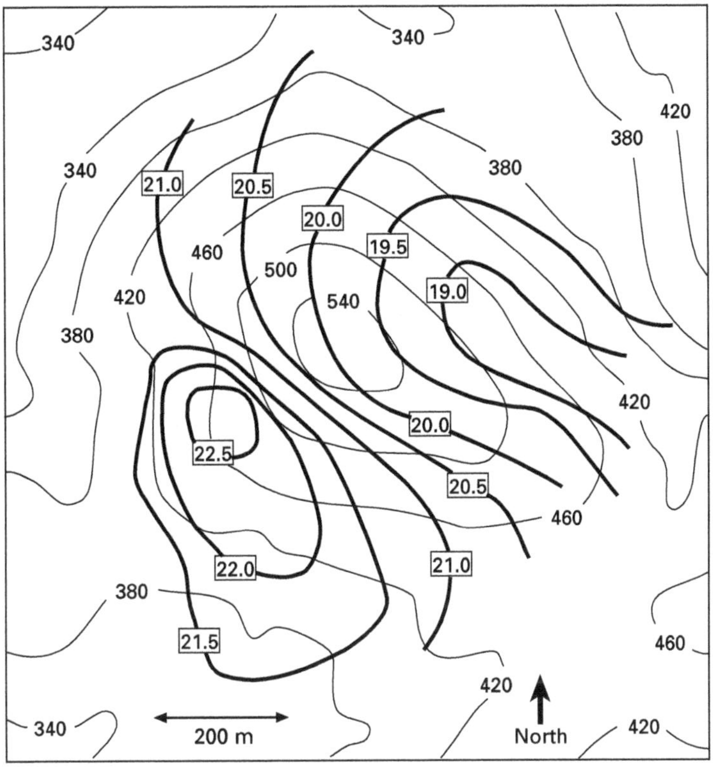

Figure 4.8. Distribution of temperatures on a sunny summer's day on a hill in Massachusetts, USA. The more southerly-facing slope has warmer temperatures than the opposite slope facing northwards. Elevation from 340 m to 540 m by 40 m. Temperature from 19.0°C to 22.5°C by 0.5°C. After Bonan.

On the equator, the sun travels a path right overhead and does not shine more on either a southern or a northern slope: in fact, it shines slightly more on east and west-facing slopes which catch additional energy from the sun around sunrise and sunset. At higher latitudes in the southern hemisphere, the sun tends to be in the northern half of the sky, so a north-facing slope will be warmer. In the northern hemisphere, south-facing slopes are warmest because the sun stays mostly in the southern half of the sky. For example, one study during a summer's day on a hill in Massachusetts found that the maximum temperature reached during the day was 3.5°C warmer on the south-facing slope than on the north-facing slope (Figure 4.8). In fact, in the mid-latitudes the sun does wander slightly into the "other" half of the sky during the early and late parts of the day during the summer; but always more energy is received from the south in the northern hemisphere, and from the north in the southern hemisphere.

Such local slope angle effects can make a difference to the ecology. A study on flowering times in a wooded valley in Indiana found that several species of wild-flowers bloomed about a week earlier on a south-facing slope than a north-facing one. This is because plants often need to be exposed to a certain amount of heat during the season before they will flower; on the warmer sunlit slope this required "heat sum" was reached sooner.

The differences with aspect tend to be most striking for types of plants which are right at the edge of their ranges, and barely able to survive in the local climate. Sometimes, they are warmer-climate plants that are at the poleward edge of their range. For example, on sand dunes on the coast of eastern England there grows a type of wild lettuce known as prickly lettuce (*Lactuca virosa*) which is at the northern edge of its distribution range in Europe. In England it will grow only on the south-facing slopes of dunes, gathering just enough energy for itself to grow and set seed. On coastlines farther south in Europe (e.g., most of France) prickly lettuce grows on both the north and south sides of dunes because the microclimate is warm enough even on the north sides, given the generally warmer air temperatures. Similarly, the stemless thistle (*Onopordum acaulon*) only grows on the south side of hills at the northern edge of its range in Yorkshire, northern England. In southern England, there is enough warmth for it to grow on both the northern and southern sides of hills.

As well as temperature, the severity of aridity differs between north and south-facing slopes. The stronger the beam of sunlight, the droughtier the conditions as more water is evaporated. In semi-arid areas of southern Europe, many "north European" plant species requiring cool damp climates only survive on north-facing slopes. I remember once walking on the steep northward-facing slope of a hill in Provence in the south of France. It was covered in beech (*Fagus sylvatica*) forest and in the shade and dampness of the understory I could for all the world have been in my rainy native land of England—a strangely comforting form of *déjà vu*. Yet, when I topped the brow of the hill to the southern side, in the space of a few meters I was back into hot, dry air, surrounded by typical open Mediterranean scrub. The influence of sun angle had made all the difference between survival of deciduous forest, and its replacement by oily brush that burns every few years.

The difference in moisture availability with aspect can even be noticeable on a more miniature scale on tree trunks; the northern side of a tree trunk in northwestern Europe tends to have a lot more mosses growing on it than the drier, hotter south-facing side.

4.1.8 Bumps and hollows in the landscape have their own microclimate

As I mentioned above, a group of rocks that provides shelter can allow a pocket of still air to form on an exposed mountain slope. Small bowl-shaped hollows in the landscape, a few meters or even just a few centimeters across, can also act as solar energy collectors (like a parabolic satellite dish which concentrates the signal into the middle), gathering heat into the center to give a warmer microclimate. In tundra—the grassy or shrubby vegetation which exists in very cold Arctic and alpine environments (Chapter 2)—the extra heat concentrated in small hollows in the landscape is crucial to the growth of certain plants, and the survival of certain species of insects. This sort of heat-concentrating effect is also very common among the bowl-shaped hollows in coastal sand dunes at lower latitudes too; temperatures can be many degrees higher on a sunny day in the hollow between several dunes than on the tops of the dunes. In an interesting variant of this heat-gathering effect, the white flowers of the "Arctic rose" *Dryas* (a widespread plant of the Arctic) also act as parabolic heat collectors concentrating light into the center of the flower. This warms up the center of the flower increasing the chances that the pollen will grow and fertilize the seeds. It also apparently warms up the bees that visit the flowers, speeding up their activity and helping them to carry pollen between the plants more efficiently.

Where there is a hollow in the landscape (caused by a small valley, or geological features such as a kettle hole or sink hole), it may have warmer sunny days because of the concentration of the sun's heat into the center, sheltered from the breeze. However, it can also have more severe winters. I lived for a while in the Appalachian fold country of east Tennessee, where small, cosy farmed valleys known as "hollers" make up some of the most beautiful countryside I have seen anywhere. My house stood perched on the somewhat cooler sloping side of a holler, and I remember how the short walk down the track to the center of the valley on a summer's day could seem like entering a furnace.

The frosts at the bottom of the holler were also more extreme; the old lady whose house was on the valley floor bemoaned the fact that her tender spring vegetables and flowers would often be hit by late frosts, while those in other people's gardens a few yards higher on the slopes survived intact. In a study of the landscape of part of the Appalachian fold country in Virginia, the date of the last frost in spring was almost a month later in small valleys that formed "frost hollows", compared with nearby areas up on the ridges. Frost hollows occur because the cold air that forms at the ground surface on the valley slopes and ridge tops during a cold night (as the ground loses infra-red radiation to space) is heavy and drains downslope as a fluid. As it enters a valley bottom, the cold air also tends to pool up to a certain level, producing a sharp transition between frosty air below and the warmer air above (Figure 4.9). One can sometimes see a "burn level" in small valleys on leafing-out deciduous trees or ferns where frosty air accumulated in a hollow, like water filling a pond; I have seen the

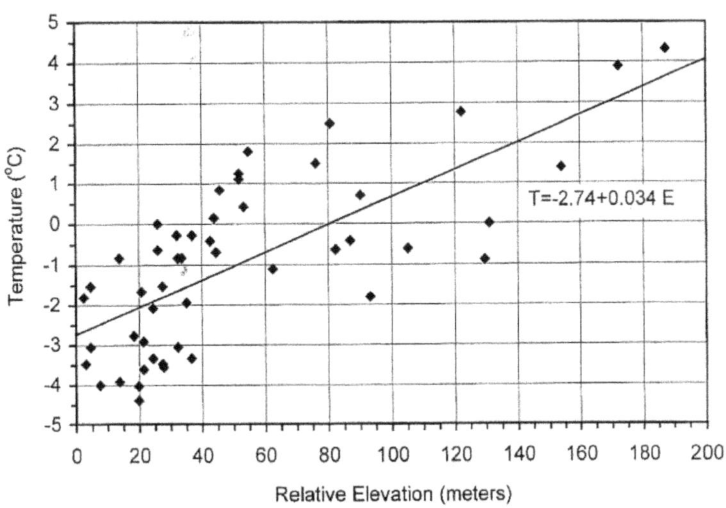

Figure 4.9. Temperature profile against height on a cold spring morning in a Pennsylvania valley that acts as a frost hollow. Sub-freezing temperatures are only present in the lowermost parts of the valley. *Source*: Bonan.

transition from no damage at all to every leaf killed in less than 50 cm vertically, all around the edge of a small valley.

Drainage of cold air also leads to more transient patterns in microclimate. On bare Mediterranean hillsides after a hot sunny day, the air near the top of a hill begins to cool after the sun goes down. Being denser it drains downslope, often forming invisible "rivers" that flow down dry stream valleys and along goat paths. Walking through the garrigue scrub in the evening one often passes through these cool rivers of air and then steps back into warm air within the space of few meters. Local people sometimes explain these patches of air as being the spirits of the dead that wander the hills, chilling or warming the people they encounter during their journeys.

Such microclimate-scale differences brought about by cold air drainage are a miniature version of the same process that can occur in large valleys, producing the mid-elevation warm belt mentioned in Chapter 1. Often, the only distinction between microclimates and mesoclimates is a matter of scale, not the fundamental processes involved.

4.1.9 Life within rocks: endolithic lichens and algae

Even in the coldest places on earth there is life, and favorable microclimates make it possible. For example, in the Antarctic mountains, where the air never gets above freezing, a north-facing rock surface on a sunny day can get much warmer. Although temperatures right on the rock surface can rise above freezing in the direct sunlight, the very dry air and rapid fluctuations in temperature prevent any form of life from growing there. Over just a few minutes, the temperature can rise above freezing and dip back down below, and this sort of instability seems to be too much for even the hardiest organisms to cope with. However, conditions are warmer and more stable a few millimeters down between the grains of a sandstone rock. The grains are transparent quartz, which allows sunlight in but provides insulation from the chilling air outside. Temperatures in the tiny gaps between these quartz grains can get much warmer than the surrounding air: to around 10°C, which is some 20°C warmer than the air outside ever gets. It is within these miniature greenhouses that specialized "endolithic" (meaning "within-rock") lichens live, photosynthesizing and growing during the few weeks each year that are warm enough, and perhaps living for centuries in a mainly dormant form.

In another of the earth's most extreme environments, lichens can also survive within rocks due to the right microclimates. Death Valley in California holds the record as the hottest place on earth. Sandstone rock faces baked by the sun and parched by lack of rain seem an unlikely place to find life, and indeed the rock surface itself has nothing growing on it. Yet, a few millimeters deep inside the rock temperature extremes are lessened and there is moisture that trickled in between the rock grains when it last rained. Endolithic lichens survive here, harvesting sunlight that reaches through the quartz grains of the rock.

4.1.10 Plants creating their own microclimate

We have considered already how forest canopies create a special environment underneath themselves, how the Arctic rose keeps its flowers warm and how cushion plants trap extra heat for themselves. There are other instances too of plants making their own microclimate.

4.1.11 Dark colors

Many algae and lichens growing on rocks in cold climates are dark-colored, even black. This helps them absorb the visible wavelengths that contain most energy from the sun (i.e., they have low albedo). It has been suggested that this dark color is a special feature evolved to cope with cold climates: the extra heating that results from this might allow better metabolism and growth. Thus, the plant modifies its own microclimate to make itself warmer. It is reasonable to suppose that in the cold, being dark might benefit the plant. However, it is not clear that dark colors have been specifically selected for in cold climates—even in the tropics algae and lichens living on rock surfaces are often dark-colored, perhaps accumulating the pigment as a defense against damaging UV light.

4.1.12 Protection against freezing

Fleshy succulent plants known as tree groundsels (*Senecio*) and giant lobelias (*Lobelia*) living on the tops of high mountains in the east African tropics have to cope with frost at night, even though the days are above-freezing. These plants seem to protect their soft, sensitive growing tips by accumulating a little "basin" of water in a rosette of leaves; this covers the growing tip with water that has heated up during the sunny days, protecting it from frost each night.

4.1.13 Internal heating

There is a widespread family of plants known as the arums—containing many thousands of species—which create heat in the flowering structures by burning up sugars, to vaporize certain chemicals that attract flies to come and pollinate the flowers. The most proficient of these self-heating plants seems to be the skunk cabbage (*Symplocarpus foetidus*), which grows in swamps in North America, creating quite a stink with the chemicals that it evaporates.The amount of heat released by a skunk cabbage can raise the temperature inside its flowering head to 35°C (almost as warm as the human body temperature), even when the air temperature is below freezing. Early in spring, when snow is still on the ground, the skunk cabbage flower heads are able to melt the snow around them, poke up and flower before any other species.

4.1.14 Volatiles from leaves

Volatile chemicals are abundant in desert scrub and Mediterranean vegetation. The scented oils evaporating from the leaves of evergreen Mediterranean scrub (such as garrigue, Chapter 2) can make a hillside smell like one big *pot pourri*, and the sagebrush of the American southwest can also smell rather strong. It is generally thought that these compounds have a protective role in making the plants distasteful or indigestible to grazers. However, some ecologists have suggested that the plants use them for an additional purpose: being kept warmer and frost-free because of the "greenhouse" heat-trapping properties of these chemicals (which strongly absorb infra-red light). However, atmospheric physicists calculate that the "heat-trapping" effect of these chemicals is probably not strong enough to make any significant difference.

4.1.15 Utilization of microclimates in agriculture

Many of the aspects of microclimates that affect plant ecology also apply to agriculture. Good farmers plan their planting to avoid unfavorable microclimates—avoiding frost pockets for sensitive crops, and allowing for the effect of aspect on temperature or water balance. They can also try to make new microclimates which will favor the plants they are growing. Shelter belts of planted trees or bushes create a drag that slows down the drying or cooling winds that blow across farmland. The effect of a shelter belt of trees on wind speed can extend across the field as far as 20 to 30 times the height of the trees.

Greenhouses and other covers in agriculture are all about forming a microclimate. It used to be thought that greenhouses worked mainly by letting visible light in and preventing infra-red light from leaving, because glass strongly absorbs infra-red light. It is now known that in fact greenhouses mostly just heat up because (having a roof on the top) they prevent heated air from rising away into the atmosphere by convection, as it would outside. However the old idea has left a legacy in science in the term "greenhouse effect" which refers to the way certain gases in the atmosphere trap heat by letting visible light in but blocking infra-red light on its way out.

The shade and coolness of the forest understory is artificially created in agroforestry, a mainly tropical practice of growing crop plants under and between rows of trees. Many plants, such as cocoa bushes, do especially well when shaded like this, because they are descended from wild plants which naturally grow in the forest understory.

Citrus growers in California or Florida sometimes resort to putting radiant heaters in the open air of the orchards if a frost is threatening; it is a very inefficient use of fossil fuel energy, but often works just well enough to keep frost off the plants. In another trick to save crops, rice farmers in northern Japan often raise the water level in their rice fields to flood the plants with a layer of insulating water to protect them against a sudden cold spell, during the most sensitive period when the rice is flowering and the grains are beginning to form.

4.2 FROM MICROCLIMATES TO MACROCLIMATES

The same factors which affect microclimates, including the plants themselves, translate into larger effects on the heat balance and moisture balance of the earth's surface. In many respects, the macroclimate (over hundreds of kilometers) is the sum total of all the microclimates across broad areas. For example, the local effect of a boreal forest canopy heating up in the sun because it has shed the snow from its branches can make a great difference to regional climate if it occurs on a broad enough scale. When an individual leaf in a rainforest canopy evaporates water and cools itself, this makes a contribution to the heat balance of the whole tree, the whole forest and the whole region. In its own tiny way it also ultimately helps to affect the distribution of heat and water vapor all around the world. So, if anything changes about the average shape of leaves, or size of trees, or the amount of bare ground around the world, this could all add up to a global change in climate.

The sort of way that changes in plant microclimates might scale up to alter global climates was neatly expressed by the "daisyworld" model of James Lovelock (see Box 4.1 and Figure 4.10).

Now imagine that instead the "sun" gets stronger, delivering too much solar energy and tending to overheat the planet. The dark plants will suffer by being

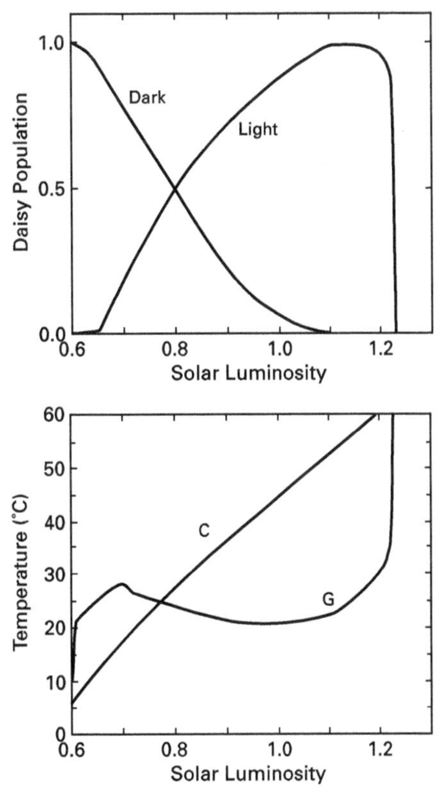

Figure 4.10. The daisyworld model of Lovelock illustrates how the microclimate effects of plants could scale up to global climates. There are black and white daisies which do better under hot and cool conditions, respectively. If the sun gets weaker or stronger, the abundance of the two types shifts according to which does better in their local microclimates (top graph). In the bottom graph, the line "C" shows how temperature would change without the daisies changing their abundance. Line "G" shows how in fact things might change if the white and dark daisies adjust their abundance to exploit the microclimatic conditions: the temperature of the whole planet is regulated by a process that is essentially controlled at the level of microclimates. After Schwartzman, from Lovelock.

overloaded with heat; they not only have to cope with the warm air temperatures but they are also absorbing a lot of sunlight which tends to heat them up even more. In this situation, the dark plants do not grow well and they get pushed out by the white plants that can keep themselves cool by reflecting back most of the sun's energy. As the white plants spread across the overheated planet, more and more of the solar radiation gets reflected back into space, and this cools the climate. Temperature is again brought down towards a more moderate level. It is as if the planet has a thermostat, regulating its temperature to prevent its climate from becoming too extreme.

Box 4.1 James Lovelock and Gaia

James Lovelock (b. 1919) is an independent scientist whose work has inspired a whole new way of thinking about the world. Of his many important contributions to science, probably the greatest has been the message that earth's environment is to a large extent controlled by life itself. Much of this book is about the ways that living organisms have seized control of climate and atmosphere. Although there have always been scientists who worked on the effects of life at the broad scale, in the last 35 years there has been an enormous expansion in this way of looking at the world. Many of the scientists who nowadays work on the effects of plants on climate or the carbon cycle attribute much of their inspiration to Lovelock's view that life has an integral role in controlling the global environment.

Lovelock honored the global system—with life at its center—with the name "Gaia" after the ancient Greek earth goddess. This choice of a label has proven controversial, and some scientists have even accused Lovelock of venturing into religious mysticism. Lovelock himself has said that the name Gaia was merely intended as an inspiring metaphor, but there is no doubt that his view of the earth system has gained a lot more attention because of his choice of this name.

Daisyworld is a hypothetical example that demonstrates a general principle: that living organisms acting in their own short-term interest on a local microclimatic scale (both responding to and changing their microclimate) might add up to very big global changes. The overall result can be a control mechanism that regulates the earth's climate. Although daisies have never changed the world, it is possible that things rather like this have actually happened in the past. Three or four billion years ago in the Precambrian, the first life on land was probably dark-colored layers and mats of algae and perhaps lichens. By altering the amount of sunlight absorbed at the earth's surface (the albedo), these plants may have brought about rather similar changes in heat balance of the world. It is thought that at that time the sun would have been fainter, and the earth in continual danger of freezing up. However, despite their possible stabilizing influence, gathering sunlight, there are signs of very severe and long-lasting ice ages before about 700 million years ago. In some of these cold phases the ice sheets spread down close to the equator, suggesting that almost the whole planet was iced up. The appearance of more complex land plants (with leaves, roots and suchlike) may have helped to impart more long-term stability on climate, keeping the earth within the band of temperatures suitable for maintaining life. However, such plants did not appear until around 400 million years ago, so they cannot be the reason in themselves why the great global ice ages ended. Nevertheless, from what we understand about the role of plant life in making climate (see Chapters 5 and 6) they may well have prevented such severe glaciations from occurring again since then.

So, in the more recent geological past, since large plants with leaves and roots first evolved, it is likely that the influence of plant microclimates in regulating broad-scale climates has become even more important. In the next couple of chapters (Chapters 5 and 6) we will explore some of these possible effects of plants on both regional and global climate.

5

The desert makes the desert: Climate feedbacks from the vegetation of arid zones

Even in the original state of the world before agriculture, forests occupied only a minority of the earth's land surface. Much of the world has dry climates that cannot support a tree cover, and their natural vegetation is open grassland, scrub or desert. There is no doubt that the location of these arid zones is largely dictated by broad-scale geography (as we saw in Chapter 1), with several factors tending to produce dry climates.

5.1 GEOGRAPHY MAKES DESERTS

The world's broad "desert belts" north and south of the equator result from the global circulation pattern (Chapter 1): equatorial air rises up into the atmosphere, heated by intense sunlight and loses its water vapor as sudden rainstorms. Eventually, this air comes back down hundreds of kilometers from the equator, and heats up as it is compressed, holding even more tightly onto what little water vapor remains within it. In such a situation, with dry air nearly always moving in from above, there is inevitably a more or less arid climate on the outer fringes of the tropics. The deserts of the Sahara, Arabia, central Asia, Australia and the USA owe their existence mainly to this process of dry air moving in from above. There are also other geographical factors that can help make a desert. Mountains can also block moist winds from the sea, forcing the air to rise, cool and drop its rain on their slopes—so that there is hardly any water vapor left to form rain on the inland side. This rain shadow effect helps to reinforce the dryness of the North American deserts, and the deserts of central Asia and Australia, combining with the descending equatorial circulation in a sort of "double whammy" of aridity.

Some other deserts are the result of cold upwelling seawater just off the coast; the cold water does not evaporate much water, and when the wind off the sea moves over hot land it holds more tightly on to what little water vapor it has. This has produced the coastal strip deserts of Peru and Namibia (Chapter 1).

5.2 BUT DESERTS MAKE THEMSELVES

In addition to all of the more traditional climatology, there is another factor whose importance is only now becoming understood. Deserts partly owe their existence to the fact that they themselves exist. The desert makes the desert, internally modifying its own climate so that less rain falls!

So the link from climate to vegetation, in Chapter 1, has been turned on its head. A fundamental fact of the earth system, that climate scientists are only now becoming fully aware of, is that vegetation can make the climate too. The mechanisms by which deserts reinforce their own dry climate was apparently first explored in the mid-1970s by Joseph Otterman, who represented his ideas in a landmark paper which inspired a whole new way of thinking. He took apart the basic physics of the local atmosphere and surface—the "mesoscale climate" that is built up from the microclimatic factors explained in Chapter 4—and he thought about what might happen if you changed the vegetation cover. One thing that he knew was important was the brightness of the surface, the albedo (roughly meaning "whiteness" in Latin).

Seen from above, green leaves look a lot darker than a bare soil surface. For example, if you look out of the window of a plane flying high above dry country, areas of dense tree and shrub cover look almost black, in contrast to the blinding brightness of patches of bare soil. The brightness of that bare soil is solar energy—sunlight—reflected straight back up into space. This is energy that might have gone into heating the surface if it had been absorbed, but instead it has been wasted. Thus, the lighter the surface, the less energy is absorbed and more is thrown right out into space. Table 5.1 shows a range of typical albedo values found for different surface types in the world.

Above bare soil with high albedo, the atmosphere is deprived of some of the warm air that would otherwise be rising up from the ground, so (other things being equal) it is cooler than it would be if it was darker. There is less convection to carry heat aloft, and the atmosphere is relatively calm and stable. The relatively shallow convection above high-albedo land tends to let air, coming in from above, descend and form a "cap" on the top, which supresses rain cloud formation. Normally, rain clouds tend to form where warm air keeps rising up from the surface carrying some water vapor and then begins cooling, causing water droplets to condense out to give clouds and then rain. Over a bare, bright surface, air tends to do the opposite thing—descending and heating as it does so. These are the sort of conditions which prevent any rain from falling (Figure 5.1). So, if we imagine two land surfaces that start off equally moist as one another, the darker low-albedo one is going to generate more rain for itself than the bright high-albedo one.

To anyone who has traveled a lot around the world, it is intuitively hard to believe that lack of vegetation cover would make the surface cooler. Arid areas of the world tend to get very hot, whereas the rainforest zones of the tropics have fairly mild temperatures most of the time. But, in fact, areas with plenty of dark vegetation are absorbing a lot more energy overall than deserts. The air temperature at ground level in these densely vegetated zones would be even higher than in a bare arid desert, except that the leaves of the forest are usually evaporating water which sucks away

Table 5.1. Typical values for various land surface types. Ice and snow, deserts, bare soil and grassland yellowed in the dry season all have substantially higher albedo than evergreen rainforest or conifer forests, deciduous forests in leaf, and water surfaces. From: A range of sources.

Surface	*Albedo*
Natural	
Fresh snow	0.80–0.95
Old snow	0.45–0.70
Desert	0.20–0.45
Glacier (dusty)	0.20–0.40
Soil (dry)	0.23–0.40
Cropland	0.18–0.25
Grassland (dry season)	0.25
Deciduous forest (in leaf)	0.10–0.15
Coniferous forest	0.05–0.15
Tropical rainforest	0.12
Water	0.03–0.10
Urban	
Road	0.05–0.20
Roof	0.08–0.35
Paint	
White	0.50–0.90
Red, brown, green	0.20–0.35
Black	0.02–0.15

heat as "latent heat of evaporation". If you need to be convinced of the difference that albedo makes to the amount of sunlight converted into heat at the surface, try walking across from a light concrete surface to dark recently-laid asphalt on a hot sunny day. The air hovering above the dark asphalt will be much hotter than that over the light concrete—often almost unbearably hot. If they were not continually evaporating water from their leaves, forests would also be even hotter than bare land.

In a desert, then, because the land surface is bare of vegetation, this tends to result in weaker upwards movement of warm air rising from the ground, which allows descending air to come in from above, and does not give rain. And, of course, without rain there can be no vegetation. So the chain of causes goes in two different directions depending on the starting position;

$$\text{Bare land} \Rightarrow \text{weakly rising air} \Rightarrow \text{no rain} \Rightarrow \text{bare land}$$

$$\text{Vegetated land} \Rightarrow \text{strongly rising air} \Rightarrow \text{rain} \Rightarrow \text{vegetated land}$$

In effect, once there is a lack of vegetation cover in an already fairly arid environment, it stabilizes its own aridity in a vicious cycle. Indeed, it may well exaggerate its own

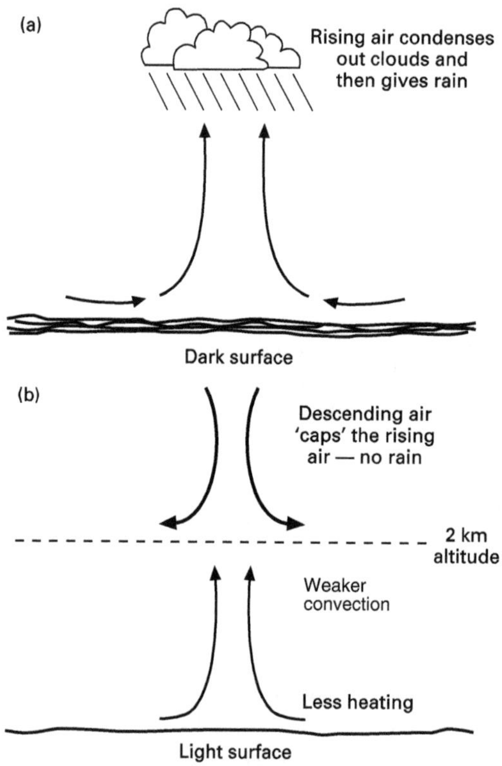

Figure 5.1. If we take two equally moist surfaces with different albedo values, the darker low-albedo one is more likely to give rain. (a) Ascending air over a dark surface rises high, then cools and condenses out water droplets that can form rain. (b) Air does not rise as high above a light-coloured surface and so is less likely to condense out water droplets.

aridity, making things more arid than they would have been to start off with. This process of reinforcement is known as positive feedback (see Box 5.1).

So, there is a positive feedback process involving surface reflectivity in deserts, and this can intensify the arid desert climate. However, the mechanism is also liable to break down sometimes and produce a sudden flip between moist and dry climates, if a slight triggering change in the environment occurs. This triggering change could be something natural, or something caused by humans.

Box 5.1 Positive feedback

In environmental science it is becoming more and more apparent that a phenomenon called *feedback* is key to understanding global processes. What do we mean by feedback? It is a process that (in some cases) "feeds off" itself, gathering momentum like a snowball rolling downhill, or conversely (in other cases) damping itself down and moderating its own effects like the central heating system of a house, controlled by a thermostat.

One example of feedback often occurs if you go to watch some awful band at the local pub. They will tend to set up the stage with the singer's microphone too close to the speakers. The mike picks up the sound of guitar and amplifies it. The more it amplifies the sound, the louder it comes out of the speakers, and the more goes back into the mike, and so on—eventually it arrives at an ear-piercing screech. This self-reinforcing process is *positive feedback*. It is a process that "magnifies" change, once the initial small triggering event (the first little noise that went into the microphone) occurs. It can be just the same with the natural environment: a slight initial change is picked up and amplified into a big shift in climate.

Or consider another type of feedback: a central heating system in a house. When the thermostat senses that the house is too hot, above the set point, it turns down the heating. If it turns the heat down too much, the thermostat senses the decrease in temperature that results, and turns it back up slightly. The temperature oscillates slightly around a fixed point. Also, if the weather outside changes, becoming warmer or hotter and tending to alter the temperature of the house, the thermostat senses the change inside the house and adjusts the heating. The operation of a central heating system is a different type of feedback: *negative feedback*. This is a process that "damps down" change. In the natural environment, negative feedback loops help keep the climate stable, resisting knocks and the destabilizing influences of positive feedbacks.

Where do we see positive feedback in nature? It is everywhere. One important example is in snow cover and ice sheet extent. An area of ice or snow cools the surface (by reflecting back most of the sun's heat and preventing it warming the surface), ensuring that more ice can form. This runaway effect helps explain why the earth went through ice ages in the past, when huge ice sheets made of solidified snow spread to cover the high latitudes of the world. Patches of snow that failed to melt could reflect sunlight and cool the air, making it easier for more snow to survive the next year, and so on. Other important positive feedbacks that involve living vegetation are explained in this chapter and the next.

So, positive feedback factors are amplifiers that intensify differences in the climate from one place to another, and also increase variability in earth's climate and environment over time. They must surely have been important in bringing about the many sudden changes in climate that have happened in the geological history of the earth, particularly during the last couple of million years (Chapter 3).

It takes a triggering event to set the positive feedback loop rolling. Particular factors that could have acted as triggers for positive feedback in the past climate include Milankovitch rhythms, which are changes in sunlight distribution due to the Earth's wobbling orbit. The change in the seasonal sunlight intensity sets a wide range of positive feedback loops rolling, involving ice, involving vegetation (Chapters 5 and 6) and involving plankton in the oceans (Chapter 7). If it moves in one direction, the Milankovitch change can trigger global cooling through all these loops. A slight shift in the other direction and the earth can suddenly warm through positive feedbacks operating in reverse.

Although positive feedbacks in climate can greatly intensify change, there will be constraints on how far things can move. A positive feedback cycle is generally "damped"—beyond a certain point the amplification loop stops responding and the system doesn't move beyond that extreme. A screeching guitar amplifier does not keep getting louder and louder until the whole building is demolished by the sound waves. It reaches its limit in terms of volume because the wattage of the speakers becomes limiting to the power of the feedback. Where a positive feedback cycle causes warming, the temperature doesn't go running away until the earth burns up completely, just because the feedback is operating to heat it up. Nor does it go down to absolute zero when a feedback that causes the cooling is in operation. All the positive feedback does is *amplify* the amount of change that occurs, up to a certain point.

If you think of it all in terms of a graph, what a positive feedback mechanism does is increase the slope of the responsiveness of the system to external changes—for example, to a Milankovitch rhythm, to random variation in regional weather patterns or to changes in the composition of the atmosphere. If you vary the amount of a certain factor (e.g., increasing the amount of greenhouse gas in the atmosphere) along the bottom axis A, there is a certain amount of response seen in terms of the vertical axis B (e.g., the temperature, responding to the increased greenhouse gases) giving a sloping line (Figure 5.2). Adding in a positive feedback changes the slope of the response: it has become much steeper. In the middle of the graph that shows the amount of response there will be a "hinge point" that remains the same, but each side of the middle the change is greater, either upwards or downwards.

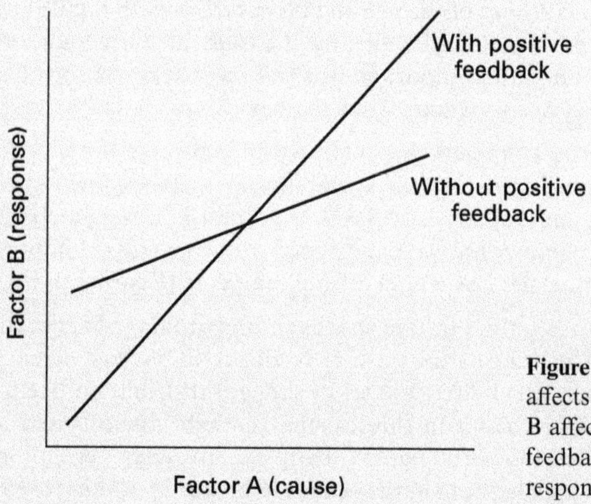

Figure 5.2. How positive feedback affects the slope of a response. Factor B affects A, and with a positive feedback working the slope of the response is steeper.

Positive feedbacks on climate can be set off by a slow, long-lasting change in background conditions. This might, for example, be a change in seasonal sunlight distribution due to the Milankovitch rhythms, moving the earth from a glacial to an interglacial state. Once the flip has occurred, this system tends to be fairly stable over thousands of years because the change in sunlight is so slow.

However, the trigger for a positive feedback could also just be a temporary set of conditions, some random event such as a much wetter than average summer or a much colder than average winter. Or it might even be too much goat-grazing in a particular year on the edge of a desert. The feedback then amplifies this initial switch until it reaches its limits, and what would have been a small temporary change has become amplified and settled into a new steady state. The new steady state is not fixed forever, though. It can always be thrown back in the other direction by a random temporary change, taken hold of by positive feedback loops running in reverse. Thus, the state of such a system is never stable, but *metastable*, liable to flip suddenly if given a small push (Figure 5.3). In a system that is metastable, both alternative states are really just as stable, and what things are like at any one time just depends what particular "peg" things have come to rest on. Such metastable states are very important in understanding the history of climates in arid regions, and many other changes over time, including some of the large global climate changes in the fossil record. As we shall see, vegetation is probably often involved in making the system metastable and liable to flip.

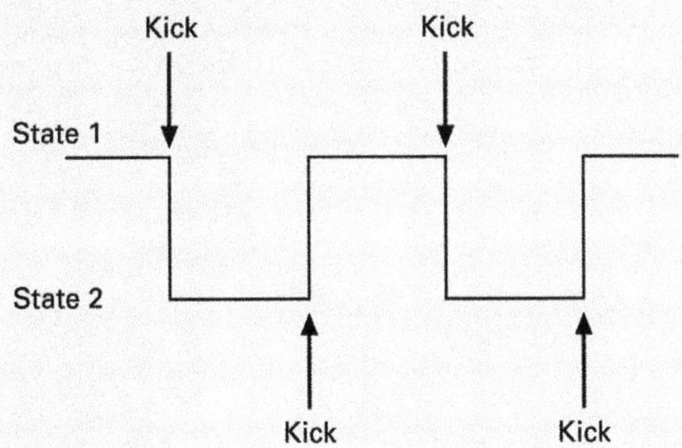

Figure 5.3. A metastable system has multiple states. It will stay stable in one state until it is "kicked" into the other one by some temporary triggering event.

5.2.1 The Sahel and vegetation feedbacks

At the southern edge of the Sahara desert is a zone of open scrub, known as the Sahel. In Arabic, the word Sahel means "shoreline", and this is a good metaphor for this long, thin strip—perhaps 200 km in depth—that borders the great desert (Figure 5.4). For thousands of years, the Sahel has been inhabited by herders and farmers, but their survival has always been made precarious by fluctuations in rainfall. In most other arid regions of the world, dry and wet years are randomly interspersed, so if one year is especially dry it is no predictor that the next year will be dry too. But, in the Sahel the climate tends to go through distinct wet or dry "phases" lasting several decades. If one year in the Sahel is drier than the long-term average, it is a good bet that the next year will be relatively dry too, and also the year after that. In the late 1800s, the region went through a moist phase, with rather good agricultural yields. Around 1900 (Figure 5.5), rainfall records show that there was an arid phase, lasting around 20 years. Then, there was a rather abrupt increase in rainfall, and rainfall

Figure 5.4. The Sahel, at the southern border of the Sahara desert. *Source*: Author.

mostly stayed high up until about 1970 when it suddenly declined, and stayed low throughout the 1970s and 1980s with a partial recovery since then. The decrease in rainfall was something like 20–35% across most of the Sahel, with some even drier phases within this.

The hardship caused by this arid phase in the Sahel was well documented by the world's media. Human populations in the Sahel had expanded in response to an abundance of food, under the high-rainfall conditions of the mid-20th century. Suddenly, there was less to eat, and nowhere for the farmers and herders to go. The whole situation was exacerbated by civil war in some parts of the Sahel, and the combination of events led to famine and many thousands of deaths.

The environmental movement, still finding its feet in the 1970s, was given a jolt when some respected climate scientists suggested that this drought was not an entirely

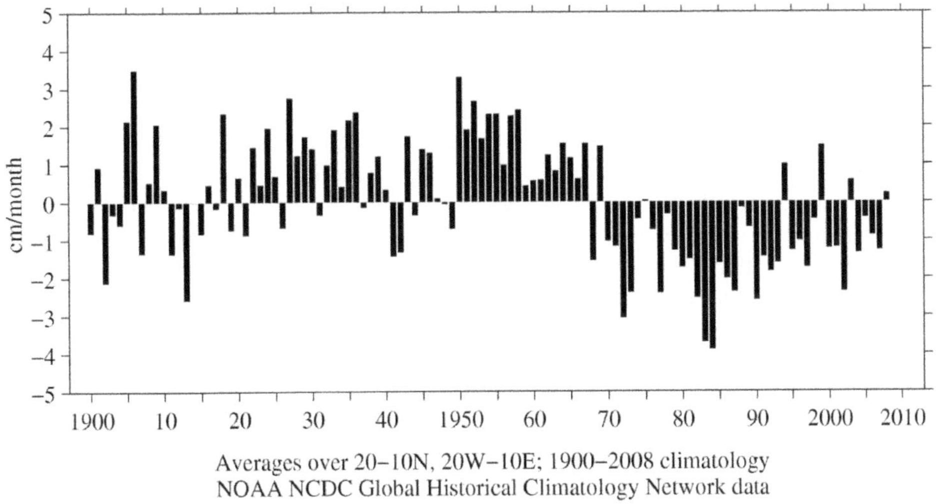

Figure 5.5. Record of rainfall in the Sahel since 1900 (JJASO mean Sahel precipitation anomalies 1900–2008). A relatively moist phase lasted until about 1970, followed by a substantial reduction in rainfall through the 1970s and 1980s. *Source*: Mitchell (2005).

natural event. Overstocking of livestock could have set off a process of positive feedback in the vegetation–climate system, which started or greatly amplified the drought. Discussion of this idea in the case of the Sahel seems to have started with that study by Otterman. He suggested that overgrazing by livestock had removed the dark vegetation cover of the Sahel, decreasing rainfall through the albedo effect mentioned above. So, with the vegetation removed there would be less upward movement of the air within the atmosphere. The lack of upwelling warm surface air would also mean that the atmosphere high above tended to stay cooler. This cool air would tend to sink gently downwards, compressing and holding its water vapor more tightly as it descended. The result would be the desert climate: a cap of descending air up above and no rain. So, essentially, the high reflectivity of the surface, caused by lack of vegetation, would produce a dry climate which would not support vegetation, ensuring high reflectivity of the bare surface, and so on ...

The idea that there is a positive feedback loop behind the rainfall cycles is a compelling one, and in fact it is not necessarily anything caused by humans. It could be that the albedo feedback operates almost independently of whatever humans do—that they and their livestock cannot affect albedo enough to bring about a lasting drought. In this case the trigger for a drought might be some sort of event imposed from outside the region. For example, there could be natural changes in wind flow patterns that produce an initial drought that is then reinforced by changes in vegetation cover. So, the albedo feedback loop involving vegetation might still be important in producing dry or wet phases in the Sahel, but humans might not be a significant part of the trigger.

After Otterman started talking about albedo, climate scientists thought about other ways in which deserts and arid zones might make their own climate. They brought in some extra factors that might alter the influence of vegetation on climate, including "roughness" which is the bumpiness of the vegetation surface compared with bare ground. The greater this roughness is, the more it tends to produce turbulence as the wind blows over it. This alters wind speed, and the vertical movement of the atmosphere. Vertical movement in the air is all-important to producing rainfall, and in transferring heat up from the surface; so, this might be another important way that vegetation modifies its own climate.

The climate modelers also thought about how evaporation might differ between vegetated areas and bare ground. Evaporation gives the water vapor that can condense out as rain again up in the sky, either over the same area or hundreds of kilometers away. It can also affect the strength of the rising convection of surface air up into the atmosphere, acting as "fuel" for the upwards motion by supplying latent heat that keeps the air warmer and thus rising. Vegetated surfaces evaporate—transpire—more water, because they catch more as rain in the root mat and in the loose soils underneath, and then let it out from the leaves. And, whereas a bare soil tends to evaporate all of its (relatively small) store of water very quickly, a vegetated area loses it gradually over a much longer period. Including such extra factors in climate models has suggested that in these respects too, vegetation tends to "make" its own climate.

So, here are two additional feedbacks on rainfall, beyond the albedo effect:

THE EVAPORATION FEEDBACK

Vegetated land ⇒ more evaporation ⇒ more rain ⇒ vegetated land

THE ROUGHNESS FEEDBACK

Vegetated land ⇒ rougher surface more turbulence ⇒ more rain ⇒ vegetated land

The more sophisticated modeling studies which included these extra feedbacks agreed with the initial conclusion that in an arid area the climate you end up with can depend very much on what you start from in terms of vegetation cover. For example, in the Sahel, if you start from a dense blanket of vegetation, this landscape generates more rain than it would if you started from sparse vegetation. In fact, in the models a more densely vegetated Sahel—such as existed in the moister phases of the last century—generates enough extra rain to sustain its denser vegetation cover. So, this is a self-stabilizing system: once set up, it can perpetuate itself (through a positive feedback loop). It is the same the other way round too—starting with a bare Sahel, the climate that the land surface makes for itself is drier, and the vegetation remains spare. So, there are two potential steady states following on from these two different sets of starting conditions.

However, these states are not truly stable; they are "metastable" (see Box 5.1), liable to flip if given a knock from external influences in climate that alter the broader atmospheric circulation across the region to give a few wetter years or drier years than normal. The sort of factors that can supply this knock are changes in ocean circulation and temperature, such as might come about in an El Niño event or during other broad climate oscillations in the climate system such as the North Atlantic Oscillation (a shift in the relative strength of air pressure systems between the north and central Atlantic). Once the surface vegetation cover has been forced to change by such strong external factors, the vegetation–climate feedback system can suddenly tumble in a different direction, to end up in the "other" state.

We see signs of the existence of these two steady states in the decadal-timescale oscillations in rainfall that are recorded by climate station records, and the famines of the Sahel. Oscillations on this longer time scale cannot be explained by external influences such as temperature shifts originating in the Atlantic (which tend to be very short-lived); they must be internally generated. So, once the Sahel gets dry it tends to stay that way for several years at least. The dry state is self-perpetuating because of the sparser vegetation influencing climate. Only when it is overwhelmed by an external change in climate (due to sea surface temperature changes forcing rainfall to come its way) does everything finally flip into a new state of greenery and higher rainfall, which is also self-perpetuating for a while. The models suggest that in fact the greener, moister state is more stable than the died-back dry one, because the bushes in the Sahel can put up with a lot of drought before they gradually give up and die. Conversely, it does not take much rain to bring about a regrowth of vegetation, dragging the system quickly out of a long drought. Observations agree with this expectation: dry phases start reluctantly and slowly as the vegetation dies back, whereas dry phases end suddenly as the vegetation responds to rain.

The variability in rainfall is also thought to have an effect on the abruptness of the boundary between desert and non-desert at the southern edge of the Sahara. A vegetation–climate model by Ning Zeng of the University of Maryland suggests that the reason a transition zone of scrub like the Sahel can exist at all is the shifting lottery of rainfall, which comes farther north in some years than others. If the rain came north to the same point every year, there would form a denser wooded cover of small trees able to exploit the moister conditions south of that point, which would suddenly give way to very sparse open scrub where the rain could no longer sustain it. The very sharpness of the transition would be amplified by vegetation–climate feedbacks from the presence of the trees themselves. In the real world, the year-to-year variability in rainfall, which is itself amplified by vegetation–climate feedbacks, produces a broad zone where moisture supply is too unstable to favor the growth of trees. The more conservative, drought-tolerating shrubs of the Sahel are able to hold on in this variable environment, and it is they that themselves favor both the variable environment and more broader gradient in vegetation! The chain of causes then goes something like this ...

Variable rainfall caused by variability in sea surface temperatures ⇒ variability is amplified by vegetation feedbacks from the Sahelian vegetation ⇒ favors Sahelian shrubs that tolerate variability ⇒ variability is amplified by Sahelian vegetation feedbacks ⇒ etc.

5.2.2 Have humans really caused the Sahelian droughts?

To return to the original question which set Otterman and others wondering: Are humans a large part of the blame for dry periods in the Sahel? Could they sometimes be the "kick" that sets the climate system tumbling towards a dry state after overgrazing takes place? The conclusion of all the models so far is that humans and their animals do not in fact have a big influence on the Sahel vegetation–climate system. While overgrazing may occur, it tends to alter details of structure and composition of the vegetation, towards thornier and less edible species, more than it affects the overall vegetation coverage. There is always at least *some* effect on overall vegetation coverage from grazing, and in this sense humans may have some small part to play in reinforcing droughts in the Sahel. However, it is an effect that is dwarfed by, and very difficult to disentangle from, the other vegetation–climate processes that cause wet and dry cycles in that region.

Interestingly, there is at least a possibility that events in the Sahel could be influenced by humans changing the vegetation *outside* the region, hundreds of miles to the south in West Africa. One model suggests that if the forests in West Africa are cleared, the loss of re-evaporation of water cuts off the supply of moisture for rainfall over the Sahel. There has already been a considerable amount of forest loss in West Africa in the last few decades and it is not certain how this might have affected the recent climate history of the Sahel. It seems that most of those working on this area presently feel that the forest removal has not had much effect, and that the variability in the Sahel is mainly due to variable sea surface temperatures in the Atlantic plus internally generated vegetation–climate feedbacks in the Sahel.

5.3 COULD THE SAHARA BE MADE GREEN?

Some models that involve both vegetation and climate have suggested the hidden potential for far more extreme changes in the climate of the Sahara than we have witnessed over the past century. These models have concentrated so far on just the western half of the Sahara desert. They tend to find that if you were to blanket the whole of the western Sahara desert in a leafy cover of grass or bushes, the climate of the region would be transformed. The low albedo, the greater roughness, the capture of rainfall and its evaporation from leaves would result in monsoon rains that normally stay to the south of the desert coming farther north. According to these models, the rain made by all this vegetation would actually be enough to sustain the vegetation cover itself and the Sahara (at least, the western Sahara, and perhaps the whole Sahara) would vanish! It would be replaced by the sort of open cover of small bushes and scattered patches of grass that we see just along the northern and southern edges of the Sahara at present, something more like "semi-desert" than the desert of the present. The higher rainfall zones to the south would also move farther north, bringing much moister climates to areas that are nowadays scrub and semi-desert. According to these models, then, the imaginary "green Sahara" and the present-day "brown Sahara" are both equally probable, equally stable in the present-day world and it is just by chance that we have one rather than the other. Some people have suggested that, given this possibility, we should set out to create a greener and more useful Sahara region for ourselves by progressively planting trees and other vegetation inwards from the edges of the desert.

However, climate modeling is a complex business and different groups' models often come up with different conclusions from one another. Some models (e.g., one put together by Hans Renssen and colleagues) set up in slightly different ways suggest that the "green" Sahara is not actually a possibility in the present-day global climate system: that even if we blanketed the whole desert in vegetation the feedbacks it set up would not manage to bring in the rains needed to keep the vegetation going. Given the uncertainties, any large-scale exercise in climate engineering that sets out to transform the Sahara through planting vegetation would risk becoming an expensive failure.

Box 5.2 Simulating climate: GCMs and mesoscale models

What changes can we expect for the Earth's climate over the coming decades, as greenhouse gases increase? Because of the importance of knowing the answer to this question, a lot of effort is going in to understanding and forecasting climate change. The world's most powerful computers (known as supercomputers) are used to calculate the effects of a given rise in greenhouse gases on global climate, using a "model": a simplified world inside the computer, complete with oceans, continents, mountain ranges and an atmosphere.

Such models are also being used to investigate how vegetation creates its own climate, and what will happen to global climate if the vegetation cover is altered. As well as looking into the future, models can be made to look backwards in time, to understand how climates in the past worked, including, for example, the effects of past vegetation changes feeding back on climate.

To get a broad global perspective, climate scientists try to simulate the circulation system of the whole planet, with what is known as a general circulation model (or GCM) (Figure 5.6). To model the entire global climate system is of course no easy task, and one which has taken a long time to get more or less right. Basically, the world in the computer is divided up into a grid covering its surface, and each grid cell is labeled as "ocean" or "land". If it is land, that surface grid cell is assigned an altitude, and also some attributes that relate to vegetation cover such as albedo and roughness. Up above the surface of each grid square, the atmosphere is represented as a stack of cubes. Each cube has its own composition and density of gases, and it exchanges energy with the cubes next to, above and below it, or (if it is at the bottom of the atmosphere) with the surface below it. Air is also exchanged sideways, and upwards and downwards from each grid cell, simulating the wind and also the process of convection. In the newest models, the ocean is also divided into stacks of cubes, much like the atmosphere except that these are under the surface and the fluid that fills them is not air but water. Heat and water move between these ocean boxes, simulating surface currents plus the sinking or

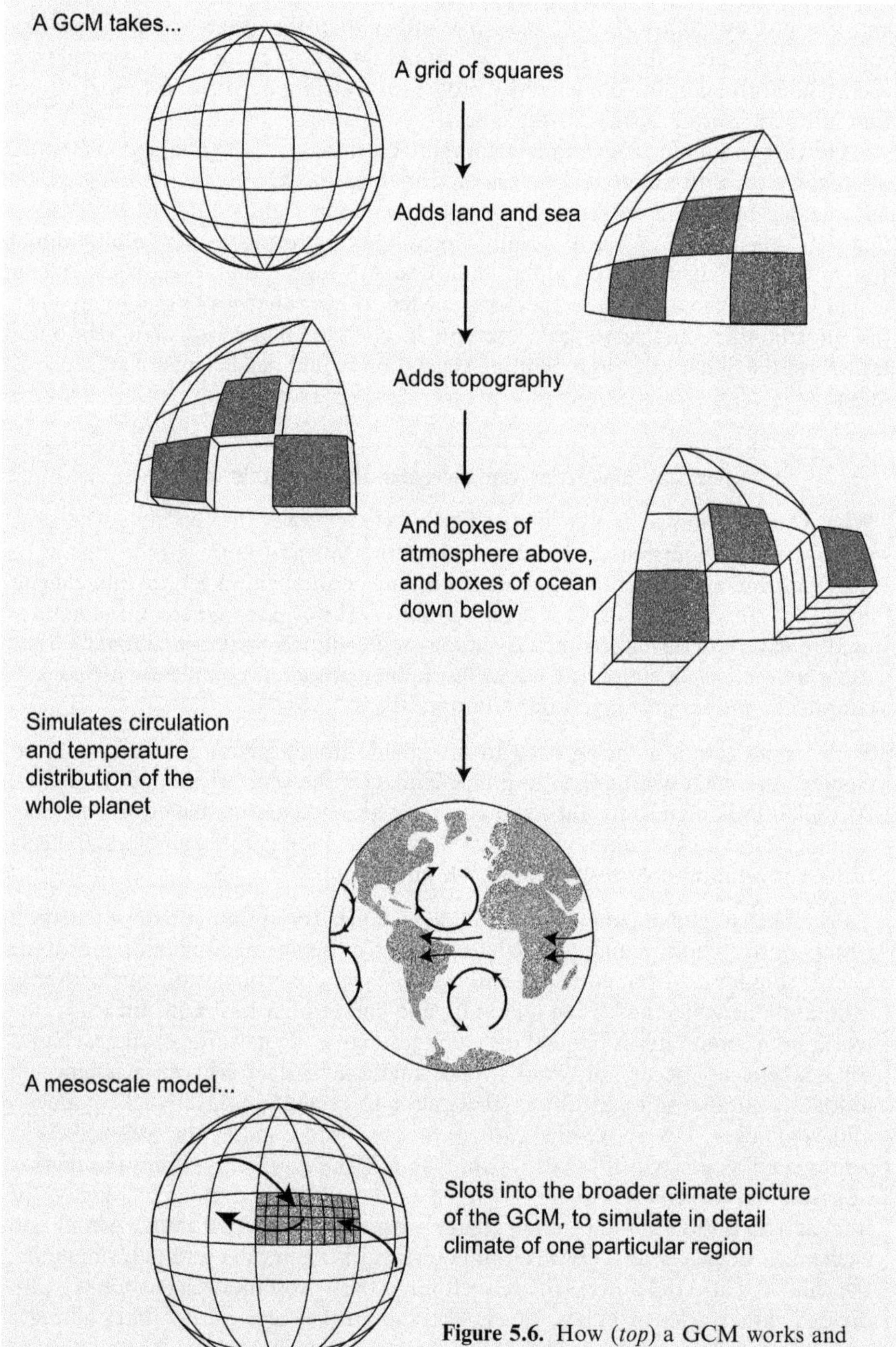

Figure 5.6. How (*top*) a GCM works and (*bottom*) how a mesoscale model slots into it.

upwelling of water. Winds and ocean currents push against one another, churning endlessly across the surface of the planet.

It is remarkable how many details of the climate system these GCMs can simulate. When a GCM is set up to run with the present-day atmospheric composition, the major wind belts and ocean surface currents can all be simulated quite accurately. Air masses form and move across the surface, colliding to give weather fronts. The patterns of average temperature and rainfall are closely similar to what we observe in the present world, and they go through their correct seasonal cycles. Furthermore, from year to year the global climate also goes through internally generated climate fluctuations that mimic those on the real earth.

When climate modelers are satisfied that their model works well for the present world, they can begin to tweak certain aspects of it to see how these will change the climate. For instance, they can add more greenhouse gases and observe the heat balance, rainfall and circulation systems changing in response. They can also change the vegetation cover and see how climate responds to this alteration in albedo, roughness and evapotranspiration. Some of the broader scale studies of vegetation–climate feedbacks use this sort of approach to reach their conclusions about the importance of vegetation cover in making climate.

Climate modeling has come a long way in the past couple of decades, as the quantity of data that computers are able to handle per unit time has increased enormously. But, there is always still room for improvement in models. A major problem in simulating the climate is still the coarseness of models—it is not possible to include every small bump or valley in the landscape, and yet such little microclimatic differences might add up to significant broad-scale effects. Many processes—such as the formation of thunderclouds—occur at a scale smaller than a single grid cell so they must be assumed to occur rather than simulated directly. Decreasing the size of the grid cells in the model increases its accuracy, but doing so magnifies the computing task enormously. Modelers have to strike a balance between the time taken to run a model, and the accuracy that it can produce. As computing power has increased, the size of the grid cells in models has decreased from $5 \times 5°$ in the 1980s, to $0.5 \times 0.5°$ in the latest models. At the time of writing, the world's most powerful computer system—located in Japan—was built especially for the task of running the most sophisticated climate models.

One way to get more fine-scale accuracy, without running up against enormous computing problems, is to focus on simulating one area of interest in detail and leaving the rest of the world outside it at a lower resolution. For this purpose, climate modelers use a special add-on model that works at a regional scale, known as a mesoscale model. A mesoscale model works in many ways like a GCM except that its grid squares and boxes are smaller and cover a more restricted area—the region simulated by such a model will be at most a few hundred or a thousand kilometers across. A mesoscale model partly creates its own climate from the sunlight that falls on it, but it slots into a broader GCM which supplies heat and water vapor in at the edges, and takes these away from the edges too when the wind blows out from the area. A mesoscale model is ideal for exploring how detailed changes in vegetation will affect a regional climate. Many of the vegetation–climate feedback studies mentioned in this book were carried out with mesoscale models coupled to broader GCMs.

5.4 A HUMAN EFFECT ON CLIMATE? THE GRASSLANDS OF THE GREAT PLAINS IN THE USA

There may be some places in the world where "climate-engineering" by humans altering vegetation cover has already occurred, albeit unintentionally. In the 1800s the grasslands of the central USA were transformed at a pace and on a scale unmatched in any other region in history. Settlers poured westwards in their millions, ploughing up the deep prairie soils to plant wheat and corn fields stretching for hundreds of miles.

Did this affect the climate? The debate about it goes back a long way, to the time when the land was still being ploughed up. In the Plains climate, a drier than average year could prove disastrous for crops, so there was plenty of interest in ensuring that the rainfall was as reliable and abundant as possible. In the 1880s, Samuel Aughey, a professor in the young state of Nebraska, suggested that ploughing a prairie soil helps it to retain water better because—with the mat of vegetation on the surface broken—water soaks in rather than running straight off into rivers. This store of water held in the crumbly soil will then evaporate, being recycled as rain which falls to earth again, instead of being lost to rivers and the sea. This idea became encapsulated by the plains farmer's adage: "Rain follows the plough." Although the idea got a lot of attention, it is now thought that ploughing actually does not have such an effect on rainfall.

Others at the time suggested that the best thing to do to ensure steady rainfall was to plant more trees. In the 1860s a US government official named Joseph Wilson pointed out that since deforestation seemed to have decreased rainfall in other parts of the world (see Chapter 6), planting trees in the Great Plains would surely increase the rainfall there. He advocated covering a third of the Great Plains in trees to ensure an adequate supply of rain. Congress was impressed enough by his arguments to pass an act that offered free land parcels to farmers who planted a certain percentage of their land with trees. However, the farmers were not motivated by these incentives—few trees were planted, and the act was eventually repealed.

More recently, aided by modern climatological knowledge and computers, scientists have been able to take a more informed look at the effects of converting the prairies to grain fields. Some modeling studies by Eastman and colleagues suggest that replacing the grasslands of the central Plains with crops caused the peak temperature reached during the afternoon to increase by between 1 and 6°C, depending on the location and time of year. The warming in the model strengthens during the growing season, and decreases as the crops are harvested. The most important factor in causing this warming is that the crops have fewer leaves per unit area than the grasslands. With fewer leaves there is less transpiration of water, and less uptake of energy in latent heat; hence, the air can get warmer over the crops.

Settlers may have affected the climate across the Great Plains even before they had managed to plough up most of the land for crops. Up until the mid-1800s, the Plains supported vast herds of bison, numbering in the tens of millions. The mass slaughter of these animals during the early phases of settlement would have greatly reduced the grazing of prairie grasses. With more leaf area accumulating uneaten, there would have been more evaporation of water from the leaves. Climate models suggest that this could have cooled summer temperatures by 0.4–0.8°C, due to extra latent heat uptake by the evaporating water. This would then have been followed by the main phase when the farmers ploughed the landscape and planted crops, which reduced leaf area to below what it had been in the grazed prairie and caused a *raising* of temperature as explained above.

In the modern Great Plains, particularly towards the western edge, farmers irrigate their crops with water from underground aquifers. What does all this extra water on the fields do to the climate? Modeling studies suggest that the uptake of heat into evaporation from irrigated crops (compared with non-irrigated crops or prairie) will cool the air and create a sort of "sea breeze" blowing outwards to nearby hotter, non-irrigated areas. Measurements comparing irrigated and non-irrigated areas of northeastern Colorado show that, as the models predict, temperatures are several degrees C cooler where there is irrigation, due to latent heat uptake, altered wind patterns and cloudiness (Figure 5.7). As irrigation in the area has expanded over the

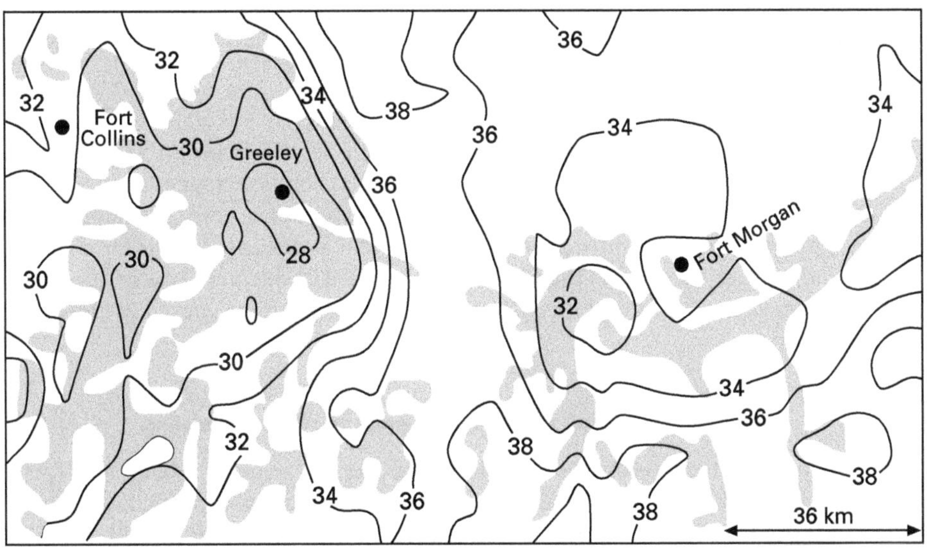

Figure 5.7. Temperature map for a warm day in northeastern Colorado. Irrigated areas such as suburbs and agricultural land (shaded) have cooler temperatures than non-irrigated areas. Surface temperature at 13:00, 1 August to 15 August 1986. Contour from 38 to 28 by 2. After Bonan.

last 45 years, there has also been a cooling trend in climate, as would be expected. The models also predict an increase in rainfall over irrigated areas as a result of both the extra water evaporated, and the movement of air that results from the temperature contrasts between irrigated and non-irrigated land. Observations from northern Texas show that extensively irrigated areas have more rainfall than otherwise similar areas that do not get much irrigation.

On the other side of the world, parts of another arid region may have been affected by climate feedbacks that result from land use change. In southern Israel over the last 50 years, intensification of farming (including increased irrigation), reduced grazing and tree-planting has resulted in lower albedo and more evapotranspiration from vegetation. Since the early 1960s there has been a dramatic increase in autumn rainfall, by as much as 200–300% depending on the location. It seems plausible that the climate change has been a result of the progressive change in land use in this area. The increased upwards movement in the atmosphere above these lands seems to suck in moist air off the Mediterranean, which gives much of the rainfall.

The Sinai desert of Egypt has cooler daytime temperatures than the adjacent Negev desert of Israel, by 3.5–5°C in the early afternoon. It seems that the key factor that makes the Sinai cooler is its lack of vegetation, due to a lot more goat and sheep-grazing and cutting of firewood. With more high-albedo soil exposed, the Sinai reflects back more sunlight and cannot heat up as much. But, doesn't this contradict what I said at the beginning of the chapter—that dark vegetated areas tend to be cooler because they evaporate more moisture? In fact, it is the exception that proves the rule that, without evaporation, dark vegetated areas would always be hotter. Conditions in the Sinai and Negev are so dry that there is no soil moisture to evaporate much of the year. So, the dark vegetation cover in the Sinai (although it is fairly sparse) merely absorbs the sun's rays but does not suck heat away into transpiration.

In slightly moister—but still arid—areas such as the Sonoran Desert in the southwest USA and Mexico, adding a bit more vegetation makes things cooler not hotter. The heavily grazed Mexican side of the border is several degrees hotter during the day than the lightly grazed US side. This is because in this case there is enough moisture in the soil for the extra leaves on the US side to have a cooling effect by transpiring more water, and this dominates over the warming caused by the darker vegetated surface.

Box 5.3 Interactive vegetation schemes in climate modeling

To simulate vegetation–climate feedbacks, it is necessary to pass back and forth between a climate model and the vegetation cover. Initially, a particular climate and a vegetation distribution are set up together. The vegetation distribution can be whatever the modeler is interested to try out, and does not need to be anything that corresponds to the actual present-day vegetation, or anything that is in balance with the climate. The purpose of the exercise is to see how the two of them—vegetation and climate—get along together. The vegetation is allowed to modify the climate (using such feedbacks as albedo, roughness and evaporation), and the modified climate is allowed to modify the vegetation (using the sort of bioclimatic relationships mentioned in Chapter 2). The two are allowed to interact, until they eventually settle down into some sort of steady state. The state that is arrived at can then be compared with what happens with a different starting point for vegetation—for example, more desert or less desert. Or it can be compared with a world in which vegetation only responds passively to climate and does not feed back to change the climate. Making such comparisons allows us to find how important vegetation is in making climate.

In the early days of modeling vegetation–climate feedbacks, this back-and-forth interaction was worked out as many separate steps. The first run of the computer would give a particular climate, and a particular vegetation distribution would now be added in. Adding vegetation would modify the climate. Then, the simulation would be stopped, and the vegetation distribution would now be changed to something which corresponded to this altered climate. The simulation would be started again exactly where it left off except with a new modified vegetation, which now had a chance to modify the climate further. The process would be repeated again and again, until eventually vegetation and climate reached a balance with one another.

Now, this rather clumsy process of stopping and re-starting the model has been replaced by interactive models. The climate and vegetation respond to one another smoothly and continuously. The key to this is to have a vegetation scheme that in effect has the plants dynamically growing or dying off as the climate around them changes. One example of such a scheme is CLIMBER, which seems an appropriate acronym because the vegetation pulls itself along in its interaction with climate.

5.5 THE GREEN SAHARA OF THE PAST

Evidence from a whole range of sources shows that only a few thousand years ago, the climate of the whole Sahara region was very different from now. Animal bones in the desert sands show that giraffes and elephants once walked where there is now no vegetation and no water. The people who lived in the central Sahara at that time even recorded the animals they saw in rock paintings and engravings, vividly illustrating just how completely this place has changed in a few thousand years. A more detailed picture of the landscape at that time comes from pollen which has ended up preserved in the dried muds of old lake beds and empty river channels. It reveals a mosaic of scrublands, open woodlands and grasslands, consisting of plant species that now only grow hundreds of kilometers farther south. Even the extremely arid core of the Sahara, which nowadays gets less than 25 mm of rainfall a year, had a dense vegetation cover capable of sustaining cattle-herding and localized wheat-growing. All the evidence shows that the moistness of the Saharan climate at that time far exceeded the alternative "green Sahara" state of the present-day world. Perhaps we should honor the memory of this remarkable phase in climate history with the upper case "Green" Sahara, to distinguish it from the merely "green" Sahara.

The picture assembled from pollen, animal fossils and ancient lake sediments shows that the ancient Green Sahara phase lasted from about 9,000 to 6,000 years ago (Figure 5.8a, b*), followed by a halting decline in rainfall to reach essentially the present state of aridity by 4,000 years ago. So, when the first organized societies began in Egypt some 6,000 years ago the landscape beyond the edges of the Nile Valley may have been entirely different from now. When the great pyramids were being constructed 4,500 years ago, the landscape they were built in was probably not yet the bare sand that exists at present. Instead, there would have been small bushes and clumps of grass dotting the landscape, spaced perhaps a few feet apart from one another. So, the landscape the pyramid builders saw around them was rather different, perhaps offering more of a contrast with the yellow rock of the pyramids themselves. Some archaeologists have speculated that the trigger that first started the phase of monument-building by the ancient Egyptian civilization was the initial drying out of climate that concentrated diverse peoples and talents into a narrow strip along the Nile Valley.

Why was the Sahara once so moist, and why did it dry out? Bearing in mind the modeling evidence showing that the present-day desert climate is very sensitive to changes in conditions, climate–vegetation scientists took on the challenge of modeling the past of the Sahara. From the models, it seems that there were several factors at work in producing this moist phase. The major one, giving most of the difference in rainfall, is the amount of sunlight the region gets in summer and does not itself depend on vegetation cover. Because of an asymmetry in the earth's orbit, there is a regular cycle of about 21,000 years in the amount of solar energy the northern hemisphere gets in summer and in winter. So, every 21,000 years there is a peak of summer input of radiation over North Africa; the sun is 7% stronger at this time than during summers at the opposite point in the cycle. Climate models show that this increased summer radiation is enough to alter the monsoon flow in the northern hemisphere. With more heating at the surface, air rises farther up into the atmosphere, and this strong convection pulls in air from the south that has picked up moisture over the tropical ocean. The moist air moving in northwards hits the ascending air from the land surface then rises and cools, generating rain. In the world as it was between 9,000 and 6,000 years ago, the monsoon rain came much farther north than it does now, because of this difference in summer sunlight.

Although the increased summer sunlight alone can explain a large part of the increase in rainfall during the Green Sahara phase, it cannot explain all of it. The combined picture from the flora and fauna of this time is that the Saharan climate was even wetter than the models can account for by the increased summer sunlight alone. Apparently then, something is missing from the calculations. Faced with the discrepancy between models and reality, the modelers added in another component. This is the vegetation itself, and the feedbacks it exerts on the monsoon rainfall that it also depends upon. The model builders took a vegetation distribution corresponding to

* See also color section.

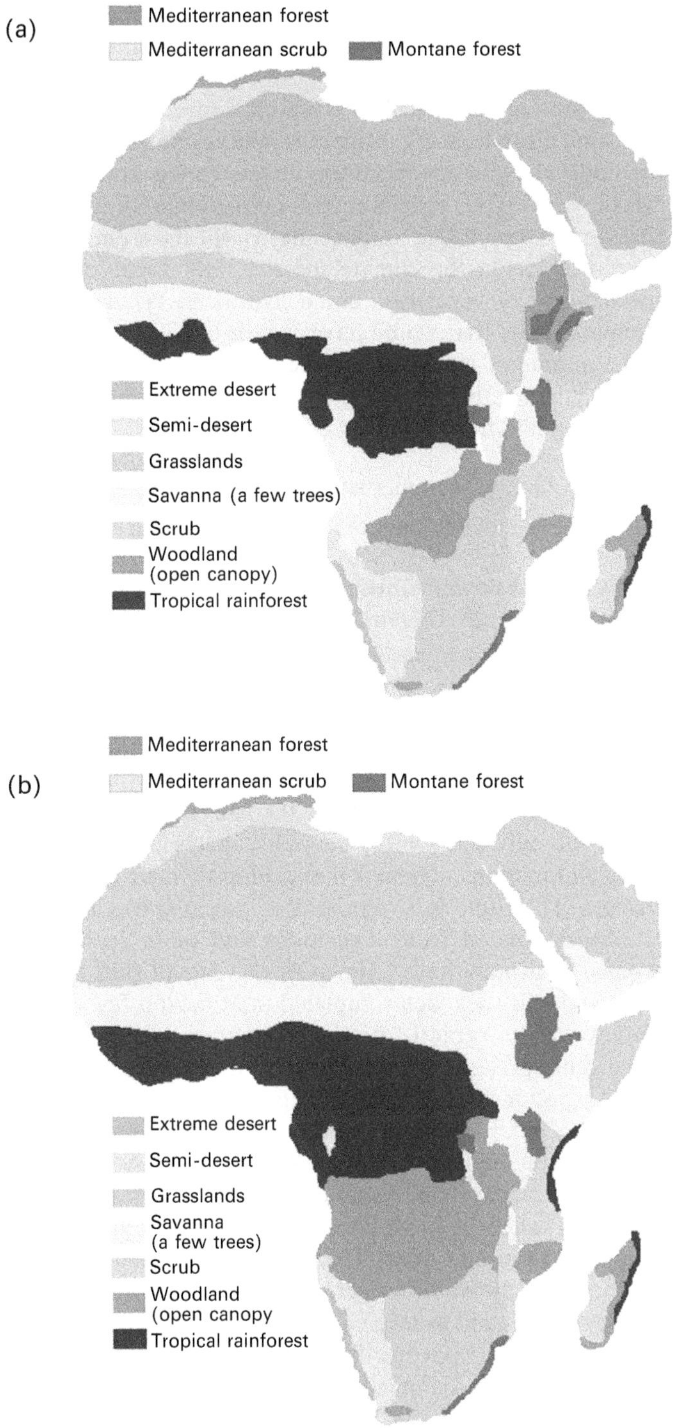

Figure 5.8. The distribution of vegetation zones of (a) the present-day and (b) the Holocene "Green Sahara" (8,000–7,000 ^{14}C years ago). Grasslands (mixed with scrub) seem to have covered the whole Sahara desert at that time. *Source*: Author.

what the fossil record suggests prevailed during the Green Sahara, and to their model they added the lower albedo, the greater roughness and the transpiration of water from the leaves of all this vegetation. The result was an even moister climate: in the model the vegetation helps to set off convection in the atmosphere, pulling in the monsoon more strongly. Moreover, the vegetation across the Sahara recycles the rain that falls, allowing the monsoon to keep going strong as it travels farther up through North Africa. The model forecasts enough rainfall to sustain the abundant vegetation that we know prevailed at that time; so, the loose ends are tied up to make a loop. The vegetation made the climate moister; and because the climate was moister, that specific type of vegetation could live there. What the vegetation did was intensify a moist climate that would have existed to some extent anyway, because of the basic underpinning of increased summer sunlight.

So, the feedback loop that made the climate "optimum" in the Sahara went something like this:

Greater summer sunlight ⇒ moister climate ⇒ more vegetation
⇒ moister climate ⇒ more vegetation

But, with this powerful feedback loop maintaining the moist climate, why did the Green Sahara end? The modern, weaker "green" (note the lower case) Sahara would be unstable, liable to end at any time given a slight push from the weather. But the models show that the ancient "Green" Sahara was far more stable, held in place by the stronger summer sunlight of the time. The only reason it ended was that the summer sunlight over the Sahara declined, to the point where the monsoon rains flickered and then died.

The final phase of drying of the ancient Sahara around 4,000 years ago apparently took only a few hundred years, much less than the sorts of timescales (thousands and millions of years) that geologists have got used to thinking of for climate changes. This rapid loss of rainfall is indicated by various forms of evidence that can be precisely dated from lake muds and other sediments. Despite any temporary reversals that may have happened, the overall shift from a lush green landscape to bare sand and rock was completed in at most a few centuries. Yet, during this time there was a slow gradual decline in summer sunlight, taking several thousand years. Since summer sunlight is really the underpinning cause of the Green Sahara, one might expect a similar gentle change in the climate of the Sahara during this period, and yet in fact it flipped relatively suddenly. Why?

The rapid end to the Green Sahara can only be explained by the way in which the vegetation system responds with its positive feedbacks. Sometimes positive feedbacks can help to stabilize a certain state, and this is what they did during a couple of thousand years (between about 7,000 and 5,000 years ago) when the summer sunlight was stronger, even though the sunlight was declining. But positive feedbacks also tend to reach a sudden breaking point, beyond which they push things in completely the opposite direction. Instead of slamming rainfall up against the top of the scale, they slammed it down against the bottom of the scale. The Sahara reached a point where the vegetation cover could no longer maintain the monsoon rains, even with its darker surface, its roughness and its abundant evaporative leaf area. The sunlight intensity could not quite ensure enough atmospheric upwelling, or enough evaporation, and the rains began to fail. Once they began to fail, the vegetation suffered and died back. And the more it died back, the more the rains failed until within a few centuries there was almost no vegetation and almost no rain (Figure 5.9).

A more detailed look at the environmental record of this critical time shows that actually the change from the Green Sahara to the brown Sahara was not a simple one-time flip. Some indicators from inland lakes and from the amount of river water coming down the Nile suggest that changes occurred in an even more sudden, chaotic manner with sudden flips and then reversals in climate each taking only a few decades. It looks like the monsoon rains flickered on and off like a failing striplight; they turned off for a few decades or a century, then back on for a few decades more, and so on, before they eventually failed completely. Some of the vegetation climate models also seem to support this detail of the picture; they predict various metastable states during the transition period that would have flipped amongst themselves rapidly, before eventually settling down into one stable barren and dry state when the summer sunlight had declined sufficiently.

5.6 COULD OTHER ARID REGIONS SHOW THE SAME AMPLIFICATION OF CHANGE BY VEGETATION COVER?

So far, relatively little modeling has been done on other grassland and desert regions of the world, but the suspicion must be that some of these also show instability in climate that is amplified by vegetation. Apart from the Sahara and Arabia, there are certainly some regions that have a history of large, repeated changes in climate over the past 10,000 years or so. One example is semi-arid northwestern China, which shows great instability in climate on the timescale of millennia. At various times, the climate on the fringes of the Mongolian desert became much moister, moist enough to grow an abundance of crops in areas that are now too dry to cultivate. Farming communities thrived in areas now barely inhabited and mostly devoid of vegetation. Fossil pollen and wood from wild plants such as trees, and changes in the iron oxide chemistry of the soils, confirms that the climate was much wetter at these times. Using a combination of different indicators, geologists have put together a general history of climate change in the region (Table 5.2).

So, it seems that there were several separate moist phases when trees and crop-growing spread out across northwest China, the main times being between about 9,400 and 7,900 years ago, and between 6,500 and 4,900 years ago. Although these broadly fall within the same phase as the moist Sahara, when summer sunlight was at its greatest, in China there are some striking fluctuations in the climate that do not occur in the Green Sahara. At these times, the climate in western China switched from much moister than present to drier than present, before later switching back again.

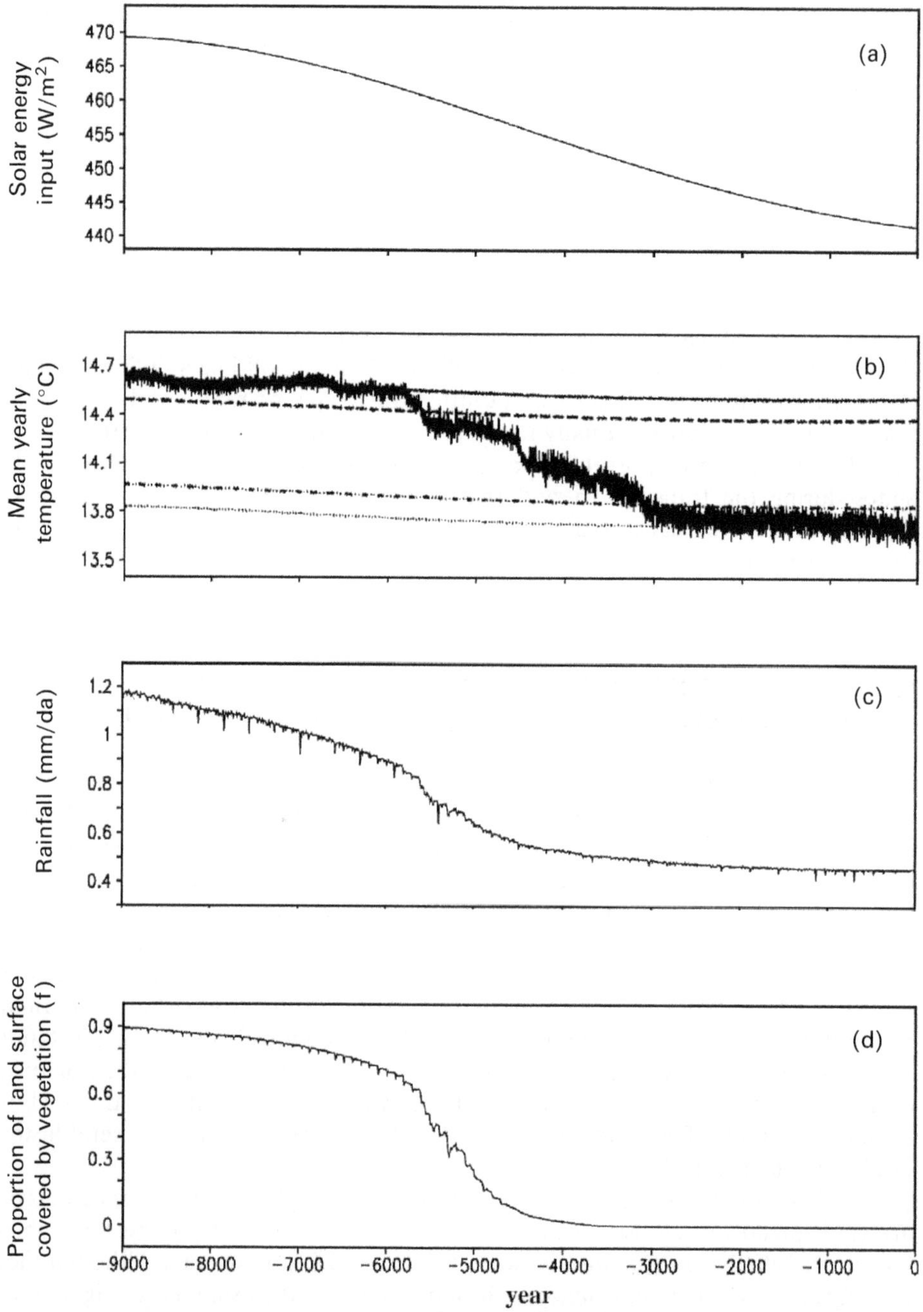

Figure 5.9. In the Sahara, during the last 9,000 years, the summer solar energy input changes slowly (a) but because of vegetation feedback the rainfall (c) and especially the vegetation cover (d) changes much faster, flipping from quite dense vegetation to virtually no vegetation. After Bonan.

Table 5.2. Climate history of northwestern China over the last 10,000 years. From: Petit-Maire and Guo.

Drier than present	9,900–9,400 yr
Moist	9,400–7,900 yr
Drier than present	7,900–6,500 yr
Moist	6,500–4,900 yr
Drier than present	4,900 yr
Moist	3,200 yr
Relatively dry phase (but still moister than present)	3,200–2,800 yr
About the same as at present (fairly arid)	Since 2,800 yr

From what we know from attempts at modeling the Green Sahara, it seems reasonable that the moist phases in northwestern China might have been accentuated by vegetation feedbacks. When the climate models are applied to this region they do indeed show that the greener landscapes would have helped to pull in more rainfall. However, while these same models produce stable moist conditions during the phase between 9,000 and 6,000 years ago in the Sahara (and instability only later, as the summer sunlight decreased), they do not predict the instability of rainfall in northwestern China. The climate instability seen in the environmental record of northwestern China is suggestive of a system where climate–vegetation feedbacks are playing an important role, but the current models at least do not show it. It may be that a future generation of more sophisticated models will show up hidden feedbacks that are occurring with vegetation.

So far then, modeling attempts in western China have not revealed any sign of multiple steady states, but the closely alternating history of arid and moist phases suggests that such instability must also exist there. Whatever the underlying cause, it is likely that part of this variability is caused by amplification of background climate changes by vegetation.

5.7 DUST

So far, we have considered albedo, roughness and evaporation of water in the feedbacks between vegetation cover and climate. Another potentially important vegetation–climate feedback comes from dust. The dust in the atmosphere mostly consists of particles of soil, fragments of the sorts of minerals that make up rocks and clays. These tiny particles tend to scatter sunlight. Dust is really a product of vegetation cover, or rather a lack of vegetation cover; areas with lots of bare soil between clumps of grass and scattered bushes tend to be the biggest contributors of dust to the atmosphere, when the wind blows across the bare surface and carries particles of dry soil aloft. In contrast, when the vegetation cover forms a continuous mat, the roots bind the soil together. With such a root mat in place, even if the soil dries out sometimes it does not crumble at the surface and get blown away. The leaves and stems of the vegetation also interrupt and slow the wind, preventing it from picking up dust from the soil surface.

As one might expect, heavily forested areas (with a deep dense root mat to bind the soil, a dense canopy to interrupt the wind and plenty of rain to keep the soil moist) contribute almost no dust to the atmosphere. Conversely, one might think that the entirely bare desert surfaces such as the central Sahara would contribute the most dust to the atmosphere—because there are no plants there to stop dust being whipped up by the wind. In fact, far more of the dust that floats around in the world's atmosphere comes from the relatively thin strips of desert margin, such as the Sahel, than from the extreme desert interiors. Why would this be? Because it is only in climates with some moisture and some vegetation activity that the rock underneath can be broken down to supply dust (a process known as weathering). In a totally dry environment, whatever dust there once was has blown away, ending up either in the sea or fixed into the soil of zones of wetter climate. There is no supply of new material and all that is left are bare stony surfaces and sand fields—the sort of landscape that makes up the Sahara. In a very moist, forested environment, breakdown of rocks can be rapid but the soil is never exposed to dry out in the open, so the clays and other minerals within it do not get picked up by the wind. A semi-arid environment such as the Sahel offers the ideal combination from the point of view of getting dust into the atmosphere: enough biological activity to chemically break down rocks into fine particles, and enough bare dry soil to allow those particles to be picked up by the wind.

By scattering some light but absorbing some too, dust has an ambivalent effect on climate. Much of the light that dust particles scatter goes back into space, so the earth is cooler because less of the warming sunlight can reach it. Also, dust that floats high in the atmosphere "wastes" some of the heat from the light that it does absorb, because it is floating above the main blanket of greenhouse gases. Without much greenhouse gas above it to trap the heat, the dust loses its heat easily back into space instead of heating the atmosphere around it and the surface below, and the earth is left cooler than it would otherwise be. On the other hand, dust itself can act like a greenhouse gas. It absorbs some infra-red radiation on its way out from the earth's surface, and then sends some of that back down to earth. So infra-red that could have been lost to space is bounced around between each dust particle and the earth's surface. This has the effect of warming the earth slightly. Also, although dust is "bright" and scatters a lot of sunlight, it is not as bright as snow and ice. If there is dust in the atmosphere above a snow-covered ice cap, the dust will actually trap more of the sun's heat than the brilliant white surface below would, and so it will help to warm the air.

The overall balance between these opposing factors determines whether dust warms or cools the earth. There is some uncertainty amongst climate scientists as to whether the cooling or warming influence is more important, and by how much.

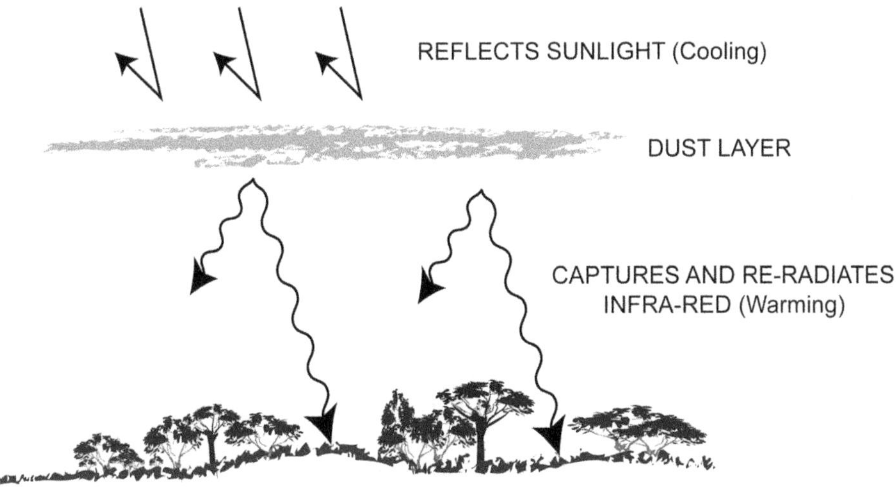

Figure 5.10. The opposing effects of dust in the atmosphere on temperature at ground level.

However, opinion favors a cooling influence for dust, in most places and at most times in the earth's history.

Dust can also have two opposing effects on rainfall (Figure 5.10). It can potentially increase rainfall, by providing condensation nuclei for water droplets. This helps the clouds to form more quickly and more abundantly, giving more rain. On the other hand, a layer of dust in the atmosphere that heats up under the sun can act as a "lid", preventing upwelling of air. This makes rain less likely. Generally, it is thought that the "rain-suppressing" influence of dust dominates over the "rain-making" influence.

So, by acting as a source of dust, semi-arid areas tend to make climates cooler and they also tend to make them drier. How extensive their influence is depends how long the dust stays aloft before it rains back down onto land or into the sea. Some climate scientists have suggested that the dust that builds up in the atmosphere in the Sahel during a drought is instrumental in intensifying and prolonging the drought. A study by Masuru Yoshioko and colleagues used a special dust model (designed to simulate how dust gets blown aloft into the atmosphere) in combination with a climate model to simulate the feedbacks from dust in the Sahel. They suggested that about 30% of the fluctuation in rainfall that occurs over a series of decades is actually due to the effects of dust from the region being whipped up by the wind into the air above. The initial "trigger" could be a change in sea surface temperature over the Atlantic that started off the drought, but the dust in the air would help to hold the climate system in a dry state for several years, with the dryness caused by the dust ensuring that more dust kept on reaching the atmosphere. So, this is a clear positive feedback, acting in parallel with other effects of vegetation cover such as albedo and evaporation. Their model is at odds with the prevailing picture, in suggesting that the dust effect is actually much more important than other vegetation feedbacks in controlling the rainfall over the Sahel. The importance of dust in drought phases in the Sahel and in other arid regions is still very much a moot point, but now we have some tantalizing clues that it might be very important.

Dust may also be able to influence climate globally on the timescale of tens of thousands of years. A geological history of the amount of dust in the atmosphere comes from coring down into ice sheets and deep ocean sediments. Dust particles that rain down out of the atmosphere end up buried in layers of snow if they land on ice sheets. If they land on the sea surface they will sink down to the sea bed and be buried in the sediment. Using sensitive techniques to study dust content in these ice and ocean bed cores, one can estimate how much dust was around in the atmosphere at particular times in the past.

The picture emerging from the last couple of million years is that the cold "glacial" climate phases tended to be much dustier than the warmer "interglacial" periods. In some areas there were tens of times as many dust particles in the atmosphere, showing up now in cores from the ice or sediment in which they were buried. Overall, the world seems to have had something like three times as much dust in the atmosphere during glacials as it has today. A big increase in the dustiness of the atmosphere is much as one might expect from comparing vegetation maps of the two types of climate phase. Glacial phases have much less forest vegetation, and far larger areas of desert and semi-desert, compared with interglacials. In fact, starting from a plausible vegetation map of a glacial phase, and using a climate model to estimate how much dust would be whipped up and carried by the wind, one can fairly accurately predict the extra amount of dust in the atmosphere during a glacial. Additional dust would also be coming from the edges of ice sheets where ground-up rock debris was being dumped and drying out. All the extra dust in the atmosphere during glacials surely had some significant effect on climate. Most likely it reinforced the cold and aridity of the time, by reflecting the sun's light back into space and suppressing the rain-giving convection of the atmosphere. The actual effect of this dust on global climate during glacials still needs to be simulated using a GCM that is ambitious enough to incorporate dust fluxes, in addition to everything else.

However, there might have been times when dust actually brought about a warming in climate. Jonathan Overpeck and colleagues at the University of Colorado pointed out that compared with the very bright surface of an ice sheet on land or floating sea ice on the ocean, dust in the atmosphere is rather darker in color. This means that if it blows over the top of a region of ice, either staying in the atmosphere or settling on the top of the ice, it will tend to mean that more of the sun's heat is absorbed. This will bring about a warming of climate, perhaps helping to melt the ice back. During ice ages, if the winds blow in certain directions and carry dust over ice sheets they might actually help to bring about the end of the glacial climate. Overpeck's group used a climate model to show that once ice sheets reach a particular size and extent, they might set off the process of their own destruction by sucking in dust.

An additional and very different effect of dust on climate may work through ocean plankton. It is thought that the growth of plankton out in the open ocean is often limited by lack of iron, and mineral dust happens to be very rich in iron. Experiments show that adding iron salts to a tank of surface ocean water will often produce a "bloom", a population explosion of algae floating in the water. It seems that, fueled by the iron they need, the algae are able to use up the additional small amounts of other nutrients—such as nitrogen and phosphorus—and multiply. Other more ambitious experiments have actually involved dumping iron salts off the back of ships traveling across the ocean. Within a few days, all along the path of the boat there is often (though not always) a bloom of phytoplankton, detectable by satellite. This burst of phytoplankton growth lasts between a few days and a couple of weeks before it disperses or is eaten up by hungry zooplankton. Given what a modest addition of iron can do, some oceanographers have wondered what effect a big increase in dust flux might have during glacial phases. Is it possible, for instance, that the increased iron input from all the dust greatly increases phytoplankton growth. The increased growth of the plankton could drag down more carbon to the deep sea, helping to decrease the CO_2 level of the atmosphere (Chapter 7). Hence, the greater dustiness of the atmosphere during glacials could be part of the cause of lower CO_2 levels. Since lower CO_2 is likely to be part of the cause of the climate and vegetation conditions that produce more dust, what we have here is a positive feedback loop that reinforces the glacial climate.

5.7.1 Sudden climate switches and dust

One of the biggest surprises to have come from studying the detailed environmental record from ice sheets and sediments, is just how quickly the world's climate can switch. Several times in the last few tens of thousands of years, the climate of large parts of the earth's surface has warmed or cooled by several degrees C over just a few decades. For example, the final warming and beginning of the meltback of the ice sheets around 11,500 years ago seems to have occurred mostly over about 70 years, and possibly much less (some indicators suggest most of the switch in climate occurred in less than 20 years). There are hints that dust fluxes may have had an important role in such sudden climate flips. High-resolution ice cores from Greenland show that in that region at least, dust fluxes from the continents switched slightly ahead of temperatures, over only about 20 years. The way that dust flux leads temperature tends to suggest that it played an important part in bringing about the sudden global climate switch. It may have acted as one of a number of amplifiers leading on from some triggering event that pushed the climate system into a different state. It is not hard to envisage how an initial warming or moistening of climate, perhaps just a random run of somewhat warmer or moister than average years, could lead to vegetation cover spreading and covering the surface. This could rapidly cut down the amount of dust blowing from the surface, and in a global climate system that is ripe for change the warming and increased rainfall could be picked up and amplified by other positive feedbacks. One important aspect of dust is that it can travel for thousands of miles, so the "signal" it sends out could rapidly affect many parts of the world simultaneously. The whole climate system would then have cascaded into a very different state, with vegetation cover effects on dust flux playing an important part in the change.

5.8 THE FUTURE

With greenhouse gases increasing in the atmosphere, it is likely that the world's climate will change significantly over the next century or two. Vegetation feedbacks will surely have a role to play in this change too—perhaps amplifying change, perhaps damping it.

There has not been much discussion in the scientific literature so far on how vegetation feedbacks in arid zones will alter the course of changing climate, but it is a reasonable guess that they will have some significant effects. Part of the problem in forecasting these future feedbacks is that different GCM simulations tend to predict quite different changes in the background climate in the drier parts of the world. So, in a particular semi-arid or arid region, there might either be less rainfall or more as the climate warms. Plus, even if there *is* an increase in rainfall, the warming caused by the greenhouse effect might increase evaporation so much that the climate actually ends up drier overall.

In a general way, some people have speculated that the greenhouse world of the coming centuries might closely resemble the warmer world of the early Holocene "optimum", between about 9,000 and 6,000 years ago, or of the early Eemian interglacial about 125,000 years ago. At both times, the Saharan, Arabian and Asian deserts were much moister than they are now. However, it is important to realize that there is a major difference from the future greenhouse world: during these past "moist" phases there was a lot more summer sunlight over the northern hemisphere and this is what helped to bring the monsoon rains in, and enabled vegetation–climate feedbacks to get going in making the climate even moister. There is nothing in the coupled vegetation–climate models that predicts a similar drastic change in Saharan climate as a result of the increased greenhouse effect alone. One of the coupled models does produce a slight increase in monsoon rainfall at the southern edge of the Sahara during the next century, and forecasts that feedbacks from the vegetation (such as decreased albedo, increased roughness and recycling of rainfall by evapotranspiration) will then amplify the amount of rainfall and result in some greening up of the southern edge of the Sahara. But this spread of vegetation and moistening of climate is not even as much as the "green" Sahara forecast by some models for the present-day state, let alone the "Green" Sahara of the past.

Something that needs to be explored further by the models is how other arid regions of the world might undergo feedbacks between climate and vegetation cover, that will either amplify or damp change. Another factor to consider is the effect of very high CO_2 levels (Chapter 8) on the feedbacks from vegetation. If CO_2 concentrations are more than double their previous background level, as they will be in another 60 years or so, this is likely to affect the way vegetation behaves. For one thing, the plants will not need to keep the stomatal pores in their leaves open for as long to get enough carbon to photosynthesize, and this means that they will tend to lose less water to the atmosphere. This means less latent heat uptake, and less recycling of rain to the atmosphere, which could surely feed back on climate. Presumably, it would tend to heat up the air near the ground (with less latent heat uptake), and also decrease the amount of rainfall (with less recycling of moisture to the atmosphere). But this scenario assumes that the types of plants growing there stay just the same. In fact, if less water is used up by each photosynthesizing leaf it might mean that small trees bearing more leaves can push their way into the vegetation, migrating in from other regions. The whole structure of the vegetation might change, and this can itself change the albedo and the roughness of the surface. Trees will also unavoidably transpire more water than smaller shrubs, counteracting the effect of increased CO_2. This could now change the climate, most likely towards increased rainfall.

Although our understanding is still evolving, the climate system of arid lands turns out to be subtly and inextricably linked to vegetation. In the next chapter we will explore the links which are also present at the other end of the moisture spectrum, involving forests.

6

Forests

Forests temper a stern climate, and in countries where the climate is milder, less strength is wasted in the battle with nature

Uncle Vanya, by Anton Chekov

Since the beginning of agriculture, 12,000 years ago, humans have had an uneasy relationship with forests. On one hand, the forests provided timber, and good hunting for game. But they also took up space where crops might be grown, and provided a refuge for malevolent creatures both real and imaginary. As farming spread out from its first heartlands in the Middle East, northern China and Central America, forests began to lose ground. Already by the time of ancient Greece 2,500 years ago, deforestation was so extensive that Plato lamented that some mountain lands that had yielded good stout timber were now "good only for bees". Evidence from pollen preserved in lake beds shows that the majority of Europe and China's natural forest was already cleared by this time. The remaining forest in both these regions continued a slow, halting decline and reached a low point some time in the last few centuries. A more recent burst of forest clearance occurred when European settlers arrived in North America from the 1600s onwards. At first, there were huge tracts of almost unbroken forest in the east, yet by the mid-1800s most of this forest had been cleared and replaced by farmland. For example, southern New England was more than 90% forested when settlers first arrived, but by 1870 there was less than 25% forest cover. In Midwestern areas such as the forested parts of Wisconsin, deforestation started later (in the 1830s) as settlers moved west, and reached a low point around 1900 with only about 10% forest cover. The character of the surviving forests was also very different. Uncut old-growth forest, which Thomas Jefferson had suggested held enough timber to last 500 years, was mostly gone by the mid-1800s and essentially disappeared in the eastern USA by the 1920s. In its place was younger, regrown forest with smaller trees and altered species composition, harvested every few decades for timber.

In the tropics, the main burst of deforestation began later—in the 20th century—and it is still under way. This big increase in deforestation started around 1950, as populations and economies of tropical countries expanded (Figure 6.1). Thus, for example, in Costa Rica the area of forest was reduced from 67% primary (meaning original, old growth forest) forest in 1940 to only 17% in 1983. Vietnam was about 45% forested in 1943, but this figure had fallen to about 20% by the mid-1980s. So far, it seems that somewhere around 12 to 15% of the primary rainforest of Amazonia has been cleared, although parts of this have now reverted back to relatively species-poor secondary forest. Over each of the last five years up to 2005, deforestation was most extensive in South America, where an average of 4.3 million hectares (10.6 million acres) were lost annually over the last five years, followed by Africa with 4 million hectares (9.8 million acres) according to figures from the UN's Food and Agriculture Organization. Although all of this represents a huge area lost each year, the rate of deforestation has declined in the last decade, giving some hope for the long-term future of tropical forests. A global average of 7.3 million hectares have been lost annually over the last five years, down from 8.9 million hectares (22 million acres) a year between 1990 and 2000.

The effects that forest loss might have on climate have been thought about for a lot longer that most people would expect. It is a surprise for modern ecologists to find a character in Chekov's late 19th century play *Uncle Vanya* already talking of the influence that forest cover might have on climate, and advocating tree-planting for the purpose of climate improvement. Yet, the ideas are far older still. Christopher Columbus—in the 1490s—suggested that the verdant forests of the Caribbean islands helped to produce the abundance of rain that fell on them. His thinking was influenced by folk knowledge of the history of the Spanish and Portuguese islands off northwest Africa: the Canaries, Madeira and the Azores. It was felt that the almost complete deforestation of these islands had resulted in a drier, less rainy climate than when Europeans first arrived.

The possible climatic effects of the rapid deforestation of the American colonies were keenly discussed by a succession of English and American scientists from the late 1600s to the early 1800s. In early colonial times, observers of nature echoed Columbus in suggesting that the humidity and frequent thunderstorms of the eastern USA in summer were a product of the abundant forest cover. Later, as the cultivated lands extended, it was suggested that forest clearance was causing rainstorms to become less frequent, and the air was becoming generally less humid than before. Another view at the time was that the climate was becoming more "moderate" as a result of deforestation, with cooler summers and warmer winters. One writer hypothesized that this was because ocean breezes could now blow further inland without the trees blocking them. Opinion on the rain-generating influence of forests on climate was by now so deeply held that in the 1790s laws were passed in the Caribbean islands to establish forest preserves. The hope was to increase rainfall, ensuring better growth of sugar cane.

During the 1800s, however, the view that forests had a significant influence on climate had both its advocates and skeptics in the scientific world. By the late 1800s it had lost favor, and mainstream scientists generally agreed that forests were unimportant in shaping climate. So, by Chekov's time this was rather an old-fashioned view that had already been mulled over and rejected by prevailing scientific opinion. However, such ideas did not entirely die out, even if they were no longer scientifically respectable. In the 1970s, for example, the idea that loss of forest in the tropics could dry out the climate over extensive areas was a major fear and a rallying point for the environmental movement.

In the last 30 years, the view that forests *are* important in making the climate has undergone a remarkable resurgence, backed up by sophisticated modeling techniques. The modern tools for understanding how forests affect climate are the high-powered computer, and complex models of atmosphere, land surface and ocean (Chapter 1), incorporating many of the microclimatic effects of vegetation cover mentioned in Chapter 4. It is looking like Columbus and the natural philosophers of the 1600s and 1700s were not too far wide of the mark, after all. Loss or gain of forest—both natural and caused by humans—may have all sorts of consequences for climate.

6.1 FINDING OUT WHAT FORESTS REALLY DO TO CLIMATE

To get very far in understanding the effects of forest cover on climate, we need to break down the complex form and behavior of the forest into simple components. These are the building blocks of a model that can include the role of forest in making climate. Several of them have already been talked about in Chapters 4 and 5, but it will do no harm to mention them again (Figure 6.2). One important basic aspect of forests is the proportion of sunlight that they absorb. Known as "albedo" (from a Latin word meaning "whiteness") this is important in determining how easily the forest can heat up in the sun. The darker the forest surface (i.e., the lower the albedo) the more solar energy is absorbed, as opposed to being reflected straight back out into space. When it is absorbed, this energy tends to heat up the leaves. Some of the heat then goes to warming the air around the top of the forest canopy. But, in fact, much of the heat energy that is in the leaves just "vanishes"; the leaves stay much cooler than you would expect from all the sun's energy that they are absorbing. The missing heat has not really vanished—it has just been stored for a while in the water vapor that evaporates from the leaves. This is known as latent heat. It is a strange thing that even though the dark forest cover is absorbing more heat from the sun—compared with a more sparsely vegetated environment—it does not show up in terms of temperature! The more open non-forested environment will nearly always be hotter during the day.

So, this brings us to a second important aspect of how forests affect climate. Transpiration, the evaporation of water out of tiny pores in the leaf surface, takes up heat. This is water that has fallen from the sky, soaked into the soil, been sucked up by roots and carried up within the tree all the way to the cells of the leaves. A single

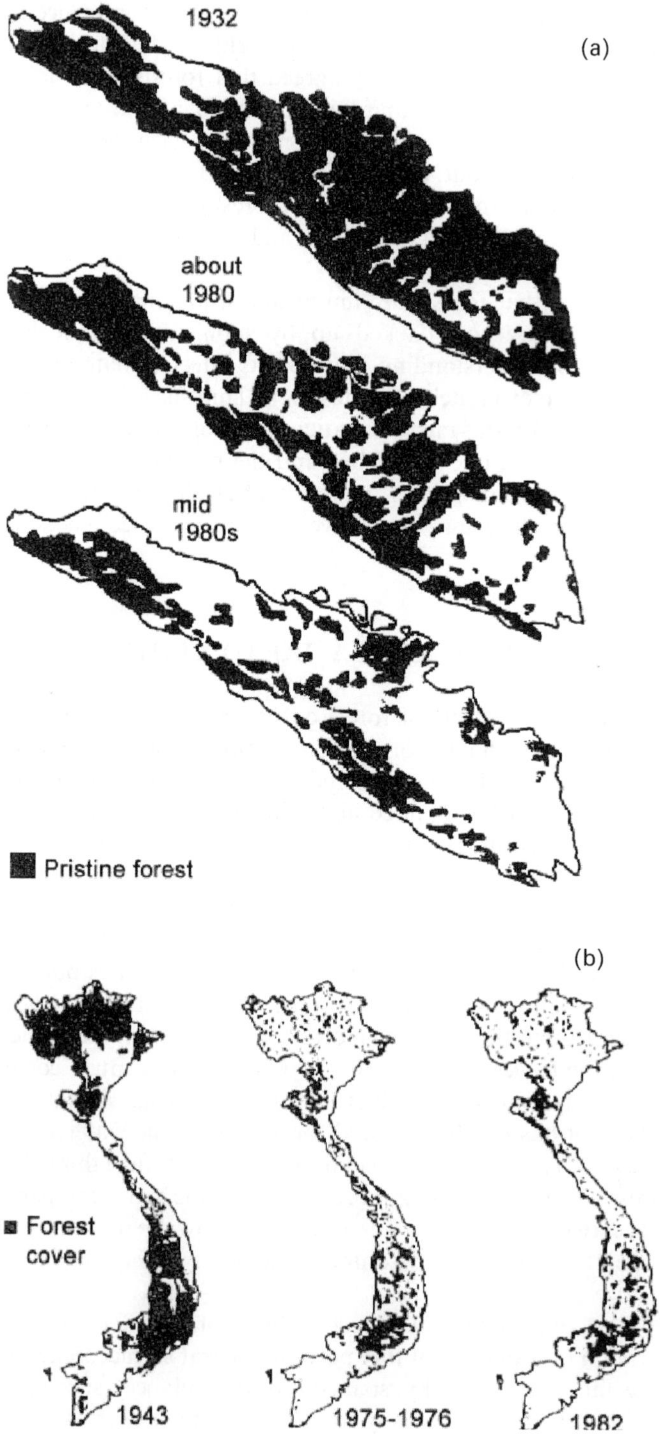

Figure 6.1. Forest loss in tropical lands. (a) Sumatra, (b) Vietnam (c) Madagascar. *Source*: Redrawn from Groombridge (1993).

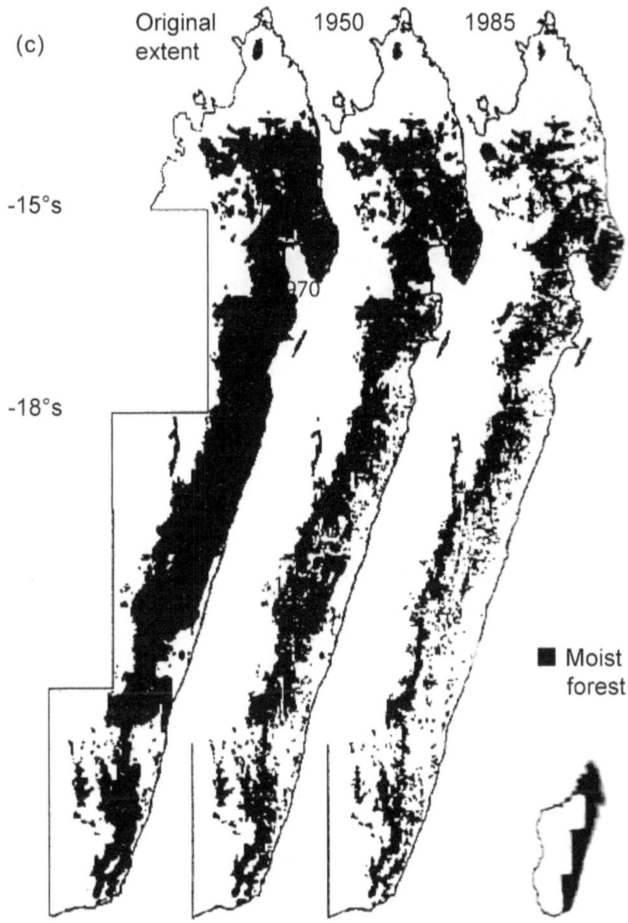

large tree can take up as much heat in evaporation as you'd get from one hundred 100 W lightbulbs burning continuously. This stored latent heat will eventually be released somewhere up in the atmosphere, thousands of meters up and perhaps hundreds or even thousands of kilometers away. What enables the heat of evaporation to be released again is the condensation of water into droplets, forming clouds and eventually rain. In addition to this, there is what is known as "physical evaporation": rainwater evaporating from the surface of the leaves or from the soil surface, without having passed through the tree. In forested areas, the greatest part of the evaporation of water going on is through transpiration from the leaves, rather than physical evaporation.

Climate scientists refer to the proportion of the heat absorbed by the forest that goes into evaporation—rather than just heating up the air—as the "Bowen ratio". This is something that varies between different forest and vegetation types, but also according to season and even with the most recent weather conditions.

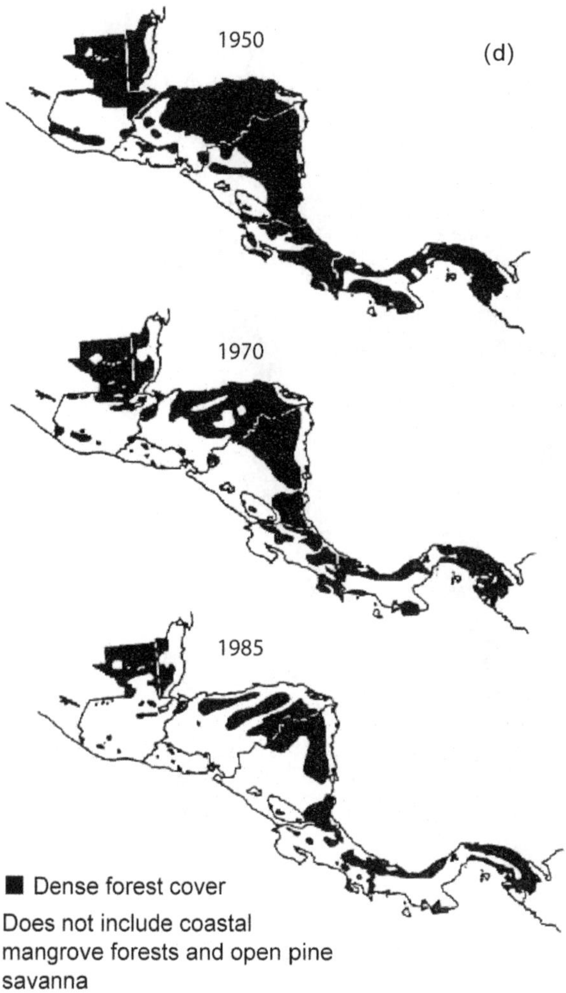

Figure 6.1 (*cont.*). Forest loss in tropical lands. (d) Central America. *Source*: Redrawn from Groombridge (1993).

In essence, forests pump heat and water out into the air above them. They do this more effectively than most other vegetation types, and far more so than bare soil. Something that also helps forests act as water pumps to the atmosphere is that they store a lot of rainwater amongst the root mass of the forest, which is rich in spongy organic matter from the decay of leaves, roots and branches of the trees. Water that would otherwise run straight off the land surface and down to the sea is instead held in the soil, to be sucked up by roots and then evaporated from leaves in the canopy. The deepest roots of many trees reach tens of meters down into the ground, and this also helps them to sustain a good rate of evaporation long after the surface soil has dried out, because the trees can continue to tap into groundwater in pores in the rocks below.

How do these two processes—heat transfer and water transfer—affect climate? On the local scale, evaporation from all the leaves in a forest canopy makes the

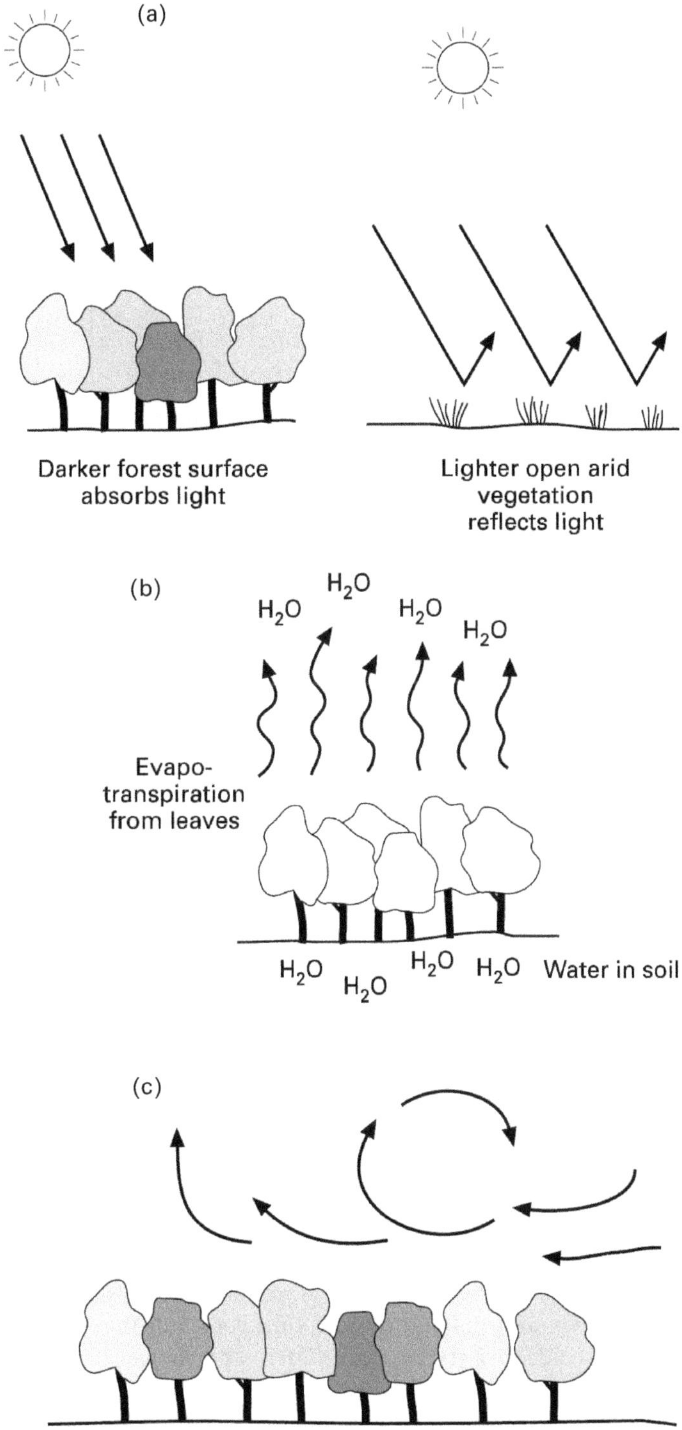

Figure 6.2. Some of the ways in which forests modify temperature. (a) Albedo: the dark forest surface absorbs sunlight, warming the air. (b) Latent heat uptake in evaporation cools the air. (c) Roughness helps to feed heat and water vapor to the atmosphere above, cooling the forest. *Source*: Author.

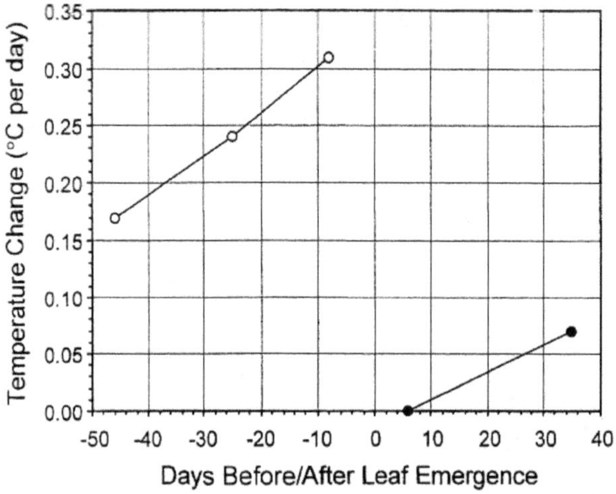

Figure 6.3. As the leaves come out, the progressive warming into spring halts for a few days because of the latent heat taken up by evaporation from the leaves. After Bonan.

surrounding air cooler than it would otherwise be (Chapter 4). The moisture from leaves also affects broader-scale aspects of regional climate. It increases the humidity—giving, for example, the sticky summer climate of the southeastern USA when plenty of heat and plenty of rainfall combine in a predominantly forested landscape. This humidity itself keeps the night warm: as the air cools in the evening, some of this water vapor condenses out yielding heat that helps prevent the air from cooling further. And the water vapor itself acts as a "greenhouse gas", trapping heat radiated by the forest canopy during the night and sending it straight back down to earth.

Leafing out in spring in the temperate forest zone has an immediate effect on temperature due to the onset of transpiration. The progressive increase in temperature into spring is halted for a few days by the transpiration from these newly formed tree leaves (Figure 6.3). These patterns seem to be paralleled more extensively in the tropics, where models and observations suggest that transpiration from forest keeps the climate cooler and rainier (see below), and less variable between night and day.

So, in at least some cases such as these, Chekov's characters were right after all: the forest does seem to moderate the climate.

Another way that forests modify atmospheric processes is through "roughness". The crowns of forest trees often look like heads of broccoli packed all against one another, with bumps on the surface and valleys between them. This uneven surface lets the wind blow down between their crowns but then find its way blocked by others in front, and it tends to set the air rolling, a type of motion known as "turbulence" (Chapter 4). The big trunks and branches also act as barriers for the wind, slowing the wind down and making it more turbulent. All this turbulence set off by the forest canopy tends to carry heat and water vapor upwards more effectively. So, the roughness of forest surfaces makes them feed water to the atmosphere more rapidly, compared with smoother surfaces such as grassland, crops or bare desert.

6.2 WHAT DEFORESTATION DOES TO CLIMATE WITHIN A REGION

What will happen if a forest is removed and replaced with much more open vegetation, such as grassland or fields of crops? In a general way, there will be two competing effects on local climate. First, albedo will be greater over the more open grassland or cropland with patches of lighter soil between the leaves. This will tend to cool down the surface because solar energy is reflected straight back to space. However, the smoother surface of a grassland or crop cover—and the smaller total amount of leaf cover—will tend to decrease evaporation of water. As mentioned above, roughness increases evaporation/transpiration, and more leaves mean more surface area to evaporate water from, so decreasing these will decrease evaporation. This decrease tends to raise the local temperature because there isn't as much latent heat of evaporation being taken up.

The balance between these two opposing effects of albedo and evaporation varies between different forest types. In tropical areas such as the Amazon, the abundant supply of rain means that the forest can feed water back up into the atmosphere very rapidly through all those leaves. The very high rate of latent heat uptake into all this evaporation means that the rainforest is cooler than more open land, despite it having a darker color (lower albedo). So, when rainforest is cleared the decrease in evaporative area of leaves and the decreased roughness can be expected to dominate, and temperature should tend to be higher following forest clearance (Figure 6.4a). In the boreal conifer forests of Canada, Siberia and Alaska, the balance is predicted to be mainly the other way around. The rate of supply of water from rain or snowmelt is less, and in the generally cooler temperatures water will not evaporate so fast from leaves. So, evaporation does not make such a difference to the temperature around the forest canopy. On the other hand, northern conifer forest is very "dark"—even darker than tropical rainforest—so clearing it makes a bigger difference to albedo. Computer climate models predict that losing boreal forest decreases the temperature, because this albedo effect dominates (Figure 6.4b). In addition, clearing boreal forest exposes snow on the ground that would otherwise lie hidden under the trees. This exposed snow provides a much bigger albedo effect, further decreasing the temperature as a result of boreal forest loss.

What about the temperate forest that sits in the mid-latitudes between the tropical and boreal forests? In this case it seems that, just as with boreal forest, the cooling effect of increased albedo is dominant (though not everyone agrees with this). Much of the year, temperatures are fairly cool so evaporation is not an important factor—the reflection or absorption of heat from the surface ends up being more significant. The temperate forests have in fact suffered more in the past from forest clearance, because their soils tend to be so good for cultivating crops, so the idea of removing them on a large scale is not just imaginary. From the computer models, it is generally thought that the deforestation that already occurred in the mid-latitudes may actually have affected the history of climate over the past few millennia. Several climate model studies have compared the effects of the original vegetation cover of unbroken forest with the present mixed cover of both forest and croplands. They tend to find that across the mid-latitudes there would be a small but significant

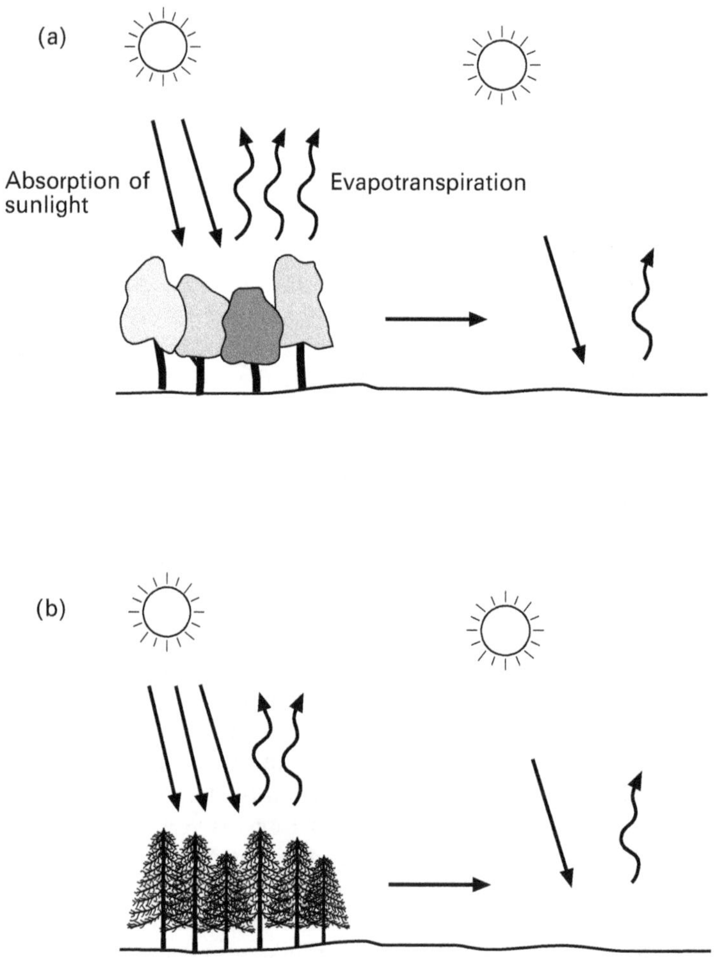

Figure 6.4. (a) In the tropical rainforest, loss of latent heat uptake and roughness dominates and deforestation is predicted to result in a regional temperature increase. (b) In boreal forest (which has less evaporation) the albedo effect dominates upon forest clearance, producing cooling. *Source*: Author.

cooling effect from the actual amount of deforestation. For example, in the eastern USA conversion of 40% of the original forest to croplands (which is about how things are at present) would both increase albedo and decrease surface roughness. According to most of the models, the albedo increase cools summer temperatures by about 0.5°C, and autumn temperatures by about 1.5–2.5°C. So, regardless of global warming or any other background trend in climate, it seems that losing part of its forest has made the eastern USA cooler than it otherwise would be.

Models by themselves are all very well, but it is best to have observations which agree in a general way with what a model predicts. This can give us some confidence that what the model says is valid. For example, it would be ideal to have good time series of temperature and precipitation data from all across the eastern USA dating back to the time of the first European settlements in North America. Then, we could compare the "before" and "after" for deforestation in the region. However, no such data series are available because no-one was collecting climate data back then, so we must look to validate the models in other ways. Fortunately, there are observations from the present day that back up parts of the general picture from models about the effect of forest cover on climate in North America. This comes in the form of an unintentional sort of "experiment": What would happen if you have one region mostly covered in crops, and another region next to it mostly covered in forest? For mid-latitude regions, such as the USA, the models would say that compared with forest the cropland area will tend to be cooler during the peak of the day, because the open cropland reflects much more sunlight back into space, preventing it from heating up so much under the sun. At night, temperatures in spring tend to cool to about the same level, because there is not much difference in the water vapor content of the air at this time of year, whatever the vegetation cover. In the tropics, the greater sun's heating combined with reduced transpiration from leaves would tend to have an opposite effect, making the cropland hotter, but in these mid-latitude areas the effect of albedo tends to be more important, Comparing observations of the daily temperature range in the Midwestern USA (where there is hardly any forest and mostly crops) with the eastern USA (where there is more forest than crops) shows that they differ in just the way that the models would predict. In the Midwest, the daily range during spring and autumn is less than in the forested east, because daytime temperatures in the Midwestern croplands do not rise as far from their night-time level. The east–west difference is greatest in the late spring when the eastern forests have leafed out but the Midwestern crop plants are still small seedlings. At this time there is a lot of bare soil in the Midwestern fields reflecting sunlight, while the eastern forests are already dark and heating up. This agrees with what one would expect from the models: clear away a temperate forest and replace it by cropland, and things will be slightly cooler on average.

Another interesting set of observations that agrees with the expectations from models is the trend in this east–west temperature difference over the last century, recorded from climate stations. Since the late 1800s, forest has been spreading back over abandoned croplands in the eastern USA. Yet, in the Midwest there is if anything less forest than there was at the turn of the 20th century, as agriculture there has intensified. The models would predict that the east–west contrast in spring temperatures would have increased as the east became more forested; and indeed that is what has happened.

Something important to realize, however, is that whether a temperate forest warms or cools the local climate overall depends on exactly what it is being compared with. Natural grasslands and meadows tend to evaporate much less water than forest, so adding temperate forest to the landscape will increase evaporation and result in an overall cooling of the climate. By contrast, planted croplands such as corn or wheat evaporate a fair amount of water from their leaves but also form a lighter coloured surface than forest for most of the year (when the crop plants are young so their leaf cover over the ground is very open or even non-existent), so adding forest at the expense of cropland is thought to warm climate. However, it is not completely certain that this relationship holds true everywhere in the temperate zone. For example, some observations in the eastern USA comparing small patches of forest and adjacent fields suggest that deciduous forest is so good at evaporating water in the summer that it cools the local climate rather more than it warms it through being darker and capturing more sunlight. If this is true, then on a local scale temperate forest is actually more like tropical forest (which cools more than it warms, compared with open land), than boreal forest (which warms more than it cools), and deforesting the eastern USA in the last few centuries could have warmed, rather than cooled the regional climate as many had assumed. A recent model by Strack and Pielke that emphasizes the cooling role of temperate forest suggests that the eastern USA—with about 60% present-day forest cover—is now nearly 0.5°C *warmer* during parts of the year than it would have been back in 1650, a time when the eastern forests were nearly intact. If so, it could be that deforesting the eastern USA actually helped bring the world out of the Little Ice Age!

All of this makes for a confusing and uncertain picture, and my guess is that actually the answer whether temperate forest warms or cools the area on balance could depend on soil type, average climate and the age of the trees. It may even depend on weather conditions in the weeks and months beforehand. For instance, in drought years in the eastern USA the crops die back, and the decrease in evaporation from the leaves can allow the parched fields to be more than 13°C warmer than the forest, which continues evaporating water brought up by its deep roots. At such times, there can be no questioning that the forest is cooling the local climate (the opposite to what it is most generally thought to do in moister years). However, even though the role of temperate forest on a local scale may be rather uncertain, there is no doubt that having it there warms our planet overall through the effect of its low albedo capturing additional sunlight—even if some of that captured energy temporarily gets taken up in the latent heat of evaporation.

The USA might also have undergone changes in *rainfall* patterns due to changes in forest cover. A modeling study by Copeland and colleagues suggests that the clearance of southeastern coastal plain (Figure 6.5) forests in the USA, and replacement by cropland since early colonial times, has shifted the peak area of rainfall southwards. Previously, the model suggests, the rainiest place in the region was the Appalachian Mountain belt. But now—both in reality and in the model—the most rainfall occurs over the northern edge of the coastal plain, at the boundary between cropland and forest. This hypothetical shift resulted from increased atmospheric upwelling over the sudden discontinuity in the landscape between forest of the mountains and rolling Piedmont country to the north, and cropland to the south. The fact that the maximum rainfall now occurs just where the model says it should is an encouraging result, which suggests that modelers are getting things about right.

The effects of tropical deforestation on climate have occupied environmental scientists for several decades now, fueled by both old traditional concerns and the new results of climate modeling. Because deforestation in Amazonia has been happening so rapidly in recent decades, much of the scientific work in observing and modeling forest feedbacks on climate has been concentrated in this area. Various projects have compared the local climate, in areas that had recently been deforested for ranch lands, with adjacent areas of intact forest. These field observations have shown that locally cleared areas tend to have an increased daily temperature range,

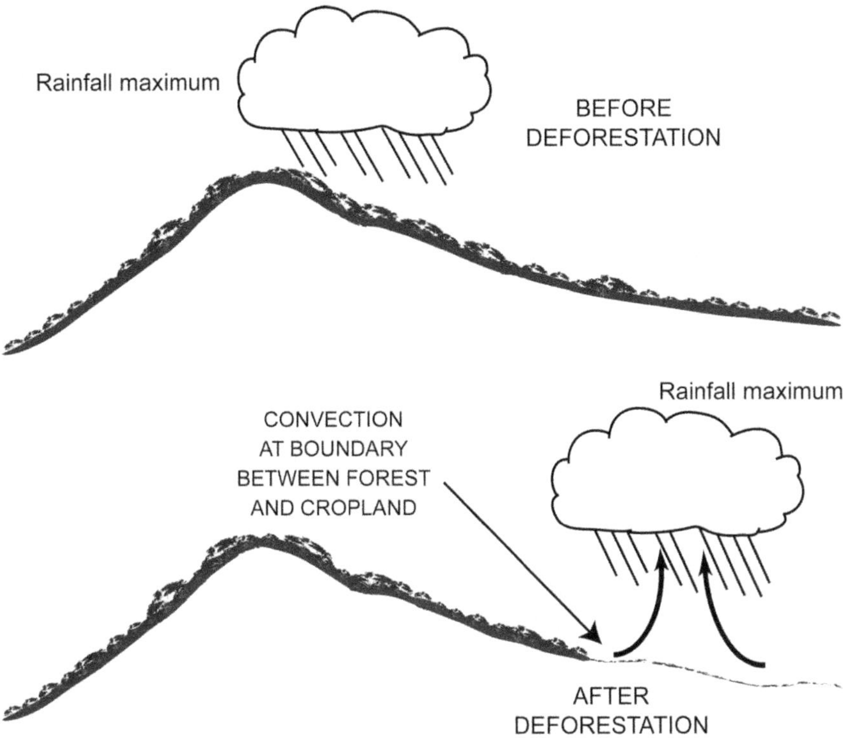

Figure 6.5. In Georgia, in the southeastern USA, models suggest that after the lowlands were deforested, the maximum rainfall area shifted away from the mountains (a) and towards the boundary between forested land and cropland (b). This is because a boundary in land surface conditions helps to create the atmospheric instability and movement which creates rain.

and an increased daily range in humidity, with a peak of temperature and dip in humidity in the middle of the day. However, the overall average temperature and humidity did not change much. These were local-scale studies, but what would happen if Amazonia was deforested on a much broader scale, with all of its forest replaced by grasslands? Computer models that have simulated this scenario of widespread destruction suggest that there would be increased temperatures, less evaporation of water from vegetation and less rainfall.

The temperature increase in a deforested Amazon, around 1.4°C on average, would be due to less latent heat being taken up into transpiration from leaves, plus evaporation from rainwater sitting on the top of the leaves. The leaves of the grassland that might replace the rainforest cannot rival a whole forest canopy as an evaporative surface, and the turf of a grassland cannot match the forest's dense meshwork of roots and spongy soil organic matter for holding water. Note, however, that even though the temperature in Amazonia would go up, the world as a whole would get cooler as a result of this deforestation. This is because the latent heat that cools Amazonia is ultimately an important source of heat to the high latitudes (see the section below, on remote effects of deforestation).

Models predict that if all the Amazon forest was cleared, rainfall inland in the Amazon Basin would decrease by about 20%, enough to make things too dry for forest in some of the more climatically marginal areas of the rainforest. The main reason for this is the loss of recycling of water vapor within the rainforest region. When rain falls on intact forest, much of it is caught by the root mat and soil, and eventually taken up by the trees and evaporated from the leaves of the canopy. The moistened air then drifts further inland, where it can once again give rain that nourishes the forest. If a forest cover had not been in place, much of this rain could have run straight off into rivers and down to the sea, and not recycled. In the interior of the Amazon Basin, a large part (around 50%) of the rainfall depends on this recycled water vapor from forests elsewhere in the Basin. A grassland cover could recycle some of this rainfall too, but not nearly as effectively as forest.

So far, no decrease in rainfall has been detected in Amazonia, although only a relatively small proportion of the region (about 12–15%) has been deforested up to the present. However, there are some disturbing decreases in rainfall in other parts of the world that look like they may be a product of deforestation. For example, in Thailand there has been a drying trend during the last 30–40 years, with a 30% decrease in September precipitation (it is now 100 mm lower). Climate–vegetation models suggest that this trend could be caused by the extensive deforestation that has occurred in Indo-China since the 1950s. Partly, the cause of this change in climate is less recycling of rainfall by evaporation from the forest canopy (so water instead runs off as streams and rivers to the sea). But, there is also another mechanism. Drier air does not promote as much convection in the atmosphere, because moisture does not condense out and release latent heat that might keep the air rising. Because of this there is less of the atmospheric instability needed to give rain. Interestingly, the drying trend in Thailand has been limited to a precise time of year; it is not present, for example, when the summer monsoon is in play during July and August. Modeling the system explains why this is so, and further implicates deforestation as the cause. During the monsoon, the regional influence of the evaporation from forest canopies is in effect flushed away by strong winds from the west that carry moisture in off the ocean. So, we only see the effect of deforestation once the system "calms down" after the monsoon, when the westerly wind has stopped blowing and rainfall comes mostly from more local sources of evaporation.

Another area where deforestation may have caused a long-term decrease in rainfall is southwestern Australia. Since the mid-20th century, rainfall in the area around Perth has decreased drastically. River inflows are about 42% less than they were previously, causing some major problems for the city of Perth that uses the rivers as a source of drinking water. A modeling attempt suggests that as much as half of this reduction is due to land cover change, with forests being replaced by croplands and pasture. In this case, apparently the main mechanism at work is the reduced roughness of the land surface. When trees were there, the uneven canopy produced greater vertical air movement, and less horizontal movement (due to the drag from the canopy). This used to favor rain production from the water vapor that blew

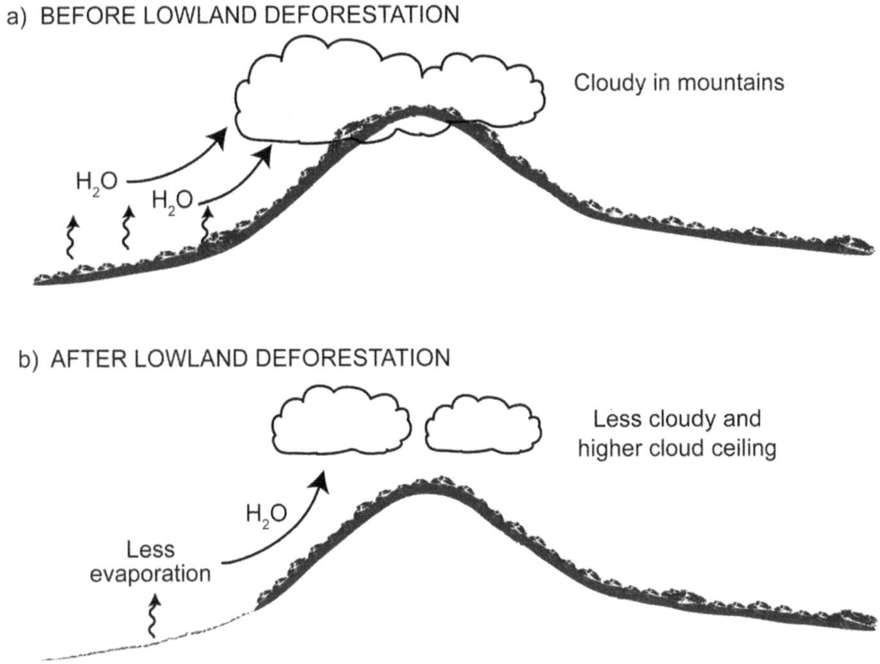

Figure 6.6. In Costa Rica, before deforestation, evaporation from lowland forests supplied abundant clouds shrouding mountaintops and supporting cloud forest (a). Now, lowland deforestation has resulted in less moisture supply to the atmosphere from the lowlands, so clouds are sparser and higher in the mountains (b). This has changed the ecology of mountain forests.

inland from the sea or evaporated from the forest, falling back over the same forested area. Now, according to the model, moist air simply moves further inland and drops its rain there, out of reach of the catchments for Perth.

Since the 1940s, Costa Rica has lost much of its lowland forest cover, and this seems to have changed the climate of adjacent mountains. Clouds now form less frequently and rainfall has decreased over the "cloud forest" zone in the mountains (Figure 6.6). The clouds also seem to form higher in the atmosphere, so they "miss" the mountain tops that they previously used to shroud and keep moist. The observations are backed up a model put together by Richard Lawton and colleagues, which seems to firmly link cause and effect: less evaporation, less convection and less turbulence over the area that was once forest has changed the distribution of clouds and rain formation up in the mountains. Clouds are predicted to form less frequently, and higher in the sky when they do form. The drying of the cloud forest is thought to have been a contributing cause in the mysterious extinction of several species of colorful tree frogs—known as harlequin frogs—that only occurred in these mountains.

The effects of removing tree cover do not just occur in areas that were previously completely forested. Generally, from the models it seems that climatic effects of adding more leaf cover "saturate" at high values; there is not much difference between a very dense and a fairly dense forest canopy in terms of what it does to climate. But, the difference between no trees at all and just a scattered open covering of trees can be far more important. Even the removal of a very open, incomplete tree cover may affect local climate. A modeling study of savannas in Brazil suggested that loss of just a fraction of the tree cover is enough to significantly decrease precipitation, increase temperature, increase wind speed and lower humidity. All these changes would tend to promote the spread of fires, further reducing the tree cover; the effect of an initial removal of trees becomes amplified so that more trees are lost.

In some areas, however, breaking up a forest cover a little might actually *increase* rainfall. In a model study by Roger Pielke and colleagues, replacing the original forest cover of the northern part of Georgia (in the southeastern USA) with the present mixture of fields and forest actually increased the rainfall. The key in this case was that the forest areas were so wet that most of the sun's energy reaching them went into latent heat, not warming of the air. With so little warming occurring, it was difficult to sustain much convection (rising air). By contrast, in the field areas there was not much latent heat uptake, so the air above the field could heat up. This caused it to rise up into the atmosphere through convection, and when it did it sucked in moist air evaporating from the adjacent forest areas. This gave storm clouds and rain when it reached high enough into the atmosphere. The key here then is that the open fields act as "focal points" for convection (Figure 6.7), and the increased convection promotes rainfall.

It is important to bear in mind that in these last two studies—as with most work on regional vegetation–climate effects—there are no relevant climate data to show

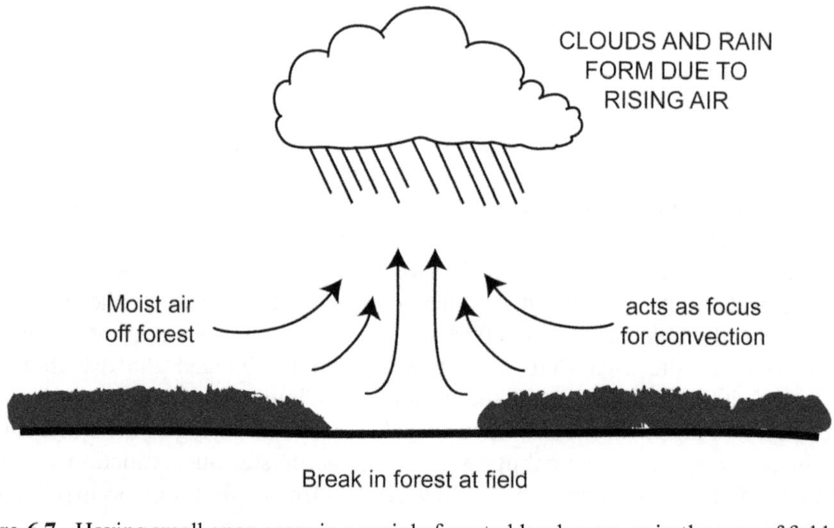

Figure 6.7. Having small open areas in a mainly forested landscape, as in the case of fields, can actually increase rainfall by providing focal points for rising air which can be carried high and condense out rain clouds.

whether nature really works like the model says it should. In most places in the world, the carefully controlled observations of climate or vegetation cover (that one needs to test a vegetation–climate feedback model by comparing the "before" and "after" scenarios) are missing. Often, this is because the "before" happened a long time ago before detailed records were being kept, or because the "after" has not yet occurred. Studies such as those in Costa Rica, Thailand and western Australia are precious gems to the world of modeling vegetation–climate feedbacks, for these provide detailed observations of climate change that correlate with a change in vegetation, and which can also be well explained by a model. Bolstered by the assurance that at least some things are well understood, modelers have the confidence to try predicting vegetation–climate feedbacks on a much broader scale.

6.3 RE-AFFORESTATION

At present, while some areas of the world are being deforested, other places are regaining the forest cover they once had. Sometimes this is due to deliberate planning and planting, as in the large areas of China and South Korea that have been re-afforested in recent decades. Other times, the return of forests is simply a by-product of social and economic forces. In the eastern USA, forests came back as a result of the collapse of the farming economy in competition with the more productive lands of the Great Plains. For example, after going down to less than 25% forest cover, the state of Rhode Island is now back to being more than 60% forested. The Low Countries of northeastern Europe have several times more forest now than they did 150 years ago. A similar process of abandonment of land and return of forest is occurring in parts of Eastern Europe at present. It would be interesting to know what these changes might be doing to climate, and the models intended to understand past deforestation effects can also be applied to the reverse process. Presumably, there would be a warming influence on climate as a result of the return of these temperate forests.

It is difficult to know what might happen in the future, but if deforestation in the tropics is eventually brought to a halt and then reversed, increasing forest cover might also have important effects on climate, presumably including decreased ground level temperatures (due to more evapotranspiration) and an increase in rainfall (due to more recycling of rainwater from the forest) in many of those regions.

6.4 THE REMOTE EFFECTS OF DEFORESTATION

Studies using GCMs (general circulation models, see Chapter 1) suggest that the influence of deforestation can extend very far afield, and the distant effects are often stronger than the local ones. Decreases in the amount of tropical forest seem likely to lead to cooler temperatures in the mid and high latitudes, especially the northern hemisphere, in winter. This is mainly because there is less evaporation of water from the tropical land surface when the forest is gone. Evaporated water is a great store of heat, and when it reaches cooler regions it can condense back out as water droplets in clouds, which releases the heat and tends to keep the air from cooling further. So, the tropical forests provide a "heat subsidy" to the higher latitudes, and when the tropical forest is reduced this long-distance source of heat also diminishes. Not only temperature is affected by the change in heat and water flux to the atmosphere.

Another remote effect predicted from deforestation close to the equator is a decrease in monsoon rainfall farther to the north. For instance, deforestation in equatorial Indonesia and Malaysia could weaken the monsoon over southern Asia (India and Bangladesh, through to Thailand and Vietnam). However, not all models predict the same effects. One model suggests that, if there was more forest in tropical Asia, Australia and east Africa, the monsoon over India would actually be weaker. The workings of the climate system on the broad scale are so complex that the slight differences in assumptions between different models can lead to entirely different results!

Such remote effects do not necessarily take the complete destruction of the rainforests to become noticeable. Even the amount of tropical deforestation that has already occurred over the last few thousand years, compared with the present natural state, is predicted by one model to have produced significant changes in climate in other parts of the world. The model—by T.N. Chase and colleagues at the University of Colorado—suggests that there would have been several degrees' centigrade cooling in winter temperature in the mid-latitudes, in such areas as Europe and North America. Much of this remote effect occurs through changes in the amount of latent heat leaving the tropics in water vapor. With less forest, there is less evaporation, and less long-distance export of latent heat to the mid-latitudes. If tropical deforestation continues, it may significantly change the climate for people living in parts of the world that seem entirely remote from these "Third World" problems. Interestingly, these cooling effects work at the same time as CO_2 released from the cleared forests (Chapter 7) is tending to warm the world, setting off a sort of "tug of war" between the two effects.

6.5 THE ROLE OF FOREST FEEDBACK IN BROAD SWINGS IN CLIMATE

Geologists have established that the earth's recent history has been dramatically unstable. On almost any timescale one looks at during the last couple of million years, there have been both regular oscillations and sudden jumps in climate. And it is turning out that forest feedbacks may have played a substantial role in these past fluctuations.

6.5.1 Deforestation and the Little Ice Age

Between 1000 and 1900 AD, forests in many areas of the northern mid-latitudes retreated under the onslaught of a growing population. The most dramatic episode of deforestation occurred after 1600 in North America, as European settlers arrived

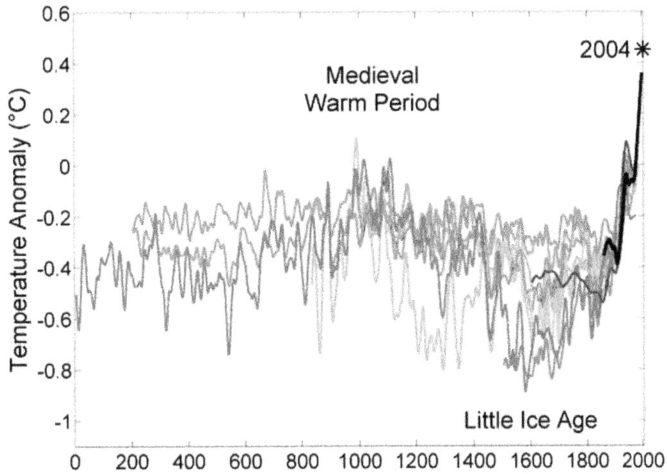

Figure 6.8. Global temperature history of the last 2,000 years from several sources of tree ring data, showing the "Little Ice Age" dip after about 1300 AD. *Source*: CDIAC.

and began clearing land for fields. By 1850, most of the forest that had covered eastern North America was gone. It is possible that the conversion of forests to fields cooled the earth's climate slightly, due to the higher albedo of the exposed soil of the fields and of the dry yellow plants as grain and hay ripens in summer. A modeling study suggests that the global cooling from change in land use in various regions after 1000 AD would have been 0.25°C overall, but concentrated in the northern hemisphere lands (which would have become 0.41°C cooler). The effect would have been especially intense in particular regions: North America, the Middle East and South-East Asia would have borne the brunt of the cooling. Europe, India and North Africa would also have been affected, to a lesser extent. Models suggest that this cooling effect of deforestation could have been strong enough to bring about a known cool period—the "Little Ice Age"—which lasted several centuries between 1000 and 1900 AD.

The Little Ice Age has been recognized from many different sources of historical and geological information (Figure 6.8*)—for example, glacier limits extending downslope, sediment isotopes, crop-planting distributions and accounts of rivers freezing over (Figure 6.9). Overall, it involved a temperature decrease similar in size to that modeled from deforestation, but it is not clear that the real climate changes corresponded in time to what the models predict. The timing of the Little Ice Age, and whether it was synchronized between different parts of the world, is still something of an uncertainty. It seems to have come on several centuries earlier than the main phase of deforestation in the Americas, although many of the coolest phases of the Little Ice Age occurred between about 1650 and 1800 when deforestation was at its peak. Whereas forest cover changes were progressive and unremitting, the Little Ice Age often reversed itself for a while; even during the coldest centuries there were brief warm phases in amongst the cool phases. This does not quite look like a simple

* See also color section.

Figure 6.9. Scene from a frozen river in Holland, 1608. During the Little Ice Age in Holland, Belgium and England the freezing of rivers and ponds often enabled large gatherings and fairs to be held on the thick ice, something that has not been possible since. *Source*: Old image, not copyrighted.

influence of forest cover on temperature. Furthermore, the Little Ice Age was already ending during the mid to late 19th century, when forest cover was actually at its minimum in eastern and Midwestern North America and when the cooling influence would be expected to be at its strongest. Thus, the timing of events in the Little Ice Age does not suggest that deforestation was the sole cause of the cooling, although it may have been a significant factor alongside multiple influences on climate that waxed and waned in importance over several centuries. Furthermore, some recent work suggests deforesting eastern North America actually *warmed* and not cooled the climate, because it decreased evaporation.

Whereas models dealing with tropical deforestation have often forecast changes in precipitation and evaporation following deforestation, the model that forecast the Little Ice Age cooling does not suggest any major effects of mid-latitude forest loss on precipitation, rather than temperature. This is because mid-latitude precipitation is controlled by the broad sweep of wind systems that blow across whole oceans and continents (an example being the wind belt known as the westerlies), not by local convection as in the tropical rainforest regions. So, the effect on rainfall in the mid-latitudes is "blurred out" by the winds, whereas the equatorial tropics that do not receive these broad wind belts are at the mercy of the aridifying influence of deforestation. However, historical deforestation in some other mid-latitude areas may have had stronger effects on precipitation.

6.5.2 Deforestation around the Mediterranean and drying in North Africa

There is a lot of evidence that the climate around the Mediterranean has gotten drier during the last 2,000 years, especially along the northern edge of North Africa. In the Atlas region of Tunisia (which was part of the Roman Empire), Roman era aqueducts lead down from hills where no streams flow now. Contemporary accounts of grain yields show that rainfall must have been much higher at that time, and there are the remains of Roman farming settlements in areas that are now too dry to farm. A model study has suggested that deforestation in the Mediterranean region from Roman times onwards was the key factor in the drying of climate in that region. It seems that the removal of tree cover would have affected the roughness, albedo and evapotranspiration from the land surface, and this may have altered the whole broad-scale circulation of the region. Even though most of the forest removal occurred around the northern edge of the Mediterranean (e.g., in Italy, Spain and southern France), the most severe effects would have been felt farther south in the Atlas Mountain region, and in the Nile Valley of Egypt farther east. It seems that, when the forests were relatively intact around the northern Mediterranean, the upwelling in the atmosphere that they promoted helped to pull in a trace of the north African monsoon, up to the southern edge of the Mediterranean. This gave just enough rain to sustain crop-growing and much denser vegetation in areas that are now on the edges of the desert. The key factor that led to deforestation seems to have been a breakdown in the traditional land ownership structure that had preserved many forest areas in Roman times. In late Roman times, large "corporations" known as *latifundia* took over, farming the land with less eye to preservation, and when Roman rule finally disintegrated after about 600 AD, the chaos that followed allowed much further deforestation.

6.5.3 Forest feedbacks during the Quaternary

The most dramatic swings in climate have occurred on longer timescales. In the last couple of million years known as the Quaternary, the world has been through numerous global "ice ages" (known as glacials) (Chapter 3). At these times, vast ice sheets built up and spread over Canada and northern Europe, New Zealand and the southern Andes, and temperature zones were pushed equatorwards. Nowhere escaped this global cooling: even the tropics were some 5 or 6°C cooler than now. As well as being colder, the glacial world was much drier. Because of the cold and aridity, broad changes in forest cover around the world occurred during the large climate swings of the last two and a half million years (known to geologists as the Quaternary), on the timescale of tens of thousands of years. At this time, before the invention of agriculture, it is thought that human influence on forest cover would have been minor and any large changes must have been due to the influence of climate itself. Many areas that are now forest-covered were completely treeless during the dry, cold glacial maxima, such as the last one that ended 15,000 years ago.

At other times there were warmer climates in the high latitudes, and more forest cover than could exist today. One example of such a warmer phase is the early

Holocene, between around 9,000 and 6,000 years ago. At that time, climates seem to have been several degrees warmer along the Arctic coast of Siberia, and forests occurred hundreds of kilometers farther north into areas that are now treeless tundra (Chapter 3).

Forest was not merely a passive participant in all of these changes in the past. There are intriguing signs that the forest cover itself has amplified many of the climate fluctuations during the Quaternary. However, it is important at the outset to emphasize that forest cover was not the primary control on the largest scale climate changes. The precise pacing of warm and cold periods shows that most of the broad swings in climate that occurred were instead brought about by the shifting orbital parameters of the earth—the Milankovitch cycles—which affect the intensity of sunlight reaching the northern hemisphere in summer (Chapter 5). The shifts in sunlight (by several percent of the total energy influx) occur on the timescale of tens of thousands of years, alternately warming and cooling the earth with their effects. The warmer summers that occur during the parts of these cycles with extra northern sunlight are key to bringing about these globally warmer times. Snow and ice have the highest albedo of any natural surfaces on earth; they can reflect back more than 95% of the solar energy that hits them. This means that they are remarkably good at preserving themselves against the heating sunlight that might melt them, and in doing so they make the whole earth cooler.

How can just a warmer summer in the north bring about a warming of the whole earth, all the year round? Well, if a more complete meltback of snow and ice occurs during a warm spring and summer, it means that more sunlight can be absorbed by the surface, instead of being reflected. This has the effect of amplifying the summer warming, further warming the climate. This can also melt more of the snow and ice. And the more the snow and ice melts, the warmer things get; and the warmer they get, the more melting there is. The large change in the temperature of this most-sensitive region alters the temperature of all the surrounding regions; it is rather like opening or shutting a door to the outside of a house on a cold night—all the rooms in the house will be affected to some extent because they all connect together. Furthermore, each of those regions has its own feedbacks which can pick up and amplify the changes brought about by the high northern latitudes. Many of these feedbacks are likely to involve vegetation in some way. In this way the "signal" from the north propagates all around the world. When the warming signal is operating most strongly, at times when the northern summer sunlight is at its greatest, the world tends to go through its very warmest phases such as the one between 9,000 and 6,000 years ago.

What northern forest cover seems to do, overall, is further amplify the changes that would have occurred anyway due to the feedback between summer temperature and ice. The warming triggered by the presence of the forest is substantial: for instance, it increases the temperature by several degrees C along the northern edge of Siberia. The most important influence of the forest is on albedo. If there is forest in a northerly climate, the tree leaves and branches tend to cover over the snow that fell through the canopy during the winter. This is especially so with the evergreen conifers that often dominate in these climates and do not need time to leaf out in spring, so the snow cover on the ground is never left exposed. Although some snow tends to accumulate on the leaves and branches, most of it tends to fall off the trees and end up under the forest canopy. If snow is covered over by the darker canopy, it cannot reflect sunlight back into space, so it cannot exert its cooling effect on the climate. The air can get warmer because of this. And furthermore, because the air is warmer due to the dark canopy under the trees it can also directly melt back the snow (it is protected against gaining heat by radiation, but not by direct conduction!). The warm air can also move beyond the forests, to open areas such as tundra, and melt back the snow there. That melting of exposed snow further amplifies the warming brought about by the forest cover. In addition, the greater the warming, the farther north the trees can grow, and the cycle continues until eventually it runs out of momentum in the farthest north where the background climate—even with positive feedbacks—is just too cold to sustain tree growth and snow melt.

Here then is the general way that forest cover seems to amplify climate change in the far north during phases of Milankovitch cycles with increased summer sunlight:

More summer sunlight ⇒ warmer temperatures ⇒ northern forest expansion

⇒ lower albedo ⇒ warmer temperatures

⇒ forest expansion ⇒ etc.

The effect of the changing forest cover thus seems to be a positive feedback on climate change. Any background change due to summer sunlight changes is amplified, producing broader swings in climate. The climate models suggest that in the Arctic there was a several degrees increase in temperature resulting from the forest cover and feedbacks that it set in place during the mid-Holocene period about 9,000–6,000 years ago. The feedbacks induced by forest itself also helped to propel the forest hundreds of kilometers farther north at that time.

When the orbital parameters shifted after 6,000 years ago, everything went into reverse. The sustaining summer sunlight that had formerly ensured the warmth now began to decline. Trees did not do so well and the canopy thinned, and snow cover increased. The northern climates cooled and forest retreated, a change once again amplified by the change in forest cover. If humans had left the climate system untouched, in several thousand years' time the earth would be ready to begin its slide into the next major glacial phase, accelerated by this forest–snowcover feedback.

It is likely that this same vegetation–climate feedback has worked in the past to help pull the world into ice ages. Models suggest that an initial cooling event about 115,000 years ago at the end of the last interglacial was greatly amplified by the loss of forest, and its replacement by tundra and snow with a higher albedo. When this extra albedo feedback is included, it turns out to cool summers in the northern lands by a massive 17–18°C, sending the world plummeting into a major glaciation.

The feedback effect between tree cover and temperature is not only relevant to understanding the distant past and the very long-term future. It might also be very important in the next few centuries. It seems likely that over the next several decades temperatures will continue to increase in the high latitudes due to greenhouse gases placed in the atmosphere by humans. As tree cover responds to the warming, expanding northwards, it is likely to further amplify the temperature increase through the same feedbacks that would have operated during the warm phase several thousand years ago.

6.6 VOLATILE ORGANIC COMPOUNDS AND CLIMATE

It is known that tree leaves evaporate many different organic compounds (VOCs) out into the air around them, especially when they are heated under a hot sun. There are several groups of compounds, including monoterpenoids and isoprene, which are thought to play some sort of protective role within the leaves, though no-one is quite sure what (it might, for example, be against insects, fungi or heat). The rate at which these chemicals are emitted depends on the particular forest type, and also the temperature conditions. Generally, the broadleaved forests of warmer climates (such as tropical rainforests) emit more isoprene, while conifer forests at high latitudes emit the most monoterpenoids.

How might these compounds affect climate? VOCs oxidize to produce a bluish natural haze in the atmosphere, and it is noticeable that many mountain regions in forested areas of the world have names that refer to this haze. A couple of examples are the Blue Ridge and the Smokey Mountains of the heavily forested southeastern USA. Analogous names occur in different languages in many different parts of the world where there is extensive forest cover. Haze tends to reflect sunlight back into space, cooling the lower atmosphere. Given that VOCs are emitted in greater quantities at warmer temperatures, if the climate warms (either naturally or due to human effects), more VOCs will be emitted, damping the warming. So far, no climate modeling has been attempted to quantify this effect, but it might turn out to be significant if studied.

VOCs may also help to make rain by forming clouds, with the oxidized particles derived from VOCs acting as nucleation centers for the cloud droplets. Observations of cloud formation over the Amazon Basin by an international team of scientists in the WETAMC study suggested that VOCs might be responsible for the formation of "shallow" clouds that are very effective at yielding rain. Thus, VOCs may help to promote rapid recycling of rainwater within a particular rainforest area, keeping the local climate wetter than it would otherwise be. It is possible that VOCs might have similar effects in other forested areas of the world, but so far there are no observations or models elsewhere that might give clues as to how important they are.

The haze from VOCs does not normally travel more than a few hundred kilometers, but VOCs can have far-reaching indirect effects on global climate because they are easily oxidized by hydroxide (OH) radicals in the atmosphere. OH is a sort of chemical vacuum cleaner that breaks down many different organic molecules in the air. Because VOCs from plants are so easily oxidized, they tend to "mop up" OH that could otherwise react with and destroy methane, an important greenhouse gas produced mainly by swamps. By in effect preserving methane from being broken down (because it produces VOC that uses up the OH), an increase in global forest area might slightly increase the warming that occurs due to the greenhouse effect. Increases in VOCs are also expected to increase the amount of ozone gas in the lower atmosphere, and as ozone is a greenhouse gas this could likewise warm the atmosphere.

Here then are some of the possible changes resulting from VOC emission by the leaves of trees. Note that 1. and 3. work in opposite directions, and it is not certain whether the cooling or warming effect predominates overall:

1. Leaves emit VOC ⇒ VOC oxidized by OH ⇒ less OH to oxidize CH_4 ⇒ more CH_4 in the atmosphere ⇒ warmer climate.
2. Leaves emit VOC ⇒ VOC oxidized to give particles ⇒ particles promote cloud formation ⇒ more rain.
3. Leaves emit VOC ⇒ VOC oxidized to give particles ⇒ particles promote cloud formation ⇒ more sunlight reflected into space ⇒ cooler climate.

It is possible that the broad changes in forest cover that followed from climate swings in the Quaternary themselves damped these changes. A study my colleagues and I carried out suggested that, due to the lower temperatures and reduced forest cover, there may have been 30–50% less VOC emission in the world under glacial conditions, soaking up less OH and thus tending to lower the methane content of the atmosphere (which is in fact what occurred, though probably due to a combination of factors such as less methane production too). Future changes in forest cover due to deforestation or forest-planting might also indirectly lead to either increased or decreased breakdown of methane, either lessening or increasing the greenhouse effect due to this gas.

6.7 FOREST–CLIMATE FEEDBACKS IN THE GREENHOUSE WORLD

It is generally agreed that the world will warm by several degrees Celsius over the next century or so, as a result of the extra greenhouse gases that humans are putting into the atmosphere. If forest cover stays just as it is now, it is likely to play a role in setting up this new climate through its own feedbacks. And if forest vegetation begins to spread in the greenhouse world, it may well further amplify changes just as it did during the Quaternary, by adding to the warming in high latitudes. For instance, a modeling study by Levis and colleagues suggested that the spread of forests over tundra in response to an initial greenhouse effect warming will eventually warm the spring climate in these northerly regions by an additional 1.1–1.6°C, on top of the 3.3°C warming expected from the greenhouse effect over the next few decades.

In some regions of the world, the influence of forests might already be altering the path of warming perceived in recent climate station records. Just measuring temperature is not necessarily a good way of assessing changes in the heat balance of a particular region. This is because a lot of heat energy is made "invisible" by getting taken up as latent heat during evapotranspiration from forest canopies. Even if more heat is being trapped by the greenhouse effect in a particular region, it might not show up because the forest simply evaporates more water, which takes up the heat and exports it to other parts of the world. If one looks at trends in total heat flux—both temperature and latent heat of evaporation—several regions of the USA are actually warming more quickly than would be expected from looking at temperature alone. By evaporating water that temporarily takes up heat, forests may be disguising the true extent of climate warming.

Recycling of rainwater by transpiration from vegetation across Europe may help to increase the amount of summer variability in both temperatures and rainfall over the region as global warming accelerates during the coming decades, according to a recent modeling study by Senevirante and colleagues. It seems that, with the strong positive feedbacks involving rainfall, evaporation and cloud cover, the summer climate will tend to flip between extreme states: either very hot and dry, or cool and very wet. Some have suggested the recent series of extreme summers in Europe shows that this pattern is beginning to manifest itself.

Another aspect not generally considered is that changes in land use over the coming decades might significantly alter the path of warming away from the trends forecasted by standard climate models. If deforestation in Amazonia continues until all its forest cover has been replaced by grassland, this will alter the time course of climate change in the region. While the greenhouse effect alone would tend to warm the climate in the Amazon Basin by about 2°C, the temperature increase from losing the forest (due to decreased latent heat uptake) would be around 1.4°C, according to a modeling study by Jonathan Foley and colleagues at the University of Wisconsin. This extra warming adds up to a considerable increase in temperature, in an already-warming tropical climate where higher temperatures tend to decrease the photosynthesis and growth of plants. It is quite possible that any remaining fragments of the rainforest ecosystem would be wiped out by this warming, and it would not be much good for the croplands either.

Conversely, cutting down the world's rainforests might actually help preserve the mid and higher latitudes against global warming. Since latent heat from the rainforests helps keep the higher latitudes warm, cutting off this heat source might help to counteract some of the warming from the greenhouse effect. However, this would be of no benefit to tropical countries which would have to sweat out the climatic consequences of losing their forest cover, and we must also consider all the extra carbon from the destroyed forests that would enter the atmosphere as CO_2, adding to the warming everywhere in the world. All things considered, the mid-latitudes might actually end up warmer rather than cooler overall if the tropical forests were cleared.

On the other hand, what if world forest cover becomes more widespread in the future? This will in itself tend to cause a warming of global climate irrespective of any greenhouse effect warming, because forest has a lower albedo than cleared land. If all the world's deforested land were allowed to return to its natural forest cover, it might warm the world overall by 1.3°C due to its effects on albedo, according to a recent modeling study by Gibbard and colleagues. Because tropical forests soak up so much heat in transpiration, this change would barely be felt in the tropics and the warming would mostly occur at higher latitudes (part of the warming there would be a result of latent heat reaching them from the increased tropical forests). On a global scale 1.3°C is a large warming—not very much less than some of the forecasts for the greenhouse effect over the next century, and larger than the warming that has occurred since the mid-1800s that has caused so many changes around the world. In addition, as well as transporting more latent heat around, more forest will mean more water vapor in the atmosphere at any one time, and water vapor is a potent greenhouse gas that might further add to the warming. This scenario of increased water vapor in the atmosphere from more forest cover is another factor that needs to be fully considered in climate modeling, if we are to carefully weigh all the options. There is at present a lot of interest in planting forests to soak up carbon dioxide, to prevent some of the warming from the greenhouse effect (see Chapter 7). Yet, as Roger Pielke and his co-authors have pointed out, with all the vegetation–climate feedbacks this increased forest cover might actually warm the planet by more than the CO_2 that it soaks up!

7

Plants and the carbon cycle

Carbon is the common currency of life. The major biological molecules are all constructed from a framework of carbon, and so living organisms need this element in especially large quantities if they are to grow and maintain their tissues. Carbon-containing molecules also serve as a store of energy for cells to work; the bonds within the molecules are broken and energy is released. For these two purposes—building bodies and fueling them—organisms are always grabbing carbon from one another, or in the case of plants, directly out of the atmosphere by photosynthesis. While many other important elements, like calcium and sulfur, are transferred too, carbon is needed in the greatest quantities and generally the most urgently.

Carbon atoms are shuttled from one molecule to another within the cells and tissues of an organism, and from one organism to another (by predation, parasitism, herbivory, decay and all the other ways in which organisms interact), until they are eventually "oxidized"—in effect, burnt—to give carbon dioxide, which goes into the atmosphere. Each CO_2 molecule may later be taken up by plants in photosynthesis, perhaps in a completely different part of the world because molecules carried by the wind can travel hundreds of miles in a few days. My friend and colleague Tyler Volk of NYU has calculated that from each individual breath a person has taken during the last several years, there are on average 60 carbon atoms present in any particular leaf taken from anywhere in the world! Having been taken up by a plant, the carbon atom can go shooting down a food chain once again. It may end up in soil, and perhaps only released by decay after many decades. Or the CO_2 molecule may drift out across an ocean, dissolve in its waters and stay there for thousands of years before it eventually wanders back out again. It might even end up in sediment on the sea floor, be buried, folded down into the earth and only released millions of years later when it gets spat out of a volcano. The global shuttling process of carbon atoms between organisms, atmosphere, oceans, rocks and soil is together known as the "carbon cycle" (Figure 7.1). Carbon has been recycled between these different compartments since the beginning of life on earth, billions of years ago. For

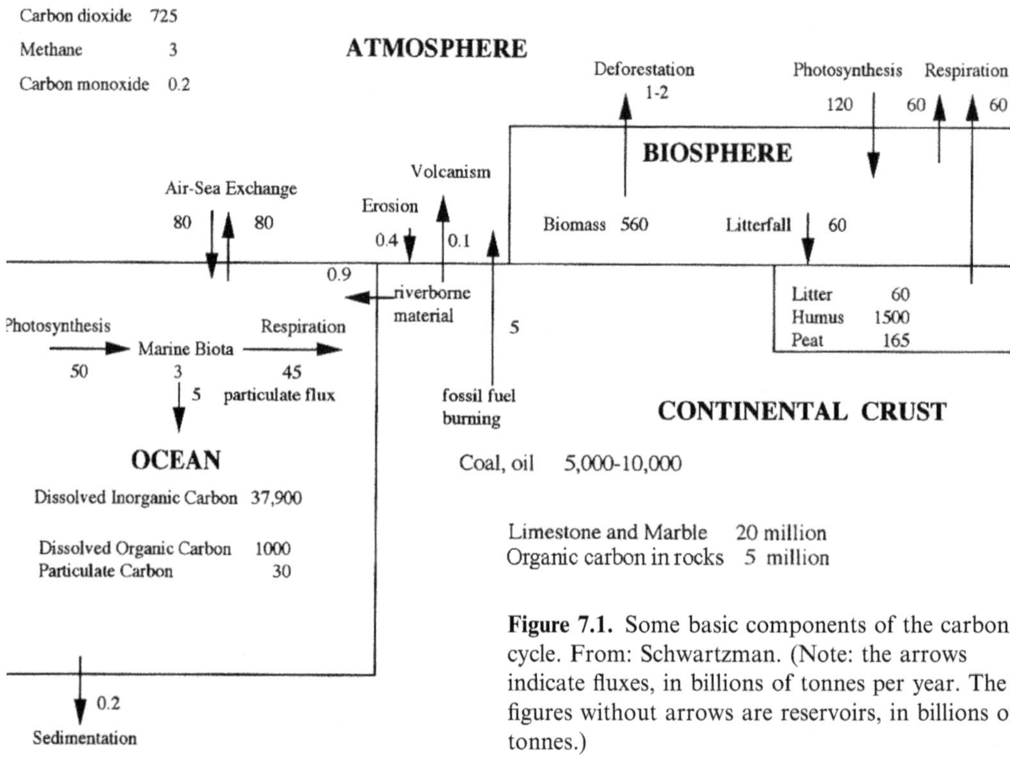

Figure 7.1. Some basic components of the carbon cycle. From: Schwartzman. (Note: the arrows indicate fluxes, in billions of tonnes per year. The figures without arrows are reservoirs, in billions of tonnes.)

sure, there are carbon atoms in each of us that once formed part of the DNA of dinosaurs.

Many people new to ecology imagine that the carbon cycle is a benevolent set-up, with organisms all helping one another out in an interlinked and cosy network. Certainly, it is interlinked, and all forms of life do depend upon it—but not at all cosy. After all, no organism in its right mind wants to get eaten outright. The carbon cycle is largely the product of a dog-eat-dog world (or, at least, a carnivore-eat-herbivore world), with organisms acting out of pure selfishness wherever there is a chance to make a living. Most of the carbon that gets transferred along a food chain is a result of misfortune for some organism or other, and not willingly given at all.

As well as providing the manufacturing base for everything else in the living world, plants excel at storing carbon. Well over 99% of the carbon in living organisms on earth is held within plants, most of this being in trees. Of the still larger "dead" store of carbon in soils, most is derived directly from breakdown of uneaten plant tissues such as fallen leaves and wood. In some ways it is a mystery why so much living green plant material manages to sit uneaten when there are hungry herbivorous insects just about everywhere. It may be that most plant tissues are just too poisonous, too poor in nutrients or too indigestible to be worth eating. Much of the world may look lush and green, but this does not mean that it is edible.

Because it is not eaten, most plant material in land ecosystems ends up falling to the ground and decaying. A lot of the early work in this is done by fungi and bacteria which can work slowly but relentlessly on the tough low-nutrient material of dead leaves and wood. Eventually, all that is left is a dark, soft material consisting of the most unreactive molecules: generally, mostly six-membered phenolic rings of carbon atoms, each ring linked to the others around it. This is what farmers and gardeners call "humus", and it is one of the major stores of organic carbon on earth. The amount of humus and other organic material that builds up in a soil depends on various factors: partly, how active the vegetation above it is sending down a rain of dead leaves, twigs and suchlike. In a moister forest climate, more carbon is likely to build up because the organisms that break it down can barely keep pace with the rate of supply. Where ground is waterlogged, oxygen is in short supply—and without oxygen microbes are barely able to break down the tough, unreactive cell walls of plants. Only partially consumed, the residue of dead plant material builds up as peat, sometimes to great thicknesses. Much of Siberia and Canada are blanketed in peat that builds up in small lakes scooped out by glaciers, on low-lying river floodplains, or over the top of permafrost layers where water cannot drain away, so the ground is always sodden when it thaws in summer. Other large areas of peat have formed in the low-lying forested floodplains of rivers in central Africa and South-East Asia. The world's peats contain perhaps 500 billion tonnes of carbon, rivaling the amount of carbon in CO_2 in the atmosphere. Despite its vast bulk, the peat of an interglacial stage such as the present usually only survives a few thousand years before the climate changes once again to conditions dry enough to allow it to decay, or too cold for the plants to grow and produce the raw materials for peat. During dry, cold glacial phases (such as the one that ended 11,500 years ago) it seems that the world generally has much less peat, and nearly all the present world's vast peat deposits have built up since the onset of the warmer, moister interglacial. Sometimes peat can survive much longer, if it is buried by other sediments before it can decay. Peats that were laid down in sinking river deltas many millions of years ago have become buried, compressed and heated to form coal, one of the major fossil fuel reservoirs which we are now burning to top up the atmosphere with CO_2 (see below).

7.1 THE OCEAN

Out in the open ocean, the store of living carbon as plants is tiny—less than in a desert on land. The floating cells of phytoplankton have lifetimes of only a few days before they sink and die or are eaten, so biomass cannot build up near the top of the ocean. The material that rains down from the surface into the deep ocean slowly rots and disperses into the water as it sinks, in a journey that may take a month. Often it clumps together as it sinks into what is aptly named "marine snow". What reaches the sea floor thousands of meters below tends to be the most inert, indigestible material that bacteria and animals find difficulty making use of. It forms a loose gelatinous material that coats the sediment surface. No-one is quite sure how much carbon is held in the oceans as this fluff on the ocean floor, or in the form of organic molecules dissolved in the seawater, but it might rival the amount stored in soils on land.

There is another much vaster store of carbon in the ocean water, which forms an integral part of the carbon cycle. This is inorganic carbon in the form of CO_2 dissolved in the water. Rather than just existing like most gases would in the form of molecules floating around in solution, CO_2 actually chemically reacts with water to form an acid, known as carbonic acid, with the chemical formula H_2CO_3.

It forms by this reaction:

$$H_2O + CO_2 \Leftrightarrow H_2CO_3$$

CO_2 also reacts with carbonates to form bicarbonate, dissolved as ions in the ocean water. So, for example, if CO_2 reacts with calcium carbonate, an insoluble substance on the sea floor:

$$CO_2 + H_2O + CaCO_3(\text{solid}) \Leftrightarrow Ca(HCO_3)_2(\text{dissolved}) \Leftrightarrow Ca^{2+} + 2HNO_3^-$$

(Note that in both cases the arrows point two ways, because the reaction is easily reversible. Both carbonic acid and bicarbonate can easily break down to yield CO_2 again if conditions shift.)

Oceanographers have chosen to call the dissolved bicarbonate and carbonate forms of carbon "alkalinity", although the term does not have much to do with pH and almost seems designed to confuse any newcomer to the subject! The oceans essentially control the CO_2 level of the atmosphere by storing most of the world's CO_2 in the form of this dissolved alkalinity. If the amount of CO_2 in the atmosphere suddenly goes up, the oceans will gradually dissolve most of it when it reacts with carbonate in the ocean and on the sea floor, forming bicarbonate so that only about one-eighth of the original amount is left in the atmosphere, like the end of an iceberg poking above the waterline when most of it is below. If, on the other hand, the CO_2 level in the atmosphere decreases, bicarbonate and carbonic acid break up to yield CO_2 and the oceans release carbon, pushing the atmosphere's CO_2 content back up again (Figure 7.2a, b). So, the oceans with their huge capacity to store and release

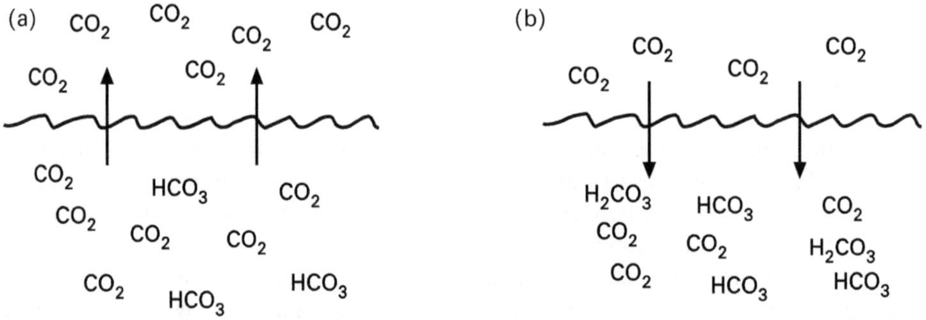

Figure 7.2. A huge amount of CO_2 is stored in the form of both bicarbonate and dissolved CO_2 in the ocean. (a) If the CO_2 concentration in the atmosphere becomes low, CO_2 will leave these reservoirs to top up the atmosphere, often depositing calcium carbonate on the sea floor as the bicarbonate breaks up to yield CO_2. (b) If CO_2 concentration in the atmosphere increases, this ocean reservoir will tend to soak up more CO_2 until most of it has been taken out of the atmosphere—often dissolving more carbonate from the sea bed to make the bicarbonate.

carbon act as a very effective buffer against any changes in CO_2 caused by living organisms, by volcanoes or by anything else. When land plants act to alter the CO_2 level in the atmosphere, they are always working against this massive buffer which rather limits how much they can change the composition of the atmosphere on the timescale of thousands of years. An increase or decrease in carbon storage in vegetation or soils may produce temporary changes lasting a few decades, but those changes will tend to be evened out by oceans taking up or releasing carbon over centuries and millennia. Only if the land plants work relentlessly over millions of years will they finally be able to overcome the effect of this big ocean reservoir and cause major changes to the CO_2 content of the atmosphere.

7.2 PLANTS AS A CONTROL ON CO_2 AND O_2

Since the beginning of photosynthetic life on earth, plants have likely had a big influence on the CO_2 level in the atmosphere. Green and (especially) blue-green bacteria, the precursors and distant cousins of modern-day green plants, began to spread through the oceans about 3.5 billion years ago. They were a source of oxygen, pouring out this highly reactive corrosive gas, which gives life to us but acts as a poison for many of the more primitive bacteria. At the same time, these photosynthesizers acted as a trap for carbon, but not in terms of standing biomass as in today's forests, for there would have been very little living carbon stored at any one time in all the green and blue-green bacteria in the world. Instead they left carbon in debris, dead cells buried in sediment that added up over all the millions of years to a huge amount of CO_2 taken out of the atmosphere. Still dispersed through the world's sedimentary rocks is a vast store of organic carbon, put there mainly by marine algae. This all adds up to an amount of carbon many times greater than the amount in CO_2 presently in the atmosphere. As the deep buried carbon reservoir increased in size over time, oxygen concentrations in the atmosphere would have risen. This is because carbon not buried in rocks tends to accumulate as CO_2 in the atmosphere, holding oxygen as well as carbon. When more carbon is buried, the oxygen is left behind. If all the dead carbon fixed by plants had quickly been able to oxidize back into CO_2, the oxygen left behind in photosynthesis could not have built up in the atmosphere—because when the dead plant cells decayed and oxidized back to CO_2, this would have taken up exactly the same amount of O_2 as was initially released in photosynthesis. Balanced by only the living biomass of plants and the dead carbon in soils at the surface, the oxygen concentration in the atmosphere would be far lower: much less than one percent. As it is, with most of the organic carbon out of reach below the surface, oxygen has accumulated to very high levels—a fifth of the atmosphere.

It is likely that the buried organic carbon reservoir in rocks has also undergone significant fluctuations, sometimes storing up extra carbon and sometimes releasing it. Unlike changes in CO_2 brought about by volcanic output and weathering (see below), this variability in the organic carbon reservoir would have been paralleled by changes in oxygen concentration, because organic carbon released from rocks will always tend to react with oxygen in the atmosphere to form CO_2. So, carbon released from rocks uses up oxygen from the atmosphere. Some calculations have it that about

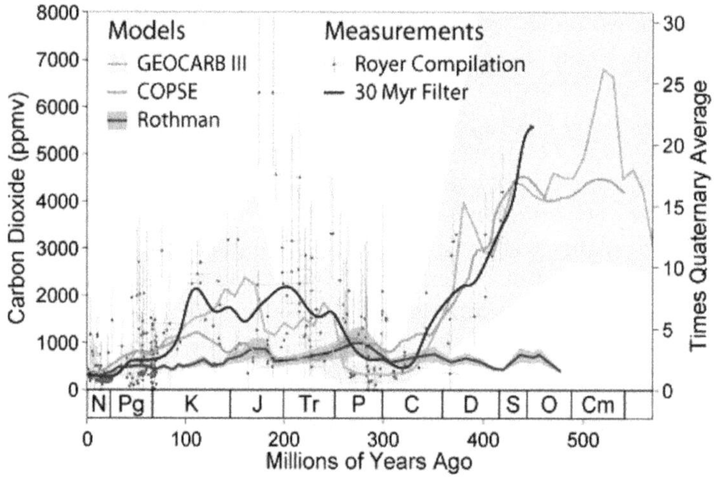

Figure 7.3. Estimated CO_2 concentrations in the atmosphere over the last several hundred million years (Phanerozoic CO_2), using different methods. From: Roger Rhode, after D. Royer.

450 million years ago the oxygen concentration in the atmosphere was only 15%, instead of the present 21%, because there was so little organic carbon held in the sorts of rocks that existed at that time. At that time the CO_2 concentration in the atmosphere would have been something like 15 to 20 times what it is now (Figure 7.3). Over the tens of millions of years that followed, land plants evolved from seaweeds and spread across the continents. Where they formed the first forests in swampy river deltas, they laid down undecayed carbon as peats. Some of these peat layers were compressed into coal, while others were washed away and incorporated as fragments into the sediments of deep ocean floors. Geologists who study the chemical balance of rocks (geochemists) suggest that the huge amount of carbon taken out of the atmosphere by undecayed parts of land plants was enough to cause atmospheric CO_2 levels to plunge down to levels similar to the present. For instance, in the environment in general at that time there was a big decrease in the abundance of the carbon-12 isotope, which is preferentially taken up by plants during photosynthesis (see Chapter 8). This suggests that plants were sucking away a lot of carbon—especially carbon-12—and that it was ending up held in undecayed organic material. In contrast, the oxygen level between around 300 and 150 million years ago might have stood at 25%, or even 30% because so much of the oxygen split off in photosynthesis was unable to rejoin with carbon in decay. Fluctuations in oxygen level would have had all sorts of interesting effects on life at the surface. For instance, they would have affected the ease with which fires could start and spread on plant material. At 15% oxygen, it is hard to sustain a fire even on dry material, whereas at 30% oxygen even moist plant tissues will burn. Some geologists have claimed to find evidence of these fluctuations in oxygen in the form of changes in the frequency of charcoal layers in rocks laid down during the past several hundred million years. For example, the coal swamp forests that existed around 350 million years ago may have at least partially burned every 3 or 4 years. Others suggest that there are too many complexities affecting the likelihood of preservation of charcoal to reach any meaningful conclusions about fire

frequency. It is also rather difficult to explain the existence of forests at times when the atmospheric oxygen level was supposedly around 30%. At this sort of concentration, a single lightning strike in even the moistest forest would cause it all to be consumed by fire, and it is rather unlikely that forest anywhere in the world would be able to grow and reach maturity. Yet, throughout the past 350 million years there is evidence of forests having existed; so, we can at least say that some of the uppermost estimates of past oxygen concentration are probably wrong. On the other hand, some independent evidence that oxygen levels were at least *somewhat* higher around 300 million years ago comes from the existence of huge flying insects, such as dragonflies with 70 cm wingspans. Calculations suggest that in our present atmosphere of 21% oxygen, such insects could not exist because they would not be able to get enough oxygen for their active lifestyle, due to the limits of how fast the breathing tubes (called tracheids) in their bodies can supply oxygen to their muscles. More oxygen around in the atmosphere would also have increased the density of the air, making it easier for such huge winged insects to hold themselves aloft.

7.3 METHANE: THE OTHER CARBON GAS

CO_2 is a sort of common currency for the carbon cycle. It is produced in abundance by all living organisms, and is the chemically stable end point for many different processes going on in the earth's atmosphere, ocean and soils. It also participates in a whole range of different processes, including photosynthesis and chemical breakdown of rocks (see below). Methane gas by contrast is only produced under special circumstances, usually where there is almost no oxygen. It is a result of the incomplete breakdown of organic molecules which would normally be burnt to CO_2 and water by oxygen-breathing organisms. Under anaerobic (oxygen-free) conditions, the energy yield is measly but better than nothing; most of the chemical energy present in the original biological molecules still remains unreleased in the methane that escapes to the surface. The oxygen-free conditions that lead to methane being produced by bacteria tend to exist in the still water and muds of swamps where the diffusion of oxygen down from the surface is slow. They also occur in the guts of animals—especially the fore-stomachs of ruminant herbivores such as cattle, and in the hind-gut of termites—which together produce a large fraction of the roughly 50 million tonnes of methane that enter the atmosphere each year.

Because methane is so rich in stored energy, it does not survive long in the atmosphere or oceans. Some bacteria at the oxygen-rich surface of swamps live by burning the methane that comes from the anaerobic layers below. Much of the methane that bubbles up from anaerobic sea floor sediments is destroyed by methane-consuming bacteria floating in the ocean waters, before it can reach the atmosphere. The methane that does get out into the air above is steadily broken down by reacting with oxidizing fragments of molecules—such as the hydroxyl radical—that exist in an oxygen-rich atmosphere. Most methane molecules are broken down to CO_2—the common denominator of the carbon cycle—within a few decades of being released to the atmosphere.

Humans are adding to the CH_4 content of the atmosphere at present by increasing the incidence of the particular environmental conditions—decay without oxygen—that lead to it being released. Wherever rice is cultivated, the oxygen-free conditions of a swamp are created artificially in the flooded field. As human populations in Asia have grown, the area of these paddy fields has expanded, and output of methane to the atmosphere has increased. Cattle, pig and sheep populations have also increased as the world's population has grown, and this too has led to a large increase in methane output. Another indirect effect of humans has been partial clearance of forest land, which has led to an increased termite population—and termites are also a potent source of methane. The net result of all these factors has been a doubling of the methane content of the earth's atmosphere in the last 200 years.

Although methane still only occurs at very low concentrations in the atmosphere—about 1,700 parts per billion—it is important because it is a very potent greenhouse gas. A molecule of CH_4 traps much more heat than a molecule of CO_2, so less of it is needed to make a big difference. At present, while CO_2 is thought to be dominating the heating up of the world due to the increased greenhouse effect, CH_4 occupies a smaller but significant second place.

While most of the methane from natural decay escapes to the air and is broken down, under some circumstances it can become trapped within the sea bed or under the ground, as a strange substance known as methane hydrate. Methane hydrate resembles ice, but it is actually a mixture of methane gas and water. It can exist under pressure at cool temperatures close to freezing, but if it is warmed or if the pressure is reduced, it will fizz until it has released a huge volume of methane gas from even a small volume of hydrate. Eventually, all that is left is water, with all the methane having escaped. Methane hydrate exists in huge quantities within the sea bed in certain parts of the oceans. The quantities are uncertain, but it could exceed the amount of carbon currently in soils and vegetation. The Gulf of Mexico is one area where it is particularly abundant, for example. In some other places methane hydrate has formed beds trapped under thick permafrost (e.g., in northern Siberia).

Most of the world's methane hydrate is many thousands—if not millions—of years old. One fear for the future is that as the deep oceans warm due to the greenhouse effect over the next couple of centuries, methane hydrate on the sea floor will start to release methane gas. This might pour methane out into the atmosphere, amplifying the warming. There are some signs that sudden global warming events in the geological past—for example, one around 55 million years ago that extended subtropical vegetation far into the Arctic Circle—were also caused by massive releases of methane hydrate. Some geologists have suggested that the sudden ending of the last ice age was boosted by methane released from hydrate layers on the sea floor and under permafrost.

7.3.1 Carbon and the history of the earth's temperature

There is a fair amount of evidence that for the first 2.5 billion years of its existence (out of roughly a 4.5 billion year history), the earth was much hotter than it is now, probably because its atmosphere was packed full of greenhouse gases such as CO_2

and CH_4. Studies on the oxygen isotopes in silica—which precipitated out of oceans and fresh waters—before about 1.5 billion years ago indicate the earth's average temperature may have been above 50°C. Although this is too hot for multicellular animals and plants, there are many types of archean bacteria (known as thermophiles) that thrive in extreme temperatures in hot springs in the present-day world. Some of them will even grow in water kept under pressure above boiling point. If the temperature reconstructions are accurate, presumably, back then the ancestors of these thermophiles were the major life forms on earth, floating in the ocean and working away in its sediments. Around 1.5 billion years ago, a long time after the time that oxygen-producing photosynthesis appeared 3.5 billion years ago, the earth seems to have undergone a dramatic cooling. This culminated in a series of massive ice ages that brought polar ice sheets right down to the tropics. The jumbled sediments that form just offshore from an ice sheet are found for example in Namibia, a subtropical country which was even closer to the equator at that time. It has been suggested that the relentless extraction of carbon by photosynthesizers in the ocean took so much CO_2 from the atmosphere that the earth's temperature dropped dramatically to around freezing on several occasions, each cold phase lasting a few million years. Then the earth would be seized by a sudden warming that melted back the ice. Exactly what brought the temperature back up again is a matter of conjecture. One idea has it that as global temperature declined to around freezing the uptake of CO_2 by rock-weathering (see below) virtually ceased, and allowed carbon dioxide added by volcanoes to build up in the atmosphere and warm the world up. Eventually, by around 700 million years ago the earth's temperature settled down to stable warmer temperatures, for reasons that are unclear. Animal life first appears in the fossil record after these great ice ages ended, suggesting that the more stable climate was more important in the evolution and survival of complex life forms.

7.3.2 Plants, weathering and CO_2

Over the last several hundred million years, plants have progressively spread out of the seas and rivers, and across the continents. This has meant more living carbon stored in vegetation, and more dead carbon in soils. It has also probably led to a dramatic increase in another route by which carbon is taken out of the atmosphere: weathering. Weathering is the gradual breakdown of the minerals in rocks that are exposed near the surface. In terms of the long-term carbon cycle, the most important chemical reactions of weathering tend to occur on igneous rocks—the products of solidified magmas from within the earth such as granites and basalts. Even in a lifeless world, weathering reactions would occur naturally where there is any water, plus carbon dioxide which acts as an acid to dissolve the silicate minerals in rocks. On the planet Mars, which is apparently without life, chemical weathering was once able to break down the rocks to yield the red iron oxide that gives the planet its characteristic color. Although weathering can occur under lifeless conditions, the signs are that the presence of organisms greatly speeds up the rate of the reactions: by tens, hundreds, even thousands of times. The products of weathering of an igneous rock—whether on Mars or on Earth—tend to be the familiar constituents of soil: clays, quartz sand

Figure 7.4. One of the thousands of species of lichens—symbiotic combinations of a fungus and alga. Lichens are thought to accelerate the chemical weathering rate by several-fold, compared with lifeless rock surfaces. *Source*: Author.

grains, iron oxide and salts. Although it varies according to the different minerals found in a rock, the general chemical process is roughly as follows:

$$\text{Rock silicate} + CO_2 + \text{water} \Rightarrow \text{clay} + \text{carbonate salts} + \text{silica} + \text{metal oxides}$$

In moist environments on earth the carbon-containing salts are dissolved in rainwater and washed by rivers down to the sea (where they form the "alkalinity reservoir", see above), although in arid regions they accumulate inland to give carbonate-rich soils and salt lakes.

As soon as any life evolved on earth, there was probably some sort of living film of microbes covering rocks on land, and it is likely that these microbes accelerated weathering by producing acids and other by-products that etched into mineral surfaces. Another step-up in the weathering rate likely occurred as the first lichens (Figure 7.4) appeared on land, perhaps 600 million years ago from the few tentative fossils. These symbiotic organisms, combining a fungus and an alga, have been shown to produce acids and chelating (ion-binding) agents that can increase the weathering rate by several times compared with a bare rock surface (Figure 7.5). From the point of view of the carbon cycle, the important thing here is that carbon is taken up into the weathering process, ending up ultimately as calcium carbonate and bicarbonate—the "alkalinity" in the oceans. The faster the rate of weathering, the faster CO_2 is removed from the atmosphere. If other sources such as volcanoes do not replenish it as quickly as it is taken up, the CO_2 level in the atmosphere will fall.

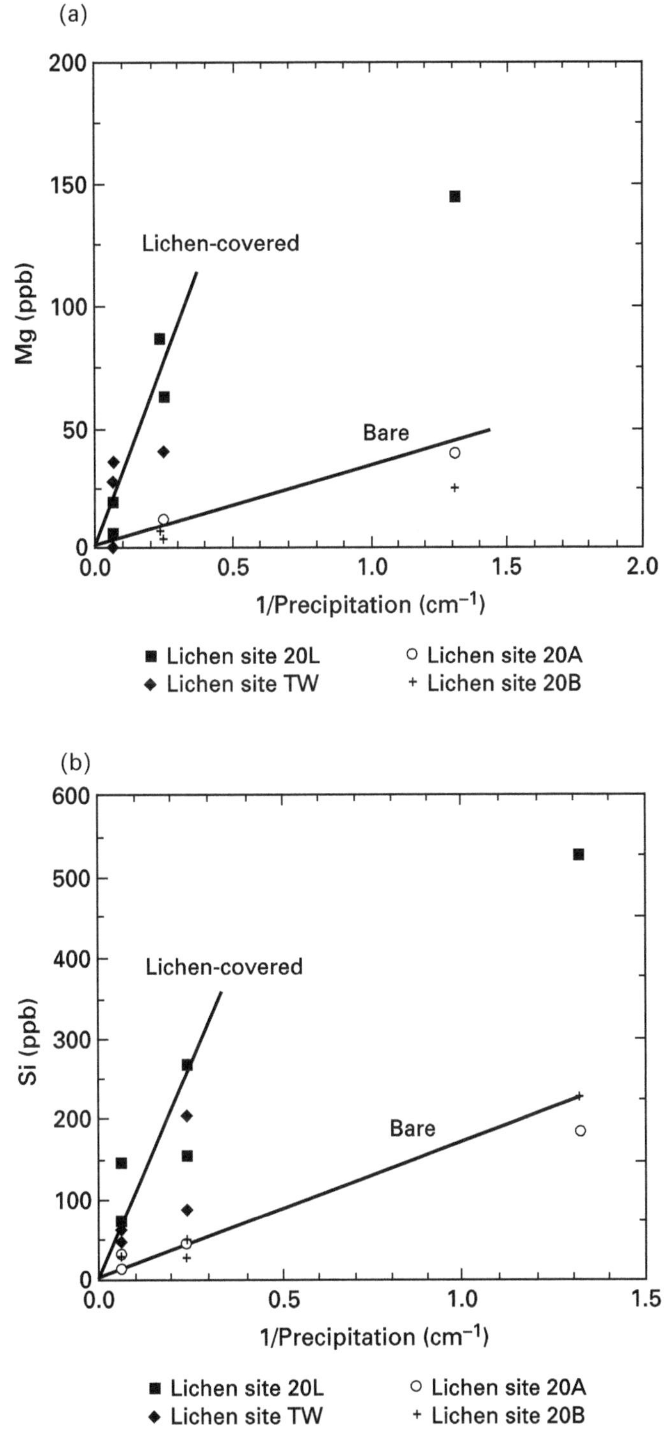

Figure 7.5. Results of an experiment that compared the amounts of salts (derived from weathering) turning up in rainwater that had run off lichen-covered rocks vs. bare rock surfaces. Weathering rate—as indicated by the magnesium ion or silicon content of the water—is several times higher on the lichen-covered areas of rock. After Schartzman.

The bigger land plants, with roots, shoots and leaves, have likely led to a further increase in weathering rate and in CO_2 uptake. Their roots are very good at insinuating themselves into cracks in rocks and between mineral grains. As well as producing their own exudates, plants employ fungi (mycorrhizae) living on their roots to help with the process of breaking down minerals and sucking in the nutrients that are released. From experiments and observations, it seems that vascular plants can increase the weathering rate ten-fold over the simpler lichens and algae which preceded them on the land surface. Mosses are also likely to be good at promoting weathering, because they can form a dense spongy mass over the rock surface, as well as accumulating dead parts that break down to release acids.

By promoting weathering, it is thought that land plants of various sorts may act as a sort of thermostat on the earth's temperature. If a burst of volcanic activity causes CO_2 levels to increase over several million years, the warming that results from it should affect the rate of growth and the mass of plant material around the world. The plants may also benefit from the direct effect of fertilization by the increased CO_2 (Chapter 8). The more vigorous the plants, the greater the rate of weathering that takes CO_2 down. The decrease in atmospheric CO_2 should tend to cool the planet, and as the planet cools the weathering rate should also ease off. This negative feedback may be part of the reason that the earth's temperatures have stayed within the general band that they have, since the origin of complex plant life on land. Without this living thermostat, fluctuations in volcanic activity would sometimes have filled the atmosphere with CO_2 gas and made it burning hot, too hot for life. By turning up its activity as the temperature increased, the weathering thermostat would have taken more CO_2 out and kept a moderate temperature. At other times, if the sun somehow became fainter or ice sheets began to spread, weathering rate would decrease allowing more CO_2 to accumulate in the atmosphere and warming the climate.

Nevertheless, at times the weathering feedback can perhaps be thrown off-balance if it is suddenly presented with too many rocks to weather. When the Himalayas, the Andes, the Alps and other mountain belts grew up almost simultaneously over the past few tens of millions of years, the huge volumes of easily weatherable igneous rocks that they exposed may have led to most of the CO_2 being sucked out of the atmosphere. While the alkalinity store in the oceans would tend to replenish it (as bicarbonate dissociates to yield carbon plus CO_2, which can leave the ocean), there are limits to how much extra CO_2 it can provide. Maureen Raymo of MIT proposed that the eventual outcome of plants weathering rocks from the Himalayas and other mountains may have been a precipitous decline in CO_2 and the drastic global cooling trend that ended in the ice ages of the last 2 million years.

The carbon taken up by weathering does not stay forever in the seawater in the alkalinity reservoir. On the timescale of millions of years, sea creatures extract calcium from the bicarbonate in sea water, to build their shells and skeletons. When they die, this calcium carbonate gets deposited in the mud on the sea floor. This becomes buried, compressed and turned into limestone and other carbonate-containing rocks. On an even longer timescale, over tens or hundreds of millions of years, the carbonate-containing rocks may be folded and heated to the point where the carbonate begins to break down and give off CO_2. The CO_2 percolates up through cracks in the rocks or gets dissolved in the molten magma of volcanoes, and returns to the atmosphere from hot springs or volcanic explosions. Thus, in addition to the carbon cycle of the surface world of plants, atmosphere, soil, and oceans there is a deeper and slower carbon cycle of rocks and volcanoes. Much of the CO_2 which enters the atmosphere from volcanoes and hot springs was previously part of the living world of plants and soils, from which it was taken up in biological weathering processes, and formed into the skeletons of marine organisms before it ended up on the sea floor. It turns out that the biological world is unexpectedly and inextricably linked to the geological world.

7.3.3 Plants, CO_2 and ice ages

Once the time of ice ages had got started about 2.5 million years ago (Chapter 1), marine plants acting on the carbon cycle may also have played an important role in altering the detailed course of both CO_2 and temperature. Analyses of bubbles of the ancient atmosphere trapped in ice cores through the Greenland and Antarctic ice caps show that there were fluctuations of around 30% in the atmospheric CO_2 level. Each time the earth slipped into a major glaciation, the CO_2 level paralleled the climate change, reaching a low point in concentration at just about the time that ice sheets were at their most extensive (Figure 7.6a, b). And each time the world warmed, the CO_2 concentration shot up just about as fast. Essentially, the only way such large and relatively rapid fluctuations in CO_2 could have occurred is if something in the oceans was storing up carbon and then releasing it again. There are various ideas to explain what was occurring, but they all involve phytoplankton as an integral part of the mechanism.

One idea is that the plankton somehow became more productive during glacial conditions, perhaps because there was more upwelling in certain parts of the oceans, bringing nutrients to the surface where they could be used. The increased rain of dead phytoplankton cells and other remains of the food chain dragged more organic carbon down into the deep ocean. There it accumulated as organic carbon in the sediments or dispersed through the water, or perhaps it was oxidized into CO_2 and became part of the alkalinity reservoir (the bicarbonate and carbonic acid combination) in the deep ocean (Figure 7.7a). Either way, part of the underlying cause of the lower CO_2 levels during ice ages would have been the continual "pumping" of carbon down into the deep ocean by phytoplankton; CO_2 was taken up in photosynthesis and (in effect) sent downwards where it accumulated. Somehow, the nutrients got separated out along the way and recycled up to the surface by upwelling, whereas the carbon remained in the abyss. Various models of ocean circulation during ice ages have been devised to explain what might have brought about this increase in plankton productivity, and at least some observations seem to support the idea that the oceans were more productive during glacials. A key area, which both the models and geological studies of the ancient ocean have focused on, is around and underneath the sea ice off Antarctica. This area is already biologically very productive in the modern-day world. If the sea ice area extended and became even more productive

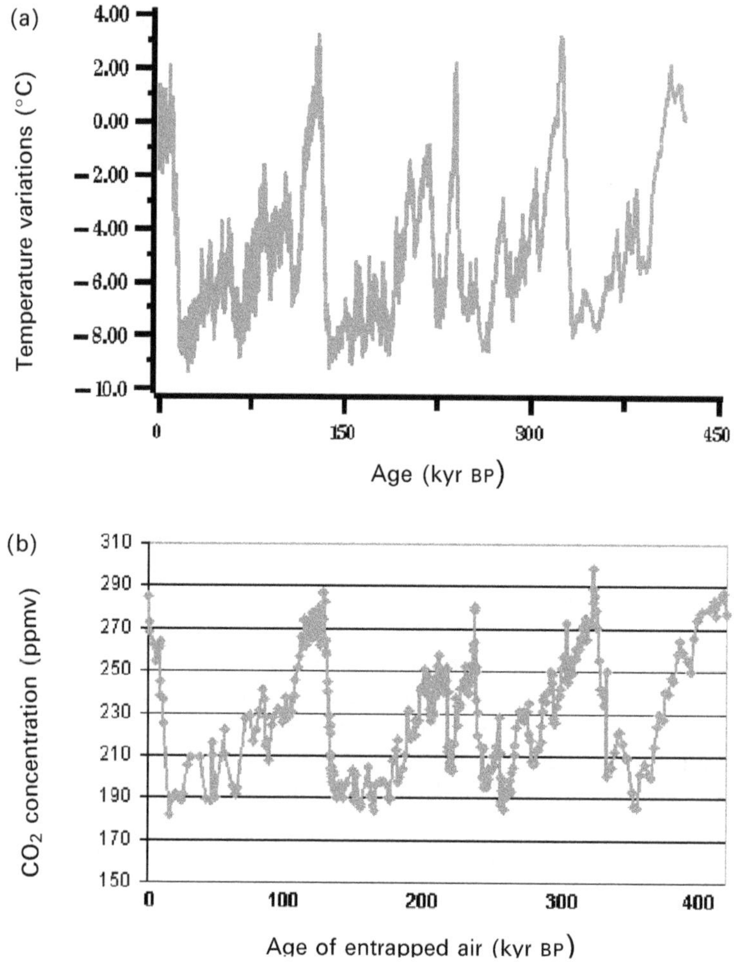

Figure 7.6. (a, b) History of temperature and atmospheric CO_2, deduced from polar ice cores. Temperature history is derived from analyzing isotopes in the ice. CO_2 history comes from analyzing bubbles of ancient air trapped within the ice. Both temperature and CO_2 show the same "sawtooth" pattern of fluctuations on approximately a 100,000 year timescale. From work by Barnola *et al.* (CDIAC).

than it is now, this could explain how extra carbon was sent to the bottom of the ocean. However, the picture from observations of ocean sediments is rather complex, with some indicators of past productivity supporting the idea of more vigorous phytoplankton growth during ice ages, and others contradicting it.

Because of the contradictory nature of the evidence, the idea that increased plankton productivity was the key to low glacial CO_2 has rather fallen out of favor. Instead, attention has tended to focus on a second mechanism that involves a slowed-down circulation in the deep ocean (Figure 7.7b). Basically, the oceans are made up of different layers of water stacked up like a layer cake, each moving in different directions, and originating in particular places where surface water folds down into the deep (see Chapter 1). Each of these layers tends to disperse upwards over time, and they also tend to well up to the surface more rapidly in particular places near coasts

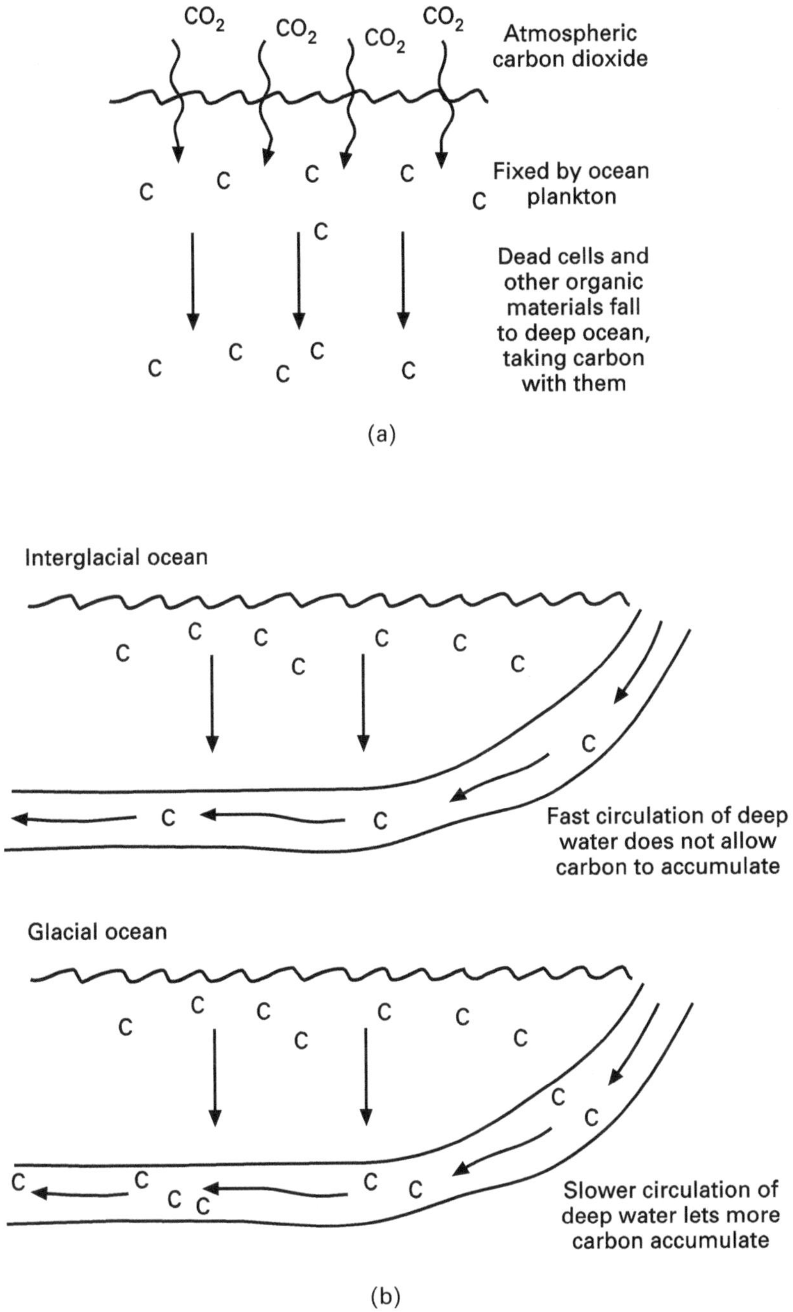

Figure 7.7. (a, b) How plankton activity may have decreased the CO_2 concentration during glacials. (a) An increased "biological pump" puts more carbon down into the deep ocean. (b) A slower circulation of the deep ocean means that more carbon can accumulate there. *Source*: Author.

(such as off the coast of Peru, see Chapter 1). As these water layers move sideways through the ocean on their journey away from the place where they originated, they accumulate a rain of carbon-containing debris from the plankton and other sea life above. The longer that these moving water layers spend in the deep, the more carbon gets loaded up within them. Although the deep waters will eventually well up or mix up into the surface waters and release the carbon they accumulated as CO_2 back into the atmosphere, if they move more slowly they will always end up holding more carbon in the deep ocean. Many questions remain about exactly what was different in the ocean circulation during glacials, but plausible models of the ice age ocean circulation tend to suggest that slower-moving deep ocean water is the best explanation for the low CO_2 of ice ages. So, in this scenario decreased CO_2 during ice ages occurs because of a basic change in the ocean circulation, but it is a mechanism that only works against a *background* of plant productivity always sending carbon down. Once again, we see plants are playing an integral role in the global carbon cycle.

However, while plants in the ocean were working in such a way as to help lower the CO_2 level of the atmosphere, plants on land seem to have been doing the opposite. During glacials when plankton were loading up the deep ocean with carbon, the world's land surfaces were much colder and drier. A vegetation map of any region of the world at that time makes the point when compared with a modern vegetation map (Figure 7.8a, b). Many areas that would now naturally be forest were scrub or grassland, and what are now grassland areas tended to be semi-desert or desert. In the glacial world, peat deposits (which nowadays make up about a quarter of the organic material stored on land) were almost non-existent. Making some reasonable guesses about how much carbon would be present in each biome in its natural state, there can be basically no doubt that there was far less carbon stored on land as vegetation and in soils during glacials, than during interglacials—such as the one we are in at present. The shift in carbon storage at the start of a glacial would have occurred gradually, as growth of new trees and productivity of new plant parts declined, while the respiration that breaks down dead plant parts and the humus in soil continued slowly but relentlessly. When breakdown of organic carbon on the forest floor and in soils is not equaled by new production, the carbon reservoir shrinks and the CO_2 floods out to other parts of the carbon cycle. So, just as the oceans were tending to drag down the CO_2 level of the atmosphere, land ecosystems were releasing carbon out into the atmosphere (Figure 7.9). The land carbon must have had the effect of keeping the CO_2 level higher overall than it would have been if the oceans had worked unopposed. The amount of carbon which left the land system during glacials, about 1,000 billion tonnes (about half the carbon that would have been there on land during an interglacial), would actually have been enough to *raise* the global CO_2 twice over. The major reason that glacial CO_2 was not higher was that most of this (about seven-eighths of it) would have dissolved in the ocean anyway as part of the alkalinity reservoir, leaving only about one-eighth above the water. However, even this amount would be enough to raise up CO_2 level by some 15%. We know of course that the CO_2 level in the atmosphere actually fell, by some 30% during glacials, because the ocean had soaked up the extra carbon. However, if the land system had not pushed out carbon it is reasonable to suppose that glacial CO_2

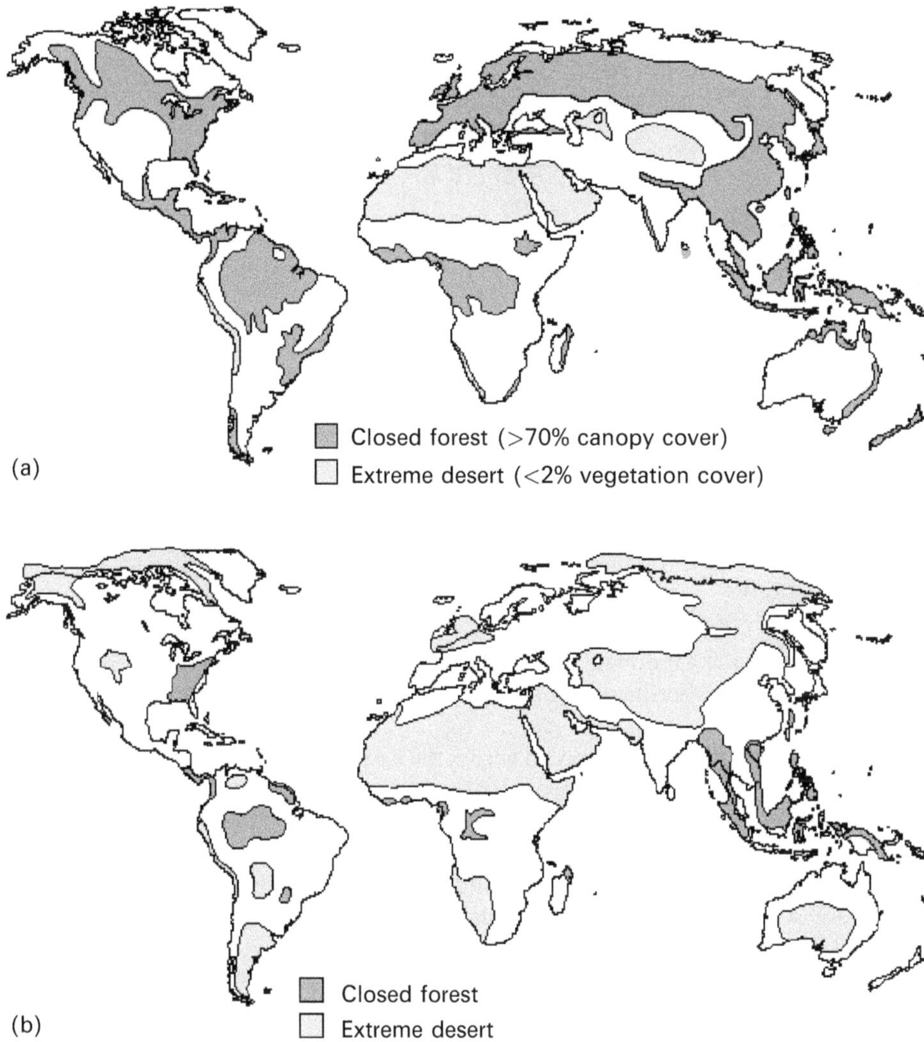

Figure 7.8. The distribution of forest and desert in (a) the present natural world and (b) the last glacial maximum or LGM (18,000 ^{14}C years ago). The LGM world is much colder and drier, resulting in less forest and more desert. *Source*: Author.

would have gone even further down, adding to the severity of the ice age through a weaker greenhouse effect. Climate models suggest that the earth is always on the edge of a runaway glaciation during ice ages, with an ice albedo feedback on temperature that would bring ice sheets down to the tropics just as happened once already about a billion years ago (see above). Part of what has kept this from happening in more recent geological history may be the readily-released reservoir of carbon in plants and soils, that prevents CO_2 from declining too far despite the best efforts of the ocean

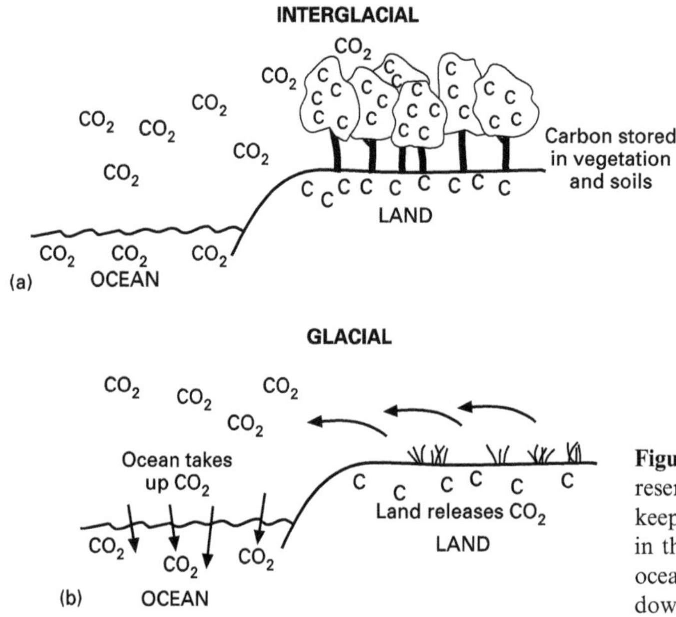

Figure 7.9. How the land reservoir of carbon may help keep up CO_2 concentrations in the atmosphere when the oceans are dragging carbon down.

carbon pumps to drag it down. This is then a negative feedback on the intensity of ice ages: as temperature declines, carbon is released from the land tending to prevent a further descent into a deeper ice age.

As life has diversified and spread across the earth's surface, the potential for such "checks and balances" within the system seems to have increased and has likely kept the climate more stable. However, it is worth bearing in mind that the land vegetation is also working in another very different way to destabilize the earth's climate during ice ages: in Chapters 4–6 we considered how vegetation affects temperature through its albedo. As vegetation cover diminished during an ice age it would not only be releasing CO_2 (which tends to keep the earth warmer) but leaving behind bare soil and exposed snow-covered ground which reflects sunlight straight back into space, cooling the world in a positive feedback. So, as the earth cools this feedback only tends to make things cooler. In the global system, nothing is ever simple; the same component—vegetation—can be participating in both positive and negative feedbacks at the same time!

7.4 HUMANS AND THE CARBON STORE OF PLANTS

No single species alters the world's habitats as much as humans do. Even in distant prehistory, the arrival of people in a particular part of the world could be heralded by an increase in burning. For example, in Australia there was a sudden jump in fire frequency around 45,000 years ago, right around the time when humans first show up in the archaeological record. From what we know of the Australian Aborigines and other hunter-gatherers when they were first contacted by Europeans, humans burned the vegetation for many different reasons. Fire was used as an aid to hunting, in driving frightened prey towards an ambush or over a cliff. It was used to clear brush to provide a clear line of sight for hunting, and to encourage green regrowth that would attract grazers to an area. Fire was also used in warfare, and as a signal, and sometimes fires were just set out of boredom. In many parts of the world, frequent burning must have altered the structure of vegetation and changed its carbon storage. In Australia, for example, the burning led to a spread of grasslands at the expense of woody scrub, and a change in the species composition of the wooded areas that remained. In areas of central Africa where humans have probably been burning the savanna for well over a million years, the fires have tended to keep back the forest while maximizing the extent of grassland.

When agriculture spread out from its several areas of origin over the last 12,000 years, the alteration of the world's vegetation became far more intense. Forests were now particularly susceptible to clearance, because humans had such an interest in the fertile land beneath the trees which can be used for agriculture. The denser populations of humans that agriculture can support also placed a huge demand on wood and other products of trees, resulting in frequent thinning and harvesting of the natural forests. Whole regions—such as Europe, China and India—had almost all their lowland forest cover removed and replaced by fields. In the uplands, where it is more difficult to farm, the forests would sometimes come back after cutting if fires were not too frequent, and if there were not too many sheep and goats to eat the seedlings. In many areas where these factors conspired against it, the forest was not able to come back and the upland landscape remained as scrub, grassland or (in very moist climates) peat bog.

These changes caused by humans occurred against a background of natural shifts in climate that also tended to alter the vegetation, giving less forest and more desert (sometimes amplified by vegetation–climate feedbacks, see Chapters 5 and 6). These natural climate changes alone would have been enough to reduce global carbon storage in forests by some 10%, and humans may have reduced the total by another 10% as they removed the forests of Europe and China during ancient times. The tens of billions of tonnes of CO_2 were released rather gradually over thousands of years, slowly enough for it to be swallowed up by the rest of the carbon cycle rather than accumulating in the atmosphere. Most of it must have gone into the oceans, into the alkalinity reservoir.

In North America, the main phase of clearance was recent enough to be recorded in the contemporary writings of naturalists. Although the American Indians had farmed there for thousands of years, in most areas their population densities were still low enough that they did not make very much impact on the landscape. When the first Europeans arrived, not only was the eastern USA mostly forested (at least 90% was forest-covered), but the physical structure of the forest was quite different too. The earliest times of settlement on the eastern seaboard are not well recorded, but later accounts as the frontier moved inland through Ohio, Indiana, Illinois and Wisconsin make it clear that big trees—very tall and of large girth—were far more abundant than they are nowadays. Particularly impressive were the chestnuts, tulip poplars, hemlock and elms which towered like the columns of a cathedral. Other common species not normally thought of as being especially large—for example, white oak, red oak and sycamore—often reached a much greater size than one would be used to seeing nowadays. Much of the reason that we do not presently see such large trees in the American forests is that there has not been time for them to reach full maturity before being cut for timber. The forests have been kept in a young state by continual harvesting of trees as soon as they became large enough to be useful. The last of the "old-growth" areas to be exploited in the east seem to have been in the Smoky Mountains, in the early 1900s. Nowadays, there are only a few small fragments that may resemble the original virgin forest, mostly in small, steep-sided valleys in the southern Appalachians.

In many areas the forest that the European settlers encountered was cut down and allowed to rot or to dry out and burn, to make way for fields. Either way, the carbon that had been held in the trees ended up being released as CO_2 to the atmosphere. The abundant organic carbon in the humus of forest soils would mostly break down to CO_2 over the first several decades of cultivation. Old-growth forests in North America and elsewhere also tended to have a lot of carbon stored in fallen branches and trunks of dead and rotting trees, with the stumps and roots of dead trees remaining below ground: a reservoir known as "woody debris". Even if the settlers did not deliberately burn this, it would have broken down and oxidized during the first decades after the forest had been cleared. In logged areas that were allowed to remain as forest, the debris would have decayed at its natural pace but would not have been replaced by new material; in a harvested forest trees do not fall over and die but instead end up at the sawmill before they can reach old age. Although much of the timber that was extracted from the original forests of North America was used in construction and not immediately allowed to rot, over centuries many of these buildings fell into disrepair or were consumed by fire. Hence, by various routes the carbon of the original forest would eventually have been released as CO_2. It is thought that the rapid clearance and exploitation of American forests in the mid to late 1700s and early 1800s would have released about 50 billion tonnes of carbon, enough to contribute to an initial up-tick in atmospheric CO_2 that occurred after 1750, beginning the rise that continues to the present (Figure 7.10).

The tropical forest areas of the world have also been exploited for agriculture and timber for many thousands of years, although in most areas the infertility of the soils prevented large-scale farming. The only major exception (at the drier margins of the tropical rainforest zone) seems to have been the Mayan civilization which grew up in the Central American lowlands, using careful cultivation techniques that recycled nutrients. Nevertheless, in the end this civilization too seems to have declined partly in response to a gradual loss of soil fertility.

It is only really in the last century or so that extensive logging and clearance of tropical forests has taken place. Initially, colonial powers began to take useful timber for export, and then (from the 1950s onwards) clearance increased enormously as populations and economies of tropical countries exploded with modern medicine and agricultural technology. The Amazonian forests of Brazil are now being cleared faster than ever, and so far around 15% of the original old-growth forest has been lost.

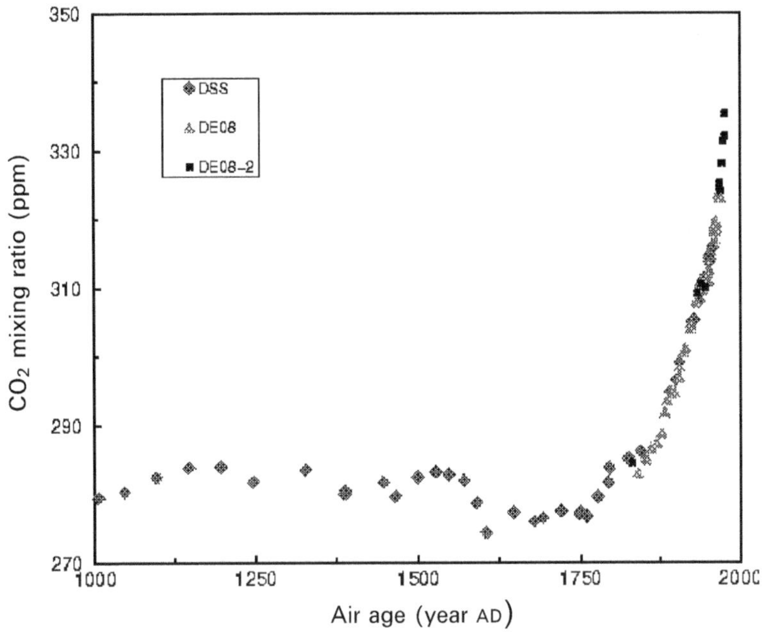

Figure 7.10. Ice core record of atmospheric CO_2 since 1000 AD, Law Dome, Antarctica ice cores. *Source*: Etheridge *et al.* (CDIAC).

Many other tropical countries—such as Vietnam, Panama, Nigeria and Costa Rica—have already lost the majority of their forest cover in the last 60 years (see Chapter 6). Many of the areas that remain forested may also have undergone a more subtle depletion of carbon storage over the last century or so, as loggers have partially exploited the forests for their best timber. For example, Sandra Brown of Winrock International has found indications that the forests of the Malay Peninsula and eastern Amazonia have had most of the really large trees extracted, except in the most inaccessible areas where these forest giants still exist, enabling the comparison. The areas that have had the big trees removed still remain as forest, but they have substantially less carbon within them.

The amount of carbon presently stored in the world's living trees alone is about 300 billion tonnes. If all the world's forest cover were to be cleared and burned, the oxidation of the trees would raise the CO_2 concentration in the atmosphere by nearly half above what it is today. Oxidation of soil organic matter from under the cleared forest land could raise it by as much again. As it is, it is unlikely that all of this carbon would ever be released. However, since deforestation contributes a major proportion (about a quarter to a third) of the CO_2 entering the atmosphere (see below), anything that alters its pace is of immediate importance.

7.5 THE PRESENT INCREASE IN CO_2

Since the late 1700s, the carbon dioxide level in the atmosphere has been increasing (Figure 7.10). The record of air bubbles trapped in polar ice shows the 18th century beginning stages of this rise, which has gently accelerated over time into a steep increase of about 1% a year. The change in CO_2 levels is also recorded in the stomatal densities of the leaves of trees preserved in herbaria; leaves collected around 1750 have a lower number of stomatal pores per unit number of epidermal cells in the skin of the leaf. This is just as one would expect from experiments that involve manipulating CO_2 concentrations (Chapter 8), where plants adjust the density of stomata on their leaves to take best advantage of the circumstances.

It is thought that the beginning of the increase in CO_2 was mostly due to deforestation in eastern North America, as settlers cleared the land for farming. Over time, more ancient sources of plant carbon from fossil fuels such as coal and oil became more important, as the industrial revolution took hold. Nevertheless, around 25–30% of the increase in atmospheric CO_2 that occurs each year is still due to deforestation, mostly in the tropics. Presently, the vast tract of forest in the Amazon Basin is the largest single source of CO_2 from deforestation, with South-East Asia following second.

However, not all forests are losing carbon. Forests areas in several parts of the world have clearly now switched from being a "source" of CO_2, to what is known as a "sink". A sink, in the language of carbon cycle science, is something that is taking up carbon and storing it. For example, in the last 150 years, forests in the eastern USA have become a carbon sink (Figure 7.11*). They made a big comeback, starting in the late 1800s as farms were abandoned as uneconomic in competition with the fertile plains lands farther west. The eastern USA is once again a mainly forested land, and its forests are still relatively young and the trees still growing, so they are storing up carbon rapidly. In China, replanting of previously deforested uplands since the late 1970s has led to a large carbon sink as the trees mature. It is likely that the large-scale movement of population to the cities, and a shift from wood-burning to coal-burning, has also helped forests to recover. Recent estimates suggest that most of the net uptake into forests is occurring in the southern half of China, while forests in the northeast of China continue to be more heavily exploited. An analogous process of forest recovery has occurred in eastern Europe, where a slump in agriculture and movement to the towns has left much land to return to forest. An important thing to bear in mind when thinking about forests as carbon sinks is that no forest can continue soaking up carbon forever. The size of the trees, the amount of fallen woody debris and the amount of organic carbon in soils underneath, will eventually reach a sort of maximum steady state. Carbon may be continually fixed by photosynthesis, and released by respiration and decay, but this is just turnover without change in the size of the carbon reservoir in the forest. Individual trees may continue to die and be replaced by new ones, but overall on the scale of the whole landscape there will be no net increase in the amount of carbon contained in the forest ecosystem (Figure 7.12). Thus, any forest carbon sink will eventually start to saturate and stop taking up carbon. However, starting from newly planted or recovering forest this steady state will only be reached after several hundred years.

* See also color section.

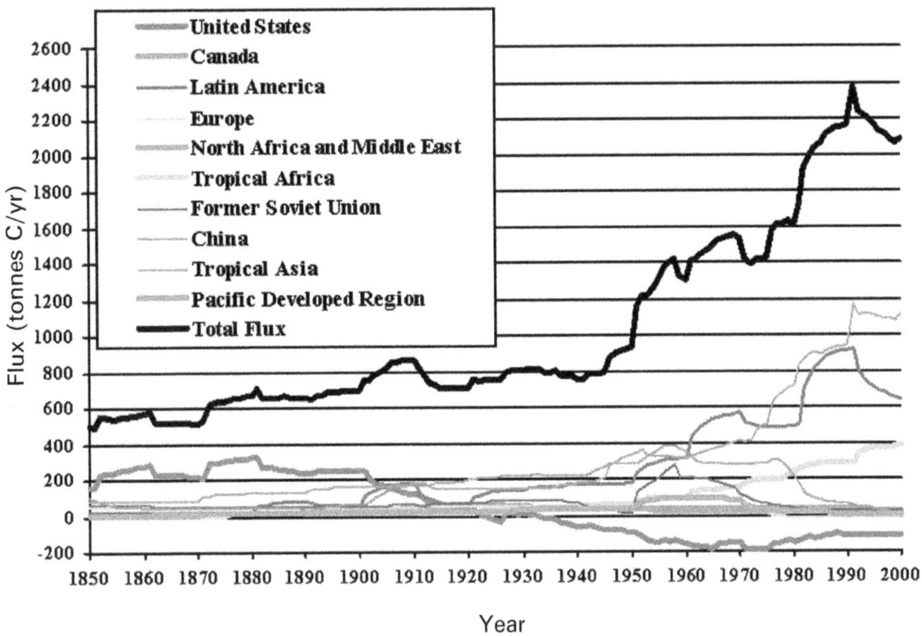

Figure 7.11. Annual net flux of carbon to the atmosphere from land use change: 1850–2000. The changing history of forests has led to some regions (such as in the USA) shifting from a net source to a sink of carbon. Other regions (such as Amazonia) have now taken over in becoming a major source of carbon. When a region goes into the negative on this graph, it is a net sink of carbon. If it goes above zero on the vertical axis, it is a net source of carbon. *Source*: Houghton and Hackler (CDIAC).

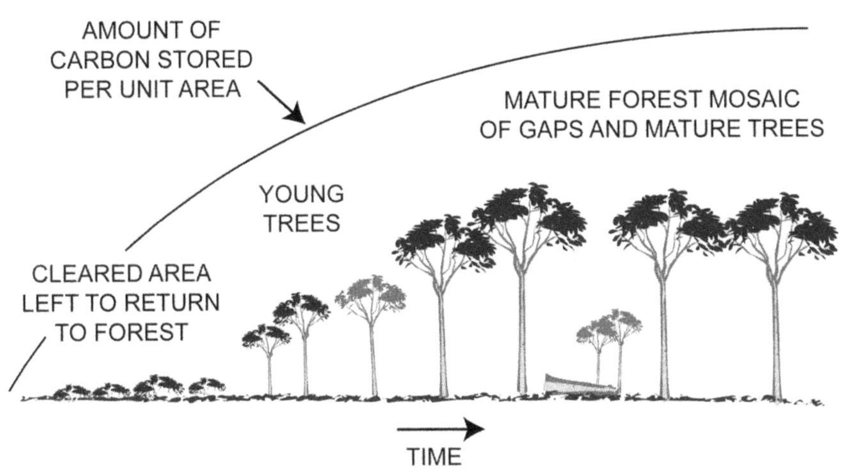

Figure 7.12. When an area of land is allowed to return to forest, there is an initial accumulation of carbon which levels off after a certain amount of time and reaches a balance between the growth and death of trees.

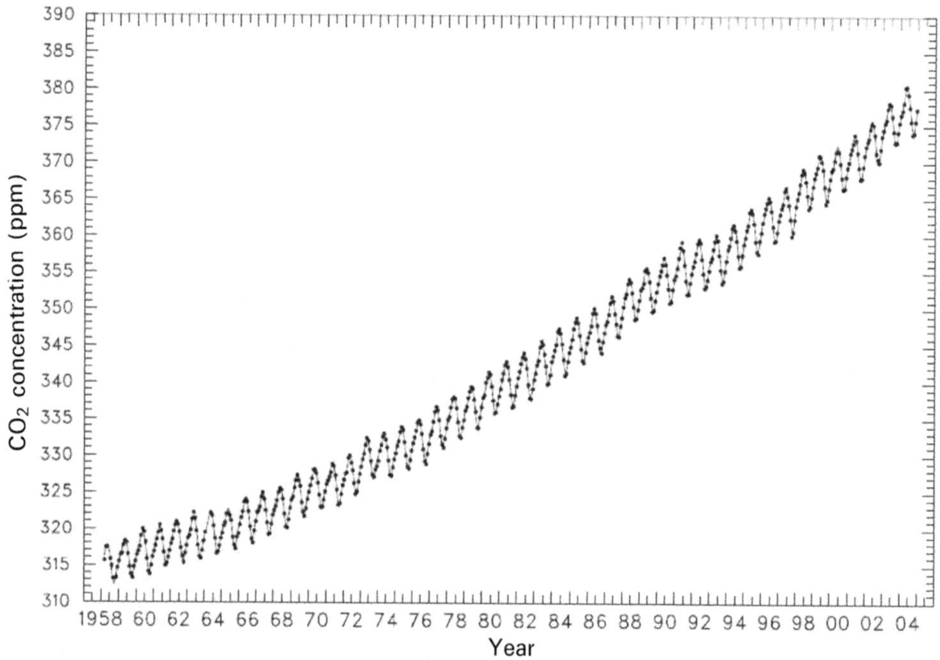

Figure 7.13. The record of atmospheric CO_2 increase since the 1950s, measured directly at the Mauna Loa Observatory in Hawaii (monthly average CO_2 concentration).

Overall, then, it is a complex picture: carbon release from the tropics and carbon uptake in the temperate zone are having competing influences on the increasing CO_2 content of the atmosphere. The loss of carbon from the tropics is large enough to win out over temperate forest uptake, raising the atmosphere's CO_2 content significantly. However, this is considerably smaller than the contribution from fossil fuel derived CO_2 increase. The two sources combined—deforestation carbon plus fossil fuel carbon—currently give an increase in atmosphere CO_2 of about 1.5 ppm/yr (Figure 7.13).

7.5.1 The oceans as a carbon sink

Meanwhile, what role are the ocean plants—the phytoplankton—playing in this story? Basically, the answer is "not much". Plants tend to grow faster with increased CO_2 levels (Chapter 8) but there is very little room for phytoplankton to benefit from this because their growth is so limited by shortage of nutrients such as nitrogen and phosphorus (and perhaps iron) in the oceans. Furthermore, the plankton cells are already bathed in high concentrations of CO_2 because the gas is so soluble in seawater, so a little more does not make so much difference to them. However, the ocean water itself *is* having a major effect in taking up carbon, independent of whatever the plants are doing. The solubility of CO_2 in water with bicarbonate present means that the oceans have the potential to take up nearly all the CO_2 we can throw at them into the alkalinity reservoir. The uptake is occurring most rapidly in particular parts of the oceans where the surface water has cooled as it has drifted up from the tropics, most especially in the North Atlantic. Here, the Gulf Stream water not only cools and takes up CO_2 from the atmosphere, but it also sinks down into the deep ocean, effectively trapping the CO_2 within it. Overall, CO_2 dissolving into ocean waters takes up about a third of the extra carbon that humans add to the atmosphere each year through fuel-burning and deforestation. On the timescale of thousands of years the oceans will eventually take up almost all of it (about seven-eighths of it, to be more exact), but for now they cannot quite keep pace and so CO_2 in the atmosphere keeps on increasing.

7.5.2 Seasonal and year-to-year wiggles in CO_2 level

As well as affecting the long-term balance of CO_2, the world's vegetation also affects the CO_2 level in the atmosphere in smaller, more subtle ways that vary with the seasons, and from one year to another. On the seasonal timescale, there is a "wiggle" in the CO_2 level (Figure 7.13), with CO_2 slightly higher in the winter and early spring than in summer and autumn. In the northern hemisphere, this seasonal variation is greatest in the far north, and it flattens out as one approaches the equator (Figure 7.14). Going south from the equator the seasonal wiggle reappears, but reversed in terms of months of the year and corresponding now to the southern hemisphere summer and winter. It is striking, however, that the seasonal wiggle is much smaller in the southern hemisphere.

What causes the seasonal wiggle in CO_2? It is the result of the way that plants change their activities with the seasons. In the spring and summer, plants in the mid and high latitudes are busy photosynthesizing—taking up carbon dioxide into sugars which is stored as starch, or built into the tough cell walls of leaves and wood. Plants in the higher latitudes take a few weeks to unfurl their leaves and start photosynthesizing rapidly, so there is a delay during May and June as leaves mature. During this stage, decaying leaf litter on the soil surface pushes out a lot of CO_2, so the atmospheric CO_2 level in the northern hemisphere reaches a peak in late spring. Then, by late June and July photosynthesis begins in earnest, and so much uptake over such huge areas is enough to produce a noticeable decrease in CO_2 level in the atmosphere across the whole hemisphere. When autumn comes, the leaves stop photosynthesizing and are dropped to the ground; trees stand leafless and herbaceous vegetation dies back. As winter cold sets in, the uptake of CO_2 from the atmosphere has essentially stopped over vast swathes of the continents. But in mid-latitudes with milder winters where the ground is not totally frozen, leaves now begin to decay on the forest floor, releasing CO_2 which joins the steady trickle of CO_2 up from the dead organic matter in soils, and the CO_2 level goes up a tad.

Why then is there much less of a seasonal wiggle in the southern hemisphere—after all, don't the plants there also experience seasons? This weaker seasonal cycle occurs because there is much less land south of the equator, and much of the land that does occur is arid and sparsely vegetated, so there is less seasonal activity of plants taking up or releasing carbon.

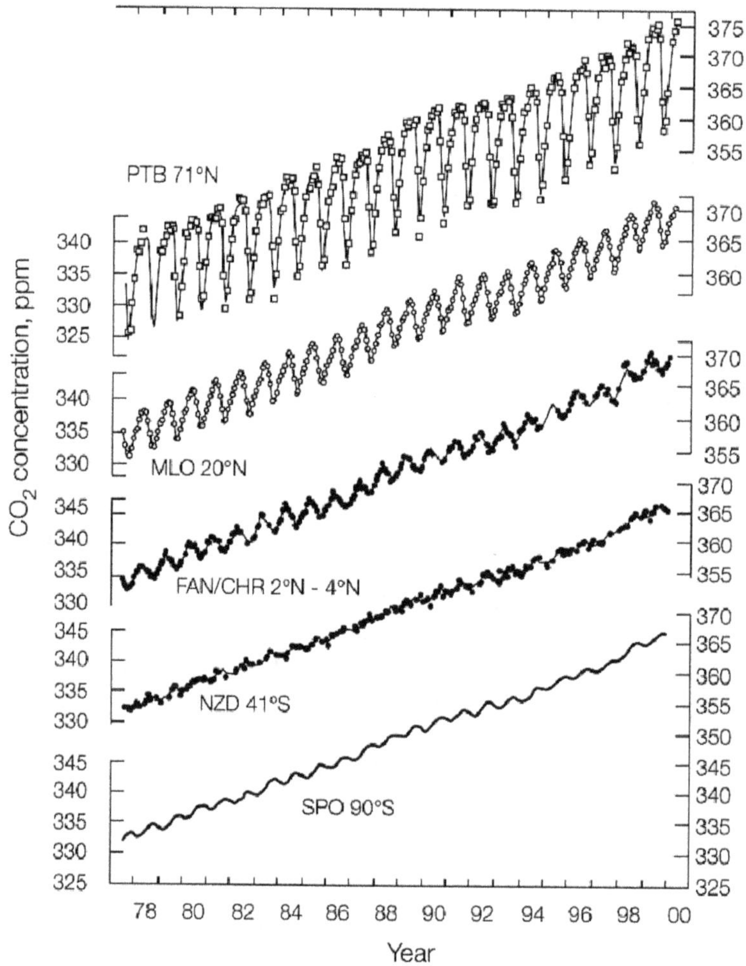

Figure 7.14. The seasonal cycle in CO_2 concentration varies with latitude. It is strongest in the far north and flattens out towards the equator. It reappears in a much weaker form in the high latitudes of the southern hemisphere. Acronyms/initials refer to the names of the CO_2 monitoring stations (e.g., MLO = Mauna Loa Observatory).

Taken from one year to the next, the CO_2 level always goes up due to the amount added by humans burning fossil fuels and clearing tropical forests. But, in some years the increase is greater than in others: as much as a two-fold difference. Human activity does not vary enough from one year to the next to explain such variations, so they must have something to do with natural processes. Partly, these year-to-year differences are due to purely physical changes in the temperature and circulation of the oceans. For example, in an El Niño year in the Pacific, upwelling of deep waters off the coast of Peru slows, and less CO_2 than usual escapes to the atmosphere. This should tend to make the atmospheric CO_2 increase a little less than usual in that year because uptake in other parts of the oceans is no longer partly balanced by this upwelling CO_2 source. Despite this predictable influence, something else clearly dominates in an El Niño year because the CO_2 level actually goes up by *more* in such years, on average. Detailed study of the way carbon isotopes in CO_2 vary from one year to the next shows that changes in the oceans are not the main cause of year-to-year differences in the rate of increase in CO_2. Year-to-year differences in CO_2 are matched by differences in the abundance of the "biological" isotope of CO_2—carbon-12—that plants are particularly good at taking up. Whenever changes in the photosynthesis or decay of plants (or other organic matter) cause a change in CO_2 levels, there is a corresponding change in the carbon-12 composition of the atmosphere (see Box 7.1 on carbon isotopes). The fact that inter-annual changes in CO_2 are paralleled by fairly big changes in carbon-12 suggests that something plants are doing is a large part of the reason CO_2 varies on this timescale.

Box 7.1 Plants and carbon isotopes

Isotopes are different forms of atoms of the same element, differing in the number of neutron particles in the nucleus of the atom. There are two main stable isotopes of carbon, and one radioactive isotope. By far the most abundant is the lightest form, carbon-12. A small percentage of any sample of carbon consists of the heavier carbon-13. Also, a very small proportion of carbon atoms are the radioactive form carbon-14, which is continuously made at the top of the atmosphere when nitrogen gas is hit by cosmic rays from outer space.

The radioactive carbon-14 form gets into plants when they take up radioactive $^{14}CO_2$ from the atmosphere. As well as entering all the living plant materials it is passed down the food chain so everything in the ecosystem picks up some of this radioactive carbon. ^{14}C disappears by radioactive decay at a very precise rate, so the level of radioactivity emanating from carbon in a sample of plant or animal material gives an accurate estimate of how old it is. This technique, radiocarbon-dating, is immensely valuable in finding out the age of samples in archeology, and also in dating ecological changes in the past (Chapter 1) from buried fragments of wood, leaves or soil organic matter. However, by about 50,000 years after the material was first fixed by plants taking in CO_2 from the atmosphere, all the ^{14}C has gone, so the technique cannot be used back beyond this age.

The two stable carbon isotopes can also reveal a lot about both past and present day ecosystem processes. When plants photosynthesize, the enzyme (called rubisco) in their cells that takes in the CO_2 tends to go preferentially for the lighter ^{12}C isotope. So, any living material is slightly enriched in ^{12}C, and depleted in ^{13}C (by about 22 parts in a thousand). This difference carries over into the animals that eat the plants, and into organic matter buried in soils and rocks.

If we study the isotope composition of the CO_2 in the earth's atmosphere over the past couple of hundred years, one thing we can tell is that it is getting "lighter"—that is, that more ^{12}C is entering the atmosphere (Figure 7.15). We can tell that this carbon is coming from a source that was once living, because it is a source rich in ^{12}C. This source could be either present-day forest and soils, or fossil carbon like oil and coal. The fact that the ^{14}C content of the atmosphere is also going down rapidly means, even though its rate of production at the top of the atmosphere has stayed constant, it is being diluted by a large portion of very old carbon. This old carbon must be coming from fossil fuels. The combined picture from both the stable and radioactive carbon isotopes tallies with general expectations from observing forest clearance and the rate of use of fossil fuels: that most of the CO_2 rise is from fossil fuels with a smaller part from tropical forest clearance.

Carbon isotopes can also reveal broad trends in the global carbon cycle going back many millions of years. In the distant history of the earth, after about 450 million years ago, there was a big decline in the amount of ^{12}C in ocean carbonate minerals. Such minerals reflect the composition of the CO_2 "left behind" in the atmosphere after some of it has been extracted by plants. What the decline in ^{12}C

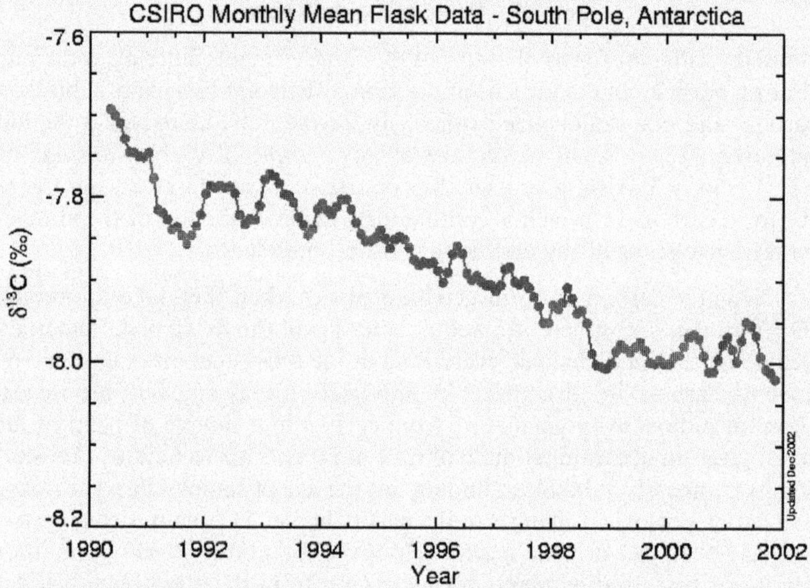

Figure 7.15. "Lightening" of the isotope composition of atmospheric CO_2 over time, measured from a monitoring station at the South Pole. CSIRO monthly mean flask data—South Pole, Antarctica. *Source*: CDIAC.

reveals is that a huge proportion of the total carbon in the atmosphere must have been taken out by plants and stored in dead organic carbon reservoirs, probably causing a major decline in atmospheric CO_2.

Other changes in the ^{12}C content of carbonates reveal catastrophic events in earth history. There were several times in the last billion years when the ^{12}C content of carbonates underwent a sudden large increase. What these reveal is that a lot of organic carbon, probably from vegetation, soils and organic-rich marine sediments, had suddenly been released as CO_2 into the atmosphere. It seems that for some reason almost all the plants in the world died and decayed, and the ecosystems that depended on them fell apart. These events are sometimes (but not always) associated with "mass extinctions", when a large proportion of the species on earth vanished. A large ^{12}C "blip" corresponds to the end of the Permian period 250 million years ago, when something like 90% of species disappeared. The cause of the end-Permian extinction is unclear (possibly an asteroid impact, possibly a phase of massive volcanic activity). It took hundreds of thousands of years for the functioning of the world's ecosystems to recover, as revealed by the time taken to return to more normal ^{12}C levels.

Other more gentle and subtle changes in the earth's ancient environment are also revealed by ^{12}C changes in buried soils. Plants with different photosynthetic systems, the C_3 and C_4 systems (see Chapter 8), concentrate the ^{12}C isotope by differing amounts. Because of the peculiar way they fix CO_2 in the leaf, C_4 plants are not so discriminating about whether they use ^{12}C or ^{13}C. So their carbon is less depleted in ^{13}C, by only around 8 atoms in a thousand as opposed to 22 atoms in C_3 plants. It is easy then to trace whether a soil has had C_4 or C_3 plants growing in it by measuring the composition of soil carbon derived from the plants that grew there. Around 7 million years ago there was a sudden decrease in ^{12}C in various soils in many parts of the world (Figure 7.16), revealing a widespread shift from C_3 plants to C_4 plants. It is thought that this switchover reflects a drying of the earth's climate, and possibly also the effects of a decrease in CO_2 levels which would also favor C_4 plants.

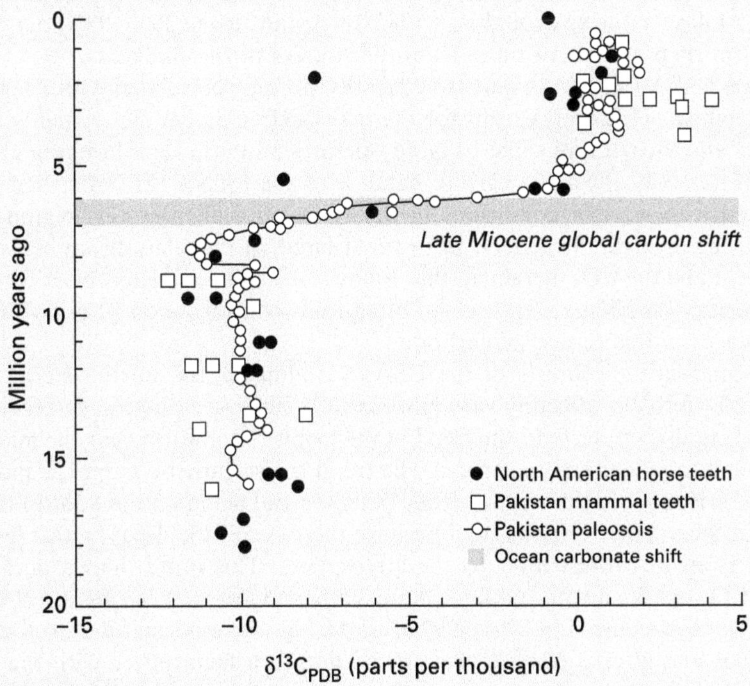

Figure 7.16. A carbon isotope shift around 7 million years ago indicates that C_4 plants suddenly became much more common. *Source*: Cerling *et al.* (1993).

7.6 THE SIGNAL IN THE ATMOSPHERE

Although its trend is always upwards, the actual amount by which the global CO_2 concentration increases tends to vary from one year to another. Years in which tropical forest regions are slightly hotter than usual tend to have a greater CO_2 increase. This suggests that in these warmer than usual years the tropical forests lose carbon through some sort of temperature-dependent process, perhaps increased respiration by the leaves, or increased rotting of dead wood and other litter in the forests. It seems that forests in South-East Asia and in the Amazon Basin particularly dominate this temperature response (Figure 7.17a, b*).

Could it be the effect of drought on the rainforest which is actually causing the big rise in global CO_2 in a hot year in the tropics? After all, the sunny, cloudless skies associated with high temperatures may also tend to be associated with lack of rain. However, when we look at the data in detail, drought in the tropical forests does not seem to have nearly as strong an effect as temperature itself; the statistical relationship of global CO_2 increment with rainfall is much weaker whereas if drought were so important we would expect it to be stronger. Nevertheless, there are exceptions—the big droughts that are associated with some El Niño events do seem to have at least some effect on carbon release in South-East Asia, where fires set by farmers often spread into tropical forests and burn across huge areas, sending smoke and haze across the region. For example, such extensive fires occurred across Indonesia and parts of Malaysia during droughts in 1982/1983, and also in 2003/2004, that they shut down airports hundreds of miles from the sources of the smoke.

Years with an El Niño event are generally strongly correlated with a large global CO_2 increment. However, except for the really extreme ones associated with strong droughts and fires, it looks like El Niño operates more through bringing about high temperatures that affect the carbon balance of the forests (El Niños are generally associated with warmer conditions in the main tropical rainforest regions), rather than causing drought. Strangely, given the strength of the relationship between an El Niño event and the CO_2 increment that follows, it is not completely clear whether it is the opposite "La Niña" event (which often follows on from an El Niño) that is the major player in this release of carbon.

The mid-latitude forests of the USA, Scandinavia and northeastern Asia also seem to play a role in affecting variability in CO_2 increase each year, but their effect is weaker. It is also opposite to the trend in the tropics: in a warm year the mid-latitude forests tend to take up more carbon. The trend is also rather complex; a particularly cool year seems to shut down the decay of leaves and wood on the floor of the boreal forests of Siberia and Canada, and because there is so little decay, much less CO_2 is released from this forest litter to the atmosphere. This more than cancels out the smaller CO_2 uptake due to reduced photosynthesis in the tree leaves in a cooler year. A year like this occurred in 1991/1992 after the big volcano Pinatubo exploded in the Philippines and altered climate around the northern hemisphere with the cloud of sulfuric acid that it pushed into the stratosphere. The northern forest zones were cooled, and with less decay the CO_2 increment in the world's atmosphere during that year was unusually small.

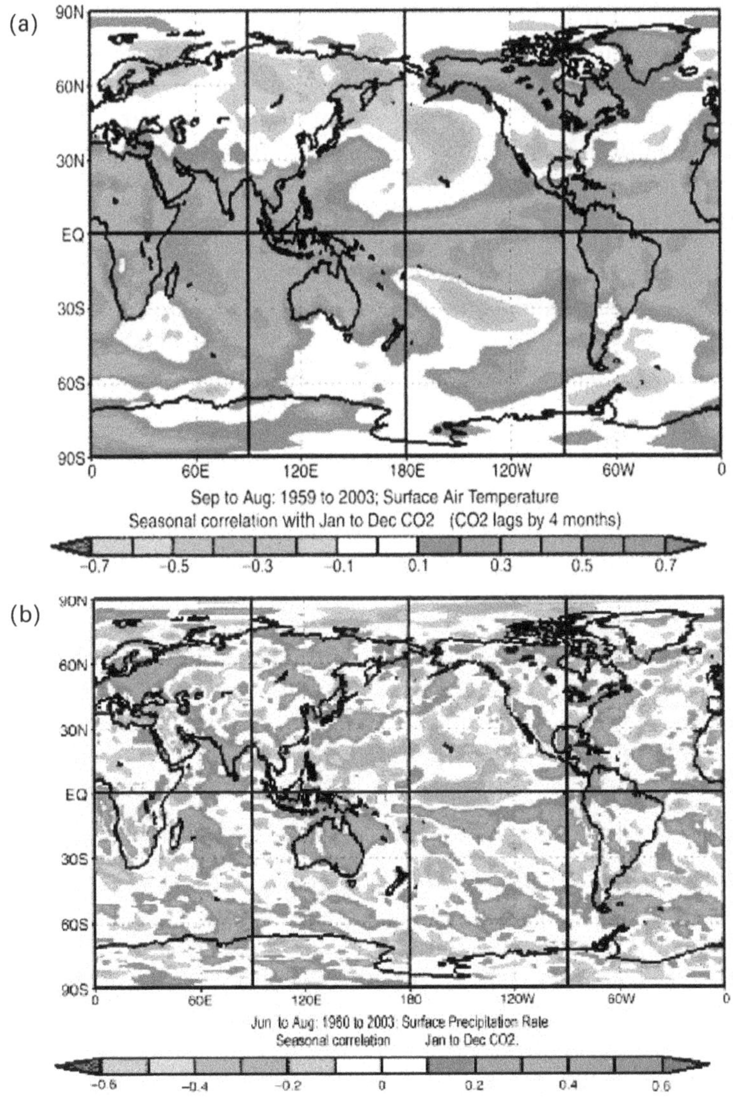

Figure 7.17. (a) This map shows the strength of correlation between temperature and global CO_2 increment each year and that CO_2 increment in a given year is correlated with mean temperature in the tropics. When temperatures in South-East Asia and Amazonia are higher, there tends to be a big increase in global CO_2 in that year (NCEP/NCAR re-analysis, NOAA/CIRES Climate Diagnostics Center). (b) A map showing the correlation between the amount of rainfall and the size of the global CO_2 increment around the world. The relationship to rainfall in forest regions of the tropics is much more scattered and weaker overall, suggesting that heat rather than lack of rainfall may be more important in producing a burst of carbon from the tropics is some years. This might be due to some combination of faster decay, poor photosynthesis and growth of trees under heat stress, or more rapid evaporation stressing trees and preventing photosynthesis. *Source*: Author, in collaboration with Gianluca Piovesan.

In the tropics there is also a weaker and rather mysterious two-year delay between a blip in temperature and a blip in their contribution to the global CO_2 increment. Compared with the "immediate" (same year) effect of temperature on CO_2 release by the tropics, the two-year lagged effect is the opposite: it takes up rather than releases more CO_2 in response to a warmer year. It is thought that this lagged response has something to do with the effect of increased temperature on recycling of nitrogen in forest ecosystems. In a warmer year more decay occurs, enabling nitrogen bound up in dead leaves and other material on the forest floor to be released as nitrates and ammonia that can then be used by the trees again. This produces a burst of growth of new leaves and wood about two years later when the trees have adjusted to the increased supply of nitrogen; and the addition of those new leaves takes up CO_2 from the atmosphere as they begin to photosynthesize. The apparent role of the tropics, and to a lesser extent the northern forests, in controlling how the CO_2 increment varies from year to year echoes their role as a consistent major sink of CO_2 on a global scale. It is not surprising then that if the sink of carbon in the tropics is greater, variability in its behavior affects the CO_2 increment the most.

As one would expect if plant and fossil fuel carbon is being burned to give CO_2, the oxygen concentration in the atmosphere is slowly declining. The amount of the decline shows a seasonal wiggle that is the opposite of the CO_2 wiggle: in the summer oxygen levels go up a bit as there is more photosynthesis producing oxygen and storing carbon in leaves. In autumn and winter the leaves decay back to CO_2 taking up oxygen, and the oxygen level goes down. Although the amount of oxygen in the atmosphere is declining, there is plenty left for us to breathe. The total amount that has been lost in the last 200 years is much less than one-thousandth of the oxygen in the atmosphere.

7.7 THE STRENGTH OF THE SEASONAL "WIGGLE" IN CO_2

The seasonal wiggle—the difference between summer and winter CO_2 concentration—also shows some variability over time. Seen from the CO_2 monitoring stations in some places (e.g., from Mauna Loa in Hawaii and from Barrow in northern Alaska), the strength of this wiggle seems to be increasing.

When it was first noticed, this increase in the seasonal oscillation in CO_2 was explained in terms of increasing CO_2 fertilization (Chapter 8) allowing more green leaves and other seasonal material to build up in the northern summers. If the CO_2 was promoting plant growth, more uptake into photosynthesis would provide a deeper summer dip in CO_2 concentration. Then, starting from autumn the increased seasonal biomass would decay, producing a larger burst of CO_2 into the atmosphere during the winter and early spring. However, this trend in CO_2 seasonality can be explained rather more simply in terms of a trend towards warmer Arctic temperatures (Chapter 3), which likewise encourages plant growth in the summer. In fact, the strength of the CO_2 wobble correlates nicely with a fluctuation in air pressure patterns known as the North Atlantic Oscillation, which gives warmer conditions in the far north (Figure 7.18).

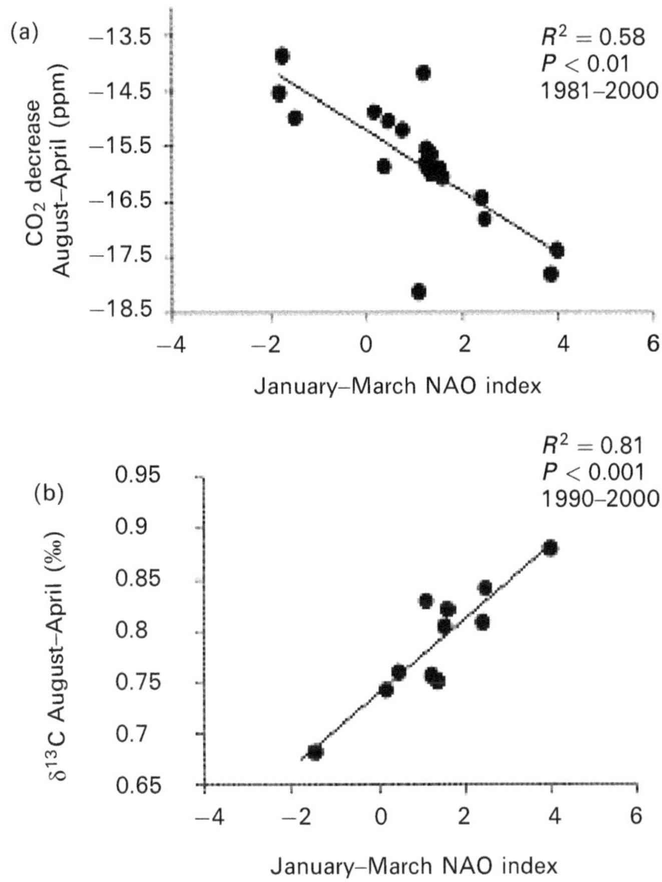

Figure 7.18. (a) The strength of the seasonal CO_2 wiggle is strongly related to the state of the North Atlantic Oscillation, a climatic fluctuation which brings warm temperatures to the high Arctic.
(b) Variation in the carbon isotope composition of CO_2 shows that it is caused by changes in plant activity, which is able to select between the isotopes.

7.8 ACCOUNTING ERRORS: THE MISSING SINK

A little over half of the carbon which enters the atmosphere each year as a result of human activities does not accumulate there. Instead, it gets taken up into one or more "sinks" which accumulate carbon. Observations of the rate at which CO_2 exchanged into ocean waters around the world suggest that of the roughly 8.5 billion extra tonnes of carbon that entered the atmosphere each year in the 1990s, 2.5 billion was being taken up into the oceans. Calculations of the rate of regrowth of young forests in the northern hemisphere suggest that 1 billion tonnes were taken up into this reservoir. Yet only 3 billion tonnes ended up as CO_2 in the atmosphere in an average year. This leaves a gap of 2 billion tonnes, a "missing" sink which is swallowing up carbon beyond that accounted for by calculations of the ocean and forest sinks.

So the balance sheet according to a summary by Skee Houghton at Woods Hole is something like this:

On the positive side:

Deforestation gives off around 2.2 billion tonnes C each year.
Fossil fuel burning produces around 6.3 billion tonnes of C each year.

Yet, only 3.2 billion tonnes C accumulates as CO_2 in the atmosphere.

On the negative side:

About 2.4 billion tonnes C are taken up by the oceans each year.
About 1 billion tonnes C are taken up into forests.

This leaves a "missing" sink of about 2 billion tonnes.

The missing sink is not quite as mysterious as it might sound; no-one believes this carbon is leaving for outer space, or being sucked deep into the earth's interior, so it must be going somewhere within either the oceans or land ecosystems. Probably, it is a result of some simple, basic errors in calculation of uptake by familiar processes in the oceans or forests. However, it might be the result of other poorly understood mechanisms, such as a direct CO_2 fertilization effect making existing forests grow faster than before and accumulate carbon (see Chapter 8). It is important to try to understand the missing sink in order to predict how it might behave in future. For instance, perhaps before long it will begin to saturate and stop taking up CO_2, or even go into reverse and start releasing carbon? For this reason there have been many studies in recent years which aim to narrow the uncertainties. These include eddy flux studies (next section).

An important clue to the missing sink comes from detailed measurement of average CO_2 concentrations around the world. The CO_2 concentration over the mid-latitudes of the northern hemisphere (Europe, Russia, China, North America and the northern Atlantic) is fairly low. This "dip" in CO_2 over the northern mid-latitudes suggests that this is where the missing carbon in the calculations is going. It looks like there is an especially strong absorption of CO_2 going on over the eastern part of the USA and Canada. Even though these regions are actually big sources of greenhouse gas due to fuel-burning, the amount of CO_2 coming off them should be substantially larger than it actually is, showing that some is going missing. Most scientists who work on this subject think that the extra CO_2 is going into maturing forests in the temperate regions, and that they are just taking it up faster than anyone had expected. It could be though that part of the unaccounted-for carbon is being absorbed in the North Atlantic, and that previous studies of CO_2 uptake in that region underestimated how fast it is dissolving in the ocean water. However, recent detailed studies of CO_2 concentration in the atmosphere—using aircraft taking measurements around the world at different heights above the ground—suggest that a lot of the CO_2 is actually being taken up into the remaining areas of tropical forests. This might be because they are recovering from past logging, increasing their carbon storage as the trees grow bigger. It might also be that the estimates of the release of carbon from tropical forest clearance are a bit too high, tending to widen a "gap" (compared with CO_2 accumulation in the atmosphere) that really was not so large in the first place.

7.9 WATCHING FORESTS TAKE UP CARBON

Because of the present and future importance of forests affecting the time course of the rise in atmospheric CO_2, there is presently a lot of work going on to understand whether—and how fast—they are taking up carbon in particular parts of the world, and how they respond to climate fluctuation. When modern ecosystem ecology first began in the 1960s, studies of forest growth concentrated on estimating the amount of wood added to all the trees throughout the forest, from the width of tree rings or the increased girth of the trunks. From this it was possible to infer the approximate amount of carbon by which a growing young forest increased its carbon storage each year, or in an old forest in equilibrium that rate at which carbon was flowing through the ecosystem (balanced by death of old trees and loss of branches). This approach has been used to estimate that the eastern USA forests are taking up carbon at the rate of several tens of millions of tonnes per year. However, it is a very broad-brush approach. It would be better if we had more details about exactly how much carbon is being taken up where, as this would enable us to predict better what will happen in the future.

In the 1980s ecologists began to consider a more ambitious and detailed approach to understanding where and how fast forests take up carbon. This relied on taking very comprehensive and precise measurements of the CO_2 concentration around the trees. The idea is, that if a forest is photosynthesizing and sucking up carbon from the atmosphere around it, this should show up as a localized depletion of CO_2 in the air just above, around and inside the forest canopy (Figure 7.19). Using

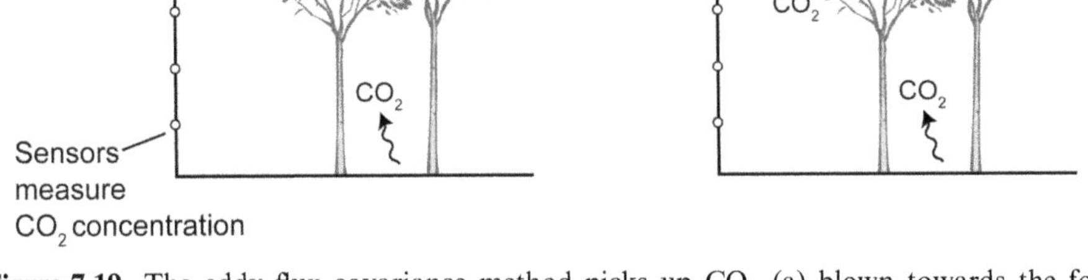

Figure 7.19. The eddy flux covariance method picks up CO_2 (a) blown towards the forest during the day and (b) compares it with CO_2 moving out of the forest at night.

large numbers of well-placed sensors to measure CO_2 concentration, it is in theory possible to estimate just how much net photosynthesis is going on during the day, and thus how fast the forest is accumulating carbon. Even though the forest ecosystem is also respiring during the day, in daytime there will normally be more photosynthetic uptake of carbon than carbon released from respiration. This estimate has to be balanced against the amount of carbon lost from the forest at night, when there is only respiration and no photosynthesis. Again this night-time assessment can be done using the sensors to measure how much the CO_2 concentration around the forest has been raised at night relative to the background level in the atmosphere. To make these estimates properly, it is necessary to estimate how fast more CO_2 is getting delivered (during the day) or taken away (during the night) by air movement. This involves a lot of complex physics and calculation. If the measurements are continued month after month, year after year, then it may be possible to infer how the balance of carbon in a sample patch of forest is changing over time.

This approach, known as the eddy flux covariance method, is compelling but also very ambitious. It requires a huge investment of labor and money to put in place the complex measuring equipment and maintain it, and analyze the data that comes out. At an intuitive level, it is easy to see that if a slight portion of the carbon loss or gain each day was not included in the accounting (e.g., because a sensor missed it) this error would accumulate over many months and might give a totally misleading picture of the direction in which the forest's carbon balance was changing. One big problem that this method has run into is that it is difficult to summarize the amount and direction of air movement over the forest during day and night. At night, especially, air movement from the forest canopy is very sensitive to local conditions and it may stay stable (stratify), or may continue to be turbulent. Turbulence in the air is generally a good thing for this methodology, because it carries extra CO_2 coming from the plants and soil out towards the sensors up on the pole, where they can detect it and correctly add it to the respiration side of the budget. Sometimes at night the cooling CO_2-rich air from within the canopy at night sinks right to the forest floor, underneath the sensors, and flows away undetected. This too gives the impression of less respiration actually occurring in the ecosystem.

Perhaps because of these problems, eddy flux covariance has sometimes given strange results that do not seem to tally with previous knowledge of ecosystem processes built up during the 1960s and 1970s. A study site in pristine Amazon rainforest seemed to be inexplicably gaining carbon so rapidly that it was set to double in carbon mass within 60 years. In Europe, forests and their soils in the north seemed to be accumulating carbon more slowly than those in the south, not because trees were growing slower in the cooler climate but because the northerly soils were breaking down carbon more rapidly. This seemed to contradict decades of knowledge about how soil respiration responds to temperature. In many other areas such as the eastern USA and Japan, the measurements seem to make much more sense in terms of previous knowledge of how climate and forest age affect carbon balance. However, there is the nagging question of whether these studies too might contain errors which are too small to clash with previous understanding of ecosystems, but scientifically important nevertheless.

Some ecologists such as Christian Koerner of Bern University have pointed out that, in addition to the measurement problems, the fact that important processes which affect forests such as tree falls, land slides, droughts and fires occur in an occasional and unpredictable way means that intensive measurements of small patches of forests may not give a particularly relevant picture of long-term trends on a broad scale, which is the sort of question these studies are basically attempting to answer.

While the eddy flux covariance method is a considerable achievement of engineering and scientific collaboration, it remains an open question as to how much it can really teach us, compared with more old-fashioned methods of looking at the carbon balance of ecosystems.

7.9.1 Predicting changes in global carbon balance under global warming

From the year-to-year variability in the amount by which CO_2 builds up in the atmosphere, it looks as if the amount of CO_2 taken up or released by the world's vegetation responds quite a lot to changes in the climate from one year to the next. Such small responses to year-to-year climate variation might give us clues to longer term trends that will emerge as the global climate warms due to the greenhouse effect.

Since CO_2 responds to climate, and climate responds to CO_2, there is the potential here for some important feedbacks. It could be that as the world gets warmer it will favor more carbon being stored in vegetation and soils, slowing the warming by taking CO_2 out of the atmosphere. This would be a negative feedback loop, tending to act against the main cause of the warming. On the other hand, in a warmer world, vegetation and soils might actually respond by releasing CO_2, adding further to the warming in a positive feedback loop.

Given what we know of the responses of forest carbon balance to year-to-year climate fluctuation, the effect that global warming might have on CO_2 uptake or release by forests is complex. It depends on the particular region, and the detailed nature of the climate shift: how big it is, and whether rainfall changes as well as temperature. The whole task of predicting what will happen tests the limits of understanding of both climate and the global carbon cycle. Attempting to model the whole system over the coming centuries requires inter-disciplinary teams of experts using some of the fastest computers available. One study by Peter Cox and his colleagues based at the Hadley Center in the UK predicted that, as the world warms, carbon will gush out of the world's ecosystems into the atmosphere, amplifying the warming in a positive feedback (Figure 7.20). The effect, then, is for an initial push of CO_2 to give warming that is amplified further by more CO_2 coming out of the world's forests. In the model, a large part of the positive feedback occurs due to more frequent and more severe El Niño events affecting the carbon balance in the tropical forests of South America, South-East Asia and Africa. Drying in the Amazon region is also predicted to occur as a result of increased temperatures in the Atlantic due to global warming.

Though such models are impressive, there are many uncertainties, and a slight error in the parameters of a model could throw the predictions way off from what will actually happen. Not all climate models predict more El Niños under global

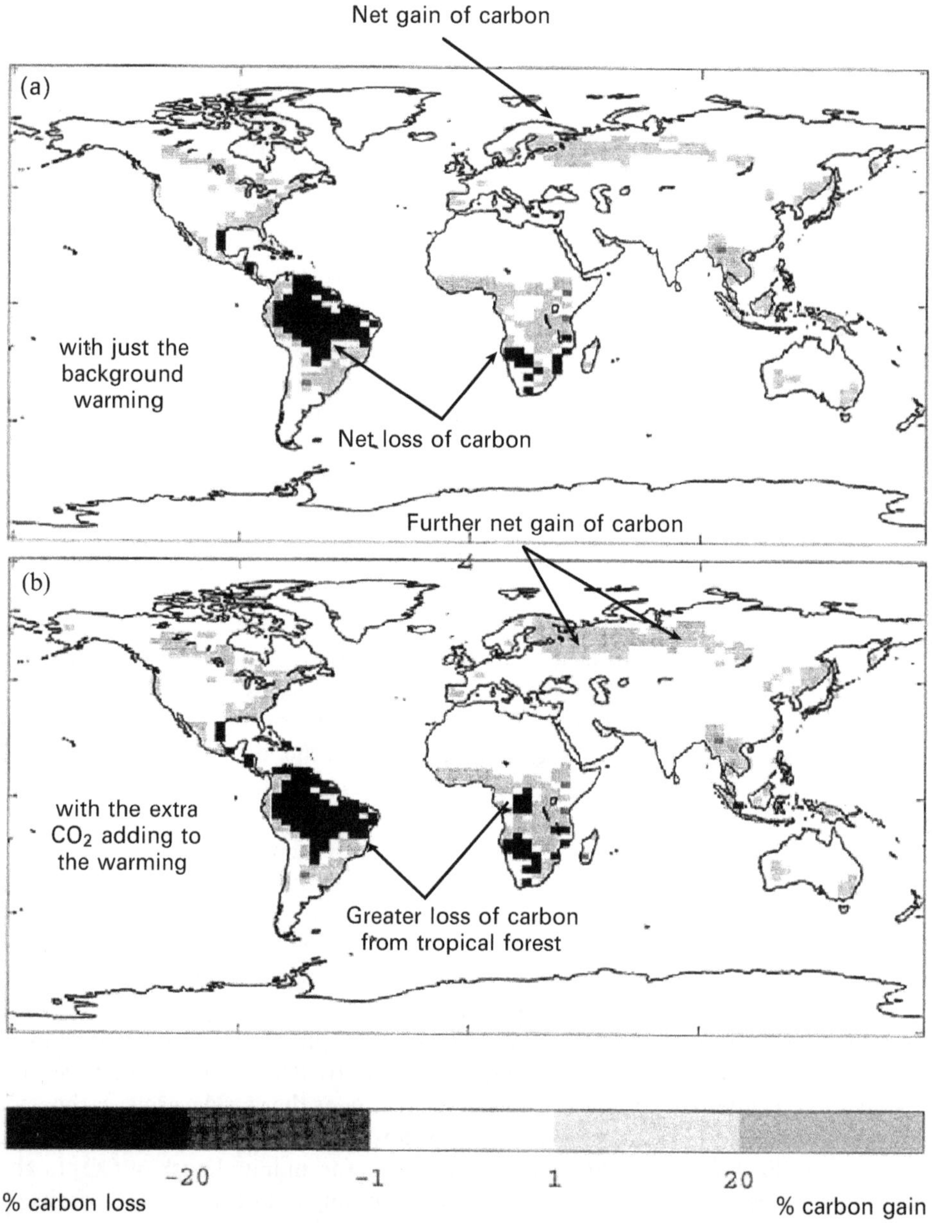

Figure 7.20. Model results with and without the "gushing out" of carbon that would result from warming affecting the carbon balance of forests. In the lower scenario, the extra CO_2 that the forests give out (in response to warming) warms the climate further and this in turn promotes more forest growth in the high latitudes. (a) Without CO_2 feedbacks. (b) With CO_2 feedbacks involving the extra CO_2 released by die-back of vegetation and decay of soil carbon, adding further to the warming and affecting vegetation in a positive feedback cycle. *Source*: Cox *et al.* (2001).

warming, and there is even some doubt whether most of the El Niño events over the past few decades have pushed more CO_2 into the atmosphere. The "blip" in CO_2 that tends to be associated with an El Niño could actually be due to the opposite "La Niña" climate episode that tends to follow soon afterwards.

There is in particular a need for more fieldwork and basic observations of how plants and soils respond to changing climate conditions. A big problem for models to predict is exactly how vegetation will respond through the direct CO_2 effect (Chapter 8). The global model for the future carbon cycle that Cox and colleagues have put together also included the influence of the direct CO_2 fertilization effect for the future, and even though they have tried to bring this in to the model, it is hard to say how this factor would modify the overall outcome in terms of carbon storage. In a more recent version of the same model, Richard Betts and his colleagues have attempted to look in more detail at this direct CO_2 effect on global carbon balance in the future: but more about that in the next chapter (Chapter 8). It is important to bear in mind, however, that the Cox *et al.* and Betts *et al.* scenarios are using just one general circulation model, the Hadley Centre Model, when there are several other ones out there. The Hadley Centre model happens to be the only one that forecasts such a widespread "Amazon Meltdown" scenario (others have the Amazon rainforest surviving the next century mostly intact, unless of course humans cut it down), so perhaps we should not take it so seriously as to exclude everything else. Another thing to bear in mind is that even quite small details of how the model of Cox and colleagues works have a big effect on the outcome of global warming in its predictions. For example, how sensitive tropical forest trees are to temperature is key to their response to warming; yet we actually lack the key data that are needed. The model itself assumes that tropical forests respond to temperature just as temperate-zone forests do. There is an urgent need for more data from experiments and field observations to show how tropical forests will actually respond to temperature increases.

Various other feedback mechanisms involving CO_2 and ecosystem carbon reservoirs are also thought possible, and not all of them have been modeled yet. One major worry for the future is how the very large reservoir of carbon in peatlands in Siberia and Canada will respond to global warming. At present, both regions are extensively blanketed in a layer of peat. This has built up from dead plant material over thousands of years under waterlogged soil conditions and cool temperatures that have slowed down decay. If temperatures increase, the slow breakdown of the peat might accelerate and result in rapid addition of CO_2 (and also the greenhouse gas methane, CH_4) to the atmosphere. CO_2 output is especially likely if the climate gets drier overall so that water tables fall, allowing oxygen to get into the peat. Under these conditions microbes, fires and also a gradual chemical reaction with the air can oxidize the peat. Digging down within the peat itself, one can often see "oxidized layers" from the past when drought caused water tables to fall and some of the peat broke down to give CO_2 gas. The fear is that under global warming this could occur on a massive scale, pushing billions of tonnes of CO_2 into the already-warming atmosphere. If all the peat in the northern latitudes were to oxidize, it would push up the CO_2 level by at least 50% beyond the present level. This is the sort of problem that models of the carbon cycle under global warming need to try to tackle. Especially important will be the field data that the models are based on, yet such data are often in short supply. A lot of the sort of basic science that enables such understanding is unglamorous and does not attract the same sort of funding as, for example, the eddy flux covariance studies mentioned above, or the FACE experiments discussed in Chapter 8. Much of what we do know is based on a long and rather esoteric tradition of the study of nature—the sort of thing that until recently funding agencies were keen to cut back on. All that we can say at present is that it is quite possible that vegetation and the soils underneath it will have a major influence on CO_2 levels once global warming gets fully under way.

8

The direct carbon dioxide effect on plants

8.1 THE TWO DIRECT EFFECTS OF CO_2 ON PLANTS: PHOTOSYNTHESIS AND WATER BALANCE

Carbon dioxide may affect plants by changing the climate, but it can have another more subtle and quite separate influence, through its direct effects on plant physiology. Since CO_2 is fundamental to photosynthesis, it makes sense that increasing the amount of CO_2 in the atmosphere will tend to allow plants to photosynthesize faster. This then is one-half of the direct CO_2 effect on plants. But there is also another less straightforward direct effect of CO_2 on the water balance of plants. Why should this be?

Ever since plants first came out of the sea to live on land, they have faced a dilemma. They must prevent themselves from losing too much water in the drying air, but they also need to take in CO_2 in order to photosynthesize. A plant could easily almost eliminate water loss by coating itself in some sort of thick waxy layer that water cannot pass through. But, at the same time this would almost totally prevent CO_2 from getting into its leaves, and it would be unable to grow. So, plants have to balance a "trade-off" between gathering enough CO_2 in order to photosynthesize, and avoiding death by desiccation. Vascular plants (those with roots, stems and leaves) have solved the problem in a satisfactory way by using tiny pores in their leaves—called stomata—which can open and close. When a plant has plenty of water, the stomata let CO_2 in to the moist interior of the leaf and the plant tolerates the evaporation of water through the stomata for the benefits of photosynthesis. When the plant has enough carbon, or when it begins to run short of water, it partially or totally closes these pores to prevent further water loss.

Much of the time, plants only open their stomata part-way, or keep them shut altogether, which limits the amount of CO_2 they can take up and the amount by which they can grow. If you add more water around the roots of the plants, they will open their stomata more fully and keep them open for longer, take up more CO_2 and grow more. If instead you add more CO_2 to the air around the plants, very often they do the opposite, keeping their stomata only part-way open or closing them after a short time. This is because at high enough concentrations CO_2 veritably pours into a leaf, even through partially closed stomata. So, without keeping stomata fully open for long, the plant has soon got all the CO_2 it needs, and has synthesized all the sugar that it can use for the time being. Having got enough CO_2, the leaf then shuts the stomata to prevent any further loss of water. Evolution has selected plants that take this conservative path, avoiding "spending" water around their roots that they might need for another day, as soon as they have enough sugars to keep them going. Thus, a plant that has more CO_2 may not actually do more photosynthesis, but instead it may avoid dying of drought because the supply of water around its roots lasts longer.

All in all, CO_2 and water are interchangeable; they are part of a trade-off for plants. More CO_2 means that a plant has more water. Giving a plant more water means that it can open its stomata and take in more CO_2, which allows more photosynthesis. So, more CO_2 can benefit plants in two ways: it can mean that they get more growing done because they can do more photosynthesis, and it can also mean that they are less susceptible to drought. An increase in CO_2 to the sort of levels that will be reached in the next century will affect plants everywhere in the world, altering their growth rate and their water balance. The only question is how large these effects will be, and what long-term consequences they will have for ecosystems and communities.

If the amount of CO_2 in the atmosphere increases, in a general way we can expect it to benefit plant activity on land. Plants can photosynthesize more, and also suffer less risk of dying of dehydration. Over the next few centuries, this "direct CO_2 effect" might well turn out to be ecologically more important than the greenhouse effect of CO_2 and other greenhouse gases. However, there is a lot of uncertainty and indeed quite a bit of mystery associated with the direct CO_2 effect. There are some good reasons for thinking that it could be very important in altering vegetation, but a frustrating lack of evidence to show whether such suspicions are right or wrong. Within the small amount of evidence that we do have, there are quite a few contradictions and paradoxes.

8.2 INCREASED CO_2 EFFECTS AT THE SCALE OF A LEAF

Some tentative clues to the effects of increased CO_2 on plants come from short-term observations of individual leaves exposed to artificially high CO_2 concentrations. It is possible to estimate how fast a leaf is photosynthesizing by measuring the uptake of CO_2 labeled with radioactive ^{14}C. The more radioactive the leaf is at the end of the experiment, the more carbon it has managed to fix by photosynthesis. Such small-scale experiments on raised CO_2 have tended to involve a doubling of CO_2 from about 350 ppm—the approximate "background" level of CO_2 during the past couple of decades—to 700 ppm, a level that CO_2 is likely to reach well before the end of this century. Short-term exposure to high CO_2 tends to result in a major increase in the amount of sugars fixed—a typical sort of change observed would be a doubling or tripling of the rate of photosynthesis. In these experiments, CO_2 tends to increase photosynthesis by proportionately greater amounts at higher temperatures. At 30°C, the relative "gain" from CO_2 fertilization is about 30% greater than at 20°C. This is because at high temperatures plants suffer quite badly from a process known as photorespiration where oxygen "gets in the way" of the photosynthetic reaction. Raising the CO_2 level helps push carbon instead of oxygen into the reaction, preventing photorespiration. Hence the greater benefit that comes at higher temperatures where the problem of photorespiration is especially acute.

8.3 MODELING DIRECT CO_2 EFFECTS

How can we predict how plants around the world will respond to raised CO_2 levels? Plant physiologists and global ecosystem modelers have put a great deal of emphasis on short-term observations of the effects of increased CO_2 on photosynthesis. They have also tended to make a lot of use of a set of principles together known as the "Farquahar Model", put together by Graham Farquahar at the Australian National University. This model reduces the complex process of assimilation of CO_2 into the plant to certain simple key components that act as bottlenecks (Figure 8.1): first, the CO_2 diffuses into the leaf through a stomatal pore and basic gas physics shows how the rate depends of the concentration of CO_2. Then, the CO_2 gets incorporated by an enzyme (known as rubisco) into organic form in the cell, at a rate that can be predicted pretty much exactly from the way a mix of the relevant components in a beaker would behave. So, CO_2 fertilization of photosynthesis comes down simply to gas diffusion, and then a chemical reaction. And, hey presto, here is a universal model to predict CO_2 fertilization effects on plants. This model has been taken up with great enthusiasm by some global modelers who have essentially used it as the core of a series of other more ambitious models, extrapolating the future effects of CO_2 on all the world's vegetation for the next several hundred years.

We will examine some of these predictions in more detail later on, but it is important to be aware that many ecologists are intuitively skeptical of any broad-scale model built upwards from biochemical principles. Ecologists are constantly reminded of the sheer com-

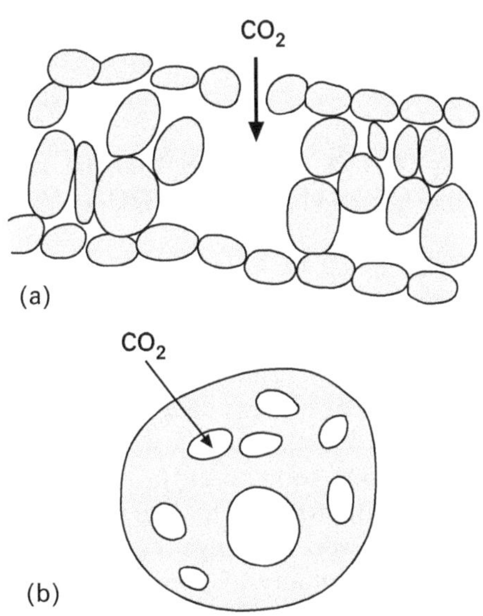

Figure 8.1. Key steps in photosynthesis which are altered by CO_2 concentrations. (a) Diffusion of CO_2 into the leaf. (b) Uptake of CO_2 by rubisco enzyme into carbon-containing molecules.

plexity of nature, and its tendency to do the opposite of whatever is expected. Considering the often frustrating experiences that ecologists have in working with the natural world, their skepticism about a model built up from molecules is understandable. In fact, it is now looking rather like their intuition is well founded: experiments which have been set up to provide a more direct indication of how plant communities will respond to increased CO_2 have shown some very complex and often unexpected results. The question of CO_2 fertilization is a classic example of the problems of scaling up in ecological systems. How things work at the smallest scale and in the short term in either a test tube or a leaf does not necessarily indicate what will happen to the whole plant over weeks or months. And what happens to the whole plant does not necessarily predict what will happen to a community of species over several years. Furthermore, what happens in a few years to the community does not necessarily indicate what will happen to the entire ecosystem over decades or centuries.

In this sense, modeling of processes that get right inside the plants, into their core metabolism and growth, is much more difficult than modeling the purely physical processes of heat exchange, turbulence and evaporation by which plants affect climate. In climate modeling, extrapolating up from the most local level (even from the individual leaf) tends to be a fairly good basis for understanding how vegetation interacts with the climate system on even the broadest scale (Chapters 5 and 6). We can tell that it can be done, from the various instances where vegetation–climate models have been tested against real changes in vegetation and found to predict climate changes more or less correctly. Modeling how living processes will be affected by CO_2 is much, much more difficult.

8.4 WHAT MODELS PREDICT FOR INCREASING CO_2 AND GLOBAL VEGETATION

Most models for predicting the broad-scale effects of increased CO_2 on vegetation tend to be based on some basic components of the photosynthetic process (from the Farquhar model, above): how fast CO_2 can diffuse into the leaf at higher concentrations, and how fast the photosynthetic chemistry in the cells of the leaf can potentially take it up at higher concentrations. Some models take a more empirical approach, seeing just how much faster on average plants tend to grow in closed-chamber experiments with raised CO_2, and then taking a CO_2 fertilization factor for the increase in growth rate, known as "beta".

What sort of things do these models predict for the coming decades and centuries? Essentially, if everything else in the world stays the same except for CO_2 increasing, the models all agree on two things. First, there will be an increase in net primary productivity (the growth rate of plants) all around the world as the supply of carbon for photosynthesis increases. Second, plants will be using the water supply around their roots more efficiently because they do not have to open the stomata in their leaves quite as much. This will be rather like an increase in rainfall as far as the plants are concerned.

Different models predict different degrees of response to any given increase in CO_2, depending on subtle details in the assumptions that they are based upon. A review of models over the past ten years or so showed that on average for a world with 580 ppm CO_2 (which will probably occur around 2050 given the current rate of increase in CO_2) compared with a baseline of 350 ppm, the models predict a 22% increase in plant productivity around the world. However, the range of estimates amongst them extends from 10 to 33%. The higher CO_2 levels go, the higher the effect on plant growth is predicted to be. In a recent review, the well-known vegetation–climate modeler Gordon Bonan suggests that above 700 ppm of CO_2, the increase in the world's plant productivity will be somewhere between 12% and 76%, averaging at 48%. If this is correct—and remember this forecast is only based on models—it would represent a transformation of the energy flow through the world's land ecosystems. And, of course, given that they are only models, if they have overlooked or misjudged some important factor, they might all be wrong.

If plants increase their productivity and also make better use of the water available to them, we can expect that there will be some changes in the structure of vegetation around the world. The distributions of biomes are often determined by water availability, and an increase in CO_2 that allows plants to use water more effectively should allow wetter-climate biomes to spread. For example, models which combine a biome vegetation scheme (Chapter 2 in this book) with a CO_2 fertilization model usually predict that there will, for example, be an increase in tropical rainforest, allowing it to spread out into zones which get less rainfall and currently have dry forest or savanna vegetation. In arid areas, plants that are able to get by on less water because of higher CO_2 will be able to spread—the deserts will become greener. There is also predicted to be a general shift around the world towards C_3 plants (see Box 8.1, p. 238), and away from C_4 plants which do not benefit so much from increased CO_2. The speed with which these changes in vegetation actually occur depends on many different factors. Even though tropical rainforest might be capable of spreading into savanna regions, it will likely take hundreds or even thousands of years for the forest trees to disperse out and grow up into dense forest in these new areas. In the meantime, more subtle shifts in the structure and composition of vegetation are also likely to occur as some of the plants that were already in place grow bigger, and shade out other species around them.

8.5 ADDING CLIMATE CHANGE TO THE CO_2 FERTILIZATION EFFECT

Over the coming century, the direct CO_2 effect will not be the only thing changing. CO_2 "together with several other gases that are currently increasing" is also a greenhouse gas, and by trapping heat these greenhouse gases will tend to warm the climate and also alter rainfall patterns (Chapter 3).

Temperature changes will also bring about changes in the water balance, whether or not the rainfall changes. All this has to be considered in relation to any forecast of global vegetation based on direct CO_2 effects, adding up to a very complicated mixture. While the direct CO_2 effect might be pulling things in one direction (towards wetter climate vegetation) by allowing more efficient water use, at the same time a decrease in rainfall or an increase in temperature—which increases evaporation—might be pulling things in the opposite direction (towards drier climate vegetation). It may be very difficult to predict which factor will dominate and in which direction the vegetation will change. There are uncertainties in both the direct CO_2 fertilization effect and in the climate simulations for the future, and the combination of both adds up to a far wider range of uncertainty than either taken by itself.

Nevertheless, it is interesting to think about what might happen as both CO_2 and climate undergo change over the next century. One model which concentrated on the USA during the next 50–100 years suggested that, initially, there will be an increase in overall forest extent and vegetation productivity due to the CO_2 fertilization effect dominating. However, the model predicts that, as the 21st century draws to a close, increasing heat and aridity from the greenhouse effect will result in a net decrease in forest extent, even though the direct fertilization effect of CO_2 is still increasing as its level in the atmosphere soars. Sitch and colleagues (2008) have tried a more extensive comparison of the various climate model scenarios, and various vegetation–climate models that also incorporate a direct CO_2 fertilization effect on the plants. What the models tend to find is that over the next century or so, the CO_2 fertilization effect is all-important in preventing the world's land ecosystems from gushing out carbon into the atmosphere as CO_2. If the direct CO_2 effect is included in the model, what was previously a scenario where vegetation and soils lose CO_2 into the atmosphere now becomes a scenario where there is a small net uptake. This is good news in terms of avoiding a runaway global feedback, but it hangs precariously on assumptions about the balance between the effect of warming (drying out vegetation, and also promoting decay of soil carbon) and the opposite effect of direct CO_2 fertilization—the latter being based on models which do not find any independent validation in the results of CO_2 fertilization experiments (see below). And furthermore, the climate models do not all paint the same rosy picture of the direct CO_2 effect saving the world. The Hadley Centre model used by Peter Cox, Peter Betts and colleagues (below) suggests that the CO_2 effect is not enough to prevent carbon gushing out of tropical rainforest in Amazonia as the result of a hotter, drier climate.

A further step is to consider how the direct CO_2 effect might set off the sorts of climate feedbacks from vegetation I mentioned in Chapters 5 and 6. If plants are opening their stomata *less* under increased CO_2 and thus losing less water by evaporation, this means slower less efficient recycling of rainwater (which allows more water to run straight off the land to rivers instead of getting evaporated from leaves). Less recycling may mean an overall decrease in rainfall, which takes away some of the benefit to the water balance of the plants from having increased CO_2. On the other hand, the increase in vegetation leaf coverage resulting from direct CO_2 effects would decrease albedo—the "lightness" of the surface. In arid areas this darkening of the surface would tend to increase rainfall by promoting convection (Chapter 5). In colder climates, the decreased albedo would also tend to warm the climate (Chapter 6). Hence, an initial boost given by the direct CO_2 effect can end up being magnified into a larger shift in vegetation. Some attempts at modeling such influences on the Sahara Desert margins over the next few decades suggest that, although the decreased evaporation from partially open stomata at high CO_2 may tend to decrease rainfall a little, the increased efficiency of water use will promote more vegetation overall and that this will then set off an albedo feedback that actually gives more rain! Richard Betts and his colleagues have suggested that this "stomata shutting" effect of raised CO_2 could make life hard for tropical forest in the Amazon Basin by the mid– 21st century. Less evaporation from the leaves of the forest in the wettest areas at the western and eastern edges of Amazonia could starve other areas farther inland of the rain that they need, by breaking down the rainwater-recycling mechanism (Chapter 5). This would be partly compensated by the reduced needs of these inland rainforests for water, due to the direct CO_2 effect allowing them to keep stomata partially closed—so different aspects of the direct CO_2 effect have both a positive and negative effect on carbon balance. Combined with the extra heat from the increasing greenhouse effect, and droughts resulting from warming the tropical oceans, the Amazon forest is forecast to begin dying back and gushing out carbon by around 2050. Some of this group's model forecasts suggest that by 2100, most of the forest of the Amazon Basin will have died. Even if global warming stops due to new policies that limit greenhouse gas emissions, the progressive dieback of the forest might continue by its own momentum, as a wave of drying climate, death of forests and collapse of the water-recycling mechanism gradually spreads across the whole Amazon Basin. It is important to emphasize, however, that this nightmare scenario is only one of a range of model forecasts. Using any one of several other different climate models, or tweaking the details of the interactive vegetation–climate scheme, predicts all or most of the Amazon forest remaining in place despite the warming climate. So the forecast for the future is all very sensitive to the detailed parameters of the model—for example, such things as the temperature sensitivity of tree leaves—which will require a lot more experimental or boots-on-the-ground observational evidence to be certain about them. Hopefully, this need will spur ecologists in the tropics to get out and make more observations that can help the models. In any case, before we dismiss the risks of an "Amazon collapse" it is worth bearing in mind that the Hadley Centre GCM, and the vegetation model used in these forecasts, is one of the more widely respected model combinations.

Clearly, it is very hard to try to model the outcome of such complex networks of interacting factors, but what the musings of modelers do show is that there is a lot of potential for changes to be magnified, in ways that we might not initially expect.

8.6 EXPERIMENTS WITH RAISED CO_2 AND WHOLE PLANTS

A leaf studied in high CO_2 concentration over a few minutes is not necessarily at all representative of nature. This statement might seem obvious, but modelers have not always been prepared to acknowledge it! Because there are many factors that could potentially change plants' responsiveness to CO_2, a good way to get a firmer idea of how wild or crop plants will behave is to do experiments on whole plants grown over weeks, months or years. For about 20 years now, plant biologists have been

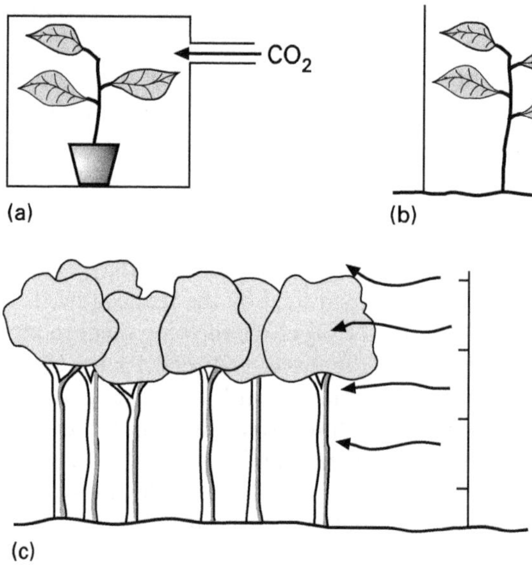

Figure 8.2. The three types of increased CO_2 experiment: (a) closed-chamber, (b) open-top chamber, (c) free air release.

experimentally raising CO_2 levels, growing plants in small-scale systems to see what effect future increases in CO_2 might have. While they are no more than isolated snapshots of the future world, these experiments at least have the advantage that they are based on actual whole plants, often growing under at least plausibly complex combinations of influences.

The earliest and simplest experiments were in closed chambers with plants growing either in compost or natural soil, adding CO_2 to the air beyond the atmospheric concentration (Figure 8.2a). This would be compared with a "control" chamber where air with the CO_2 concentration of normal outside air (ambient air) was piped in. The trouble with these closed chambers was that they always seemed rather artificial. There was not the exchange with the outside world that might allow insects, herbivores and fungi to move in and out: the plants were effectively living in a "sterile" environment. And you couldn't grow big trees in these chambers, only small plants.

To help deal with these limitations, more sophisticated experiments were developed to look at the effects of raised CO_2. Open-top chambers (Figure 8.2b) of translucent plastic were used out in fields and natural vegetation; CO_2 was piped in to make a "double-CO_2" atmosphere. Because CO_2 continually leaked out of the top, you had to use more CO_2 than in a closed chamber, making them rather more expensive to maintain (the cost of all the pure CO_2, maintained over months or years, adds up to quite a lot). The trouble with these open-topped chambers has been that they tend to be warmer inside, which complicates the conclusions (in an experiment it is always best to start by holding all things constant except for the one factor you are studying the effects of). Nevertheless, if their internal temperature is controlled with some sort of cooling system, such chambers do not necessarily have to be very different from the outside. Another problem with open-top chambers is that herbivores like deer cannot get in to nibble the plants, making the situation a bit artificial in ecological terms; although at many of the more ambitious CO_2 experiments mentioned below, grazing may also be very limited simply because of low densities of large herbivores in the areas studied. For studying how trees respond to increased CO_2, they have also been limited to seedlings or very young trees, because it is so difficult to build a chamber for a big tree.

The latest generation of raised CO_2 experiments uses open air release of CO_2 in fields or natural vegetation—an apparently more "realistic" situation which does not involve artificial enclosures (Figure 8.2c). These are called "Free Air CO_2 Experiments" or FACE. Some of these FACE set-ups are very large-scale, involving areas of forest. Other FACE experiments on desert scrub or agricultural grassland are much smaller, scaled down to correspond to the smaller size of the plants. Generally, the FACE experiments use a ring of towers that reach just above the height of the local vegetation, and release CO_2 at various points along their height, at rates carefully calculated to produce an atmosphere with double the normal amount of CO_2 (Figures 8.3, 8.4). Each individual tower only releases CO_2 part of the time, when the wind is blowing past it towards the plants within the ring. When the wind is blowing in the other direction it switches off, to avoid wasting CO_2. Nevertheless, a lot of CO_2 must be thrown around in such an experiment, much more than in open-topped chambers, and this adds greatly to the costs. Regular deliveries of tankers loaded with pure liquid CO_2 are necessary to keep the supply up. Because the experimental equipment and running costs for simulating future CO_2 concentrations are so high, few countries outside the USA, Europe and Japan have conducted any such work.

The classic FACE set-up of a ring of CO_2 release towers does not seem to be suitable for all vegetation types. Mature forests of heavy trees with thick branches do not tend to be very amenable to CO_2 release from towers, because the trees themselves block the movement of the gas and create too much turbulence that mixes the CO_2-enriched air in with normal air, in rather unpredictable ways. For this reason young forests of thin, straight trees have normally been studied using the FACE system. The only exception is an experiment by Christian Koerner and colleagues in Switzerland, on a mature mixed oak forest (Figures 8.5, 8.6*). They used a branching series of pipes that released CO_2 into the crowns of the trees in precisely calculated ways, simulating a uniformly CO_2-enriched atmosphere.

In addition, there is one interesting example of a "natural experiment" in the form of a local CO_2-rich atmosphere near some hot springs in central Italy. The springs release carbon dioxide from within the earth, much of it derived from heating of ancient carbonate rocks. The growth of nearby clumps of evergreen oaks that are exposed to CO_2 levels two or three times the background level is compared with other clumps further away that experience normal background levels of CO_2. These observations have continued for more than 35 years, much longer than any of the FACE experiments.

* See also color section.

Figure 8.3. The Tennessee FACE site showing the towers used to release CO_2 into the forest. *Source*: Rich Norby/ORNL FACE.

8.6.1 The sort of results that are found in CO_2 enrichment experiments

Looking at whole plants rather than just isolated leaves, we find that the actual growth response is generally weaker with whole plants. The strongest initial responses to raised CO_2 in terms of growth rate are usually found in well-fertilized crop plants with the "normal" C_3 photosynthetic metabolism (see Box 8.1, p. 238). C_4 plants—like corn and sugar cane—tend to show only a weak response to CO_2 levels, and even this only when water is in especially short supply. When they are kept well-watered, the response is negligible.

In "wild" C_3 plants grown at "natural" low soil nutrient levels, there is often at least a temporary increase in plant growth by about 30–100% as CO_2 is doubled from the present concentration. CO_2 fertilization initially has more or less the same effect

Figure 8.4. Aerial view of the Tennessee FACE experiment showing rings of towers (see arrows). *Source*: Rich Norby/ORNL FACE.

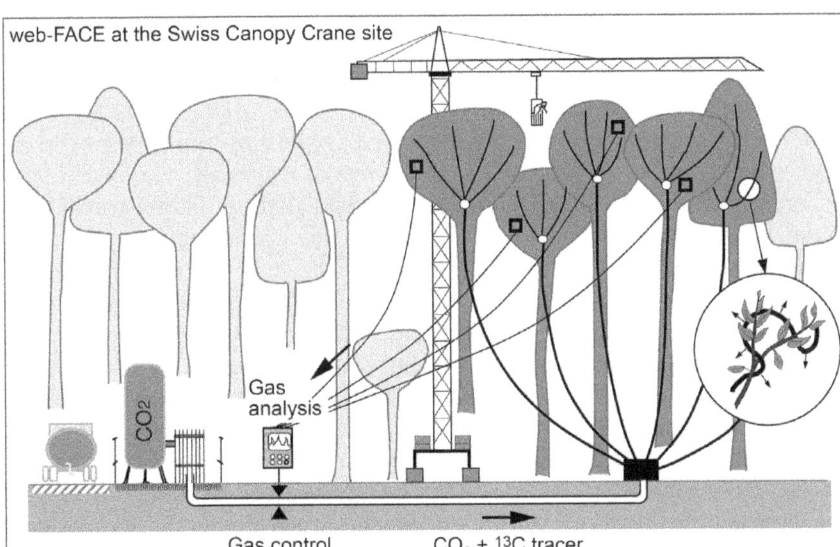

Figure 8.5. The Swiss FACE site on mature mixed temperate forest uses a network of tubes twisting up the branches to deliver CO_2 in the right places and right quantities to simulate a higher CO_2 atmosphere. *Source*: Christian Koerner.

Figure 8.6. Scientists at the Swiss FACE site inspect the forest canopy for direct CO_2 effects using a crane. *Source*: Christian Koerner.

as watering more in most ecosystems, since watering also enables greater carbon uptake through the opened stomata.

The wide range of open-topped chamber CO_2 fertilization experiments in ecosystems around the world suggest a broad range of responses, varying widely from one species and locality to another. When several slightly different treatments are used on the same species, it is often found that the growth response of the plants varies according to precise nutrient conditions in the soils. For example, take a look at the range of responses in tropical forests. Experiments with tropical tree seedlings—grown to two meters in height under increased CO_2 (around 550 ppm)—showed no response when grown out in full sunlight in the normal tropical forest soil. Yet there was a dramatic response in growth rate when nutrients were added to the soil (Koerner, 2004). What this suggests, then, is that in tropical forest there is not going to be much effect of increasing CO_2 because nutrient shortage prevents the plants growing any faster. On the other hand, when tropical rainforest seedlings grown in shade are fertilized with CO_2, there can sometimes be a quite dramatic response. Tree seedlings in the shade of the forest floor grew up to 50% faster over more than a year, an interesting and important finding because survival in the seedling stage is key to determining which tree species make it to the canopy (Koerner, 2004). Tropical forest vines seem to respond much more strongly than forest trees to CO_2, sometimes doubling their growth rate, which might explain why Oliver Philips and colleagues are noting an increase in the abundance of vines in tropical forest plots across Amazonia during the last few decades (although, strangely, at levels above 550 ppm of CO_2 the vines seem to grow slower again—something hard to explain). The fast-growing "secondary" forest trees, which normally grow in high-light conditions in disturbed areas of forest, also show a stronger response to CO_2 than the canopy trees of older undisturbed forest—unsurprisingly perhaps because they are normally growing in more nutrient-rich soils when they are fed additional CO_2.

8.6.2 A decline in response with time

Generally, the effect of high CO_2 levels shows a strong initial effect on growth, which diminishes over time. Although there is always at least some decrease in response, the amount of this decline varies a lot. In most natural and semi-natural ecosystems, there is an initial response in growth rate of the plants above ground, followed by a major decline back almost to a "normal" growth rate. In the case of woody plant ecosystems, the enhancement of growth rate generally seems to keep up better in young forest that is growing on relatively fertile soils (such as the example of the North Carolina FACE site, see below). Amongst the whole range of vegetation types that have been looked at so far, the most consistent responses are found in well-fertilized pastures and crop systems, which maintain more of the enhanced growth rate. However, even these agricultural systems do show some decline after an initial burst of response to CO_2.

8.7 TEMPERATURE AND CO_2 RESPONSES INTERACTING

Over the next century or so, CO_2 itself will not be the only environmental factor changing. Climate is likely to become substantially warmer in most parts of the world, because of the greenhouse effect of the CO_2 itself, as well as other greenhouse gases. So, to know how vegetation will really change, it may be helpful to try altering both factors simultaneously and see how they interact.

There have been a few short-term (one or two year) open-top chamber experiments in the temperate zone which artificially warmed plants growing in natural settings, while they were being fertilized by CO_2. Usually, this was done with either under-soil heating, or with infra-red lamps. For example, a study carried out in Tennessee by Rich Norby and colleagues studied young red maples (*Acer rubrum*) under this combination of conditions. The results were compared with controls at normal temperatures and CO_2, or under either only increased CO_2 or only increased temperatures. What studies such as this have tended to find is that there can be an additive effect in terms of growth rate of increasing both temperature and CO_2: the warming increases growth somewhat (at least in temperate climates, which tend to be heat-starved part of the year), and CO_2 also increases growth some more. Some studies seem to show a synergistic effect: the acceleration of growth under both

CO_2 and warmer temperatures is greater than the sum of the two separately. This might be, for example, because the benefit of high CO_2 in reducing wasteful photorespiration is greater at higher temperatures, so there is more to gain. In the warmer greenhouse world, this might then mean that biota will respond strongly to CO_2.

8.8 A FEW EXAMPLES OF WHAT IS FOUND IN FACE EXPERIMENTS

FACE studies (above) have now been carried out on a fairly diverse array of ecosystems. Rather than try to give a full account of the results from every one in turn, I will focus mostly on two very different types of ecosystems—forests and arid lands—to give some idea of the sorts of complexities that are found in the results.

8.8.1 Forests

Forests have been studied quite closely over time in relation to raised CO_2, using the FACE method. Where stands of young temperate deciduous and conifer trees growing in forest soils have been exposed to raised CO_2 over a year or two, they tend to show a strong initial increase in growth rate. However, after a few years, there is often a major decline in the effect of CO_2. I'll take one particular case study to look more closely at the complex and contradictory results of such experiments. At a FACE experiment in Tennessee on young sweetgum (*Liquidambar*) forest, wood growth was 35% greater in CO_2-enriched plots than in control plots in the first year of the experiment. In the second year, the growth response was reduced to 15% and was no longer statistically significant, with further reductions in the third year and so on (Figure 8.4). However, some other longer lasting changes have been found. For example, below ground the rate of production and turnover of fine roots has remained higher throughout the experiment. In fact, there has been so much increase in growth rate and turnover of fine roots that the rate of primary production at the site has stayed about constant; it has simply shifted underground. Bear in mind, though, that the trees that have been exposed to increased CO_2 remain a bit larger than the untreated ones because they got a "head start" a few years back, even if they are not responding much any more. The increase in NPP is just about at the average of what several different models of CO_2 fertilization have predicted, at around 23% (the models predict a 22% increase, though the range amongst them is quite broad). However, the form of the increase in NPP is not quite what the modelers would expect; it goes into making small roots which grow and then rot quickly, and not into long-lived wood. Some plant physiological modelers have found ways to try to explain all of this, in the context of familiar expectations of the CO_2 fertilization models. However, others such as the redoubtable Christian Koerner point out that these arguments, and the selection of parts of the data, involve a fair amount of gymnastics to make the results fit with the expectations of the models. My own view is that the confusing and conflicting results of CO_2 fertilization experiments put the grand global models in doubt. I am not saying that the models are necessarily wrong: it is just that we have no strong reasons at present for thinking that they are right.

A drop-off in the increased wood production at high CO_2 does not always occur. A strong response in terms of NPP (about as much as at the Tennessee site over the other side of the Smokey Mountains) was found at a FACE experiment on pine forest in North Carolina. After 10 years there is still a growth enhancement of around 20%, but in this case much of it seems to be going into extra wood and not fine roots. Among the various CO_2-fertilized plots of forest at the North Carolina site, those that are on nitrogen-rich patches of soil tend to be responding more than those on nitrogen-poor soil. In fact, most of the strong, sustained response to CO_2 comes from a single plot where the soil is very rich in nitrogen.

In the experiment in Switzerland, the mature mixed oak forest exposed to raised CO_2 (530 ppm) at first grew several percent more than an untreated area of forest adjacent to it, but then the effect diminished rapidly. By the fourth year of the experiment, the raised CO_2 forest was not putting on additional wood any more rapidly than the forest exposed to ambient CO_2 levels. In this case, root changes were not studied, so it is not known if the increased CO_2 instead affected roots as in the Tennessee experiment.

So far no FACE experiments have been run for tropical forest, which reflects the logistical and monetary challenges of setting up such a complex piece of work. The closest thing to a tropical FACE experiment so far has been the very localized and short-timescale release of extra CO_2 onto branches of various tropical tree species, using a canopy crane for access. These experiments showed that an expected "benefit" of CO_2 for tropical trees, reduced evaporation of water through the stomata, did not materialize. Instead, the stomata stayed stubbornly open for as long as usual, suggesting that the trees wouldn't economize on water under enhanced CO_2 (Koerner, 2004). This particular piece of knowledge looks like bad news for the accuracy of global change models, such as that used by Peter Cox and colleagues, to predict the loss of water recycling from tropical trees and a collapse of part of the rainfall system of the Amazon Basin. It is, however, perhaps good news for the Amazon rainforest, which would not then be expected to suffer as a result of the loss of this water-recycling effect.

Because the FACE experiments have not yet been run for several decades, we do not know whether the raised CO_2 response will produce any effects that only emerge as the trees grow bigger. Only the "natural" experiment at the hot springs in Italy has so far run for this length of time. In this particular study, the evergreen oaks were studied after they were cut and started regenerating from stumps (as coppice). The oaks close to the CO_2 source initially grew much faster, but they do not now grow any quicker than the nearby "control" individuals under normal ambient CO_2 levels that were cut at the same time. However, because they got a head start during its first few years of regrowth from stumps, the high CO_2 trees are quite a bit bigger than the normal CO_2 area.

The overall conclusion—from the various temperate forest types that have been studied using CO_2 fertilization—is essentially that there is "no conclusion". In some experiments there is a lasting response, but in other cases the response seems to have vanished. In some the response is mainly above ground, in others it is below ground in the roots. At some sites the total increase in primary productivity is much as global

CO_2 fertilization models would predict, but the nature of the response is rather odd (e.g., going into fine roots, not wood). It is difficult to know whether to take such results as supporting the models, or refuting them. Also, it is important to bear in mind that there are many forest types in the world that have not yet been studied using FACE experiments, including boreal conifer and tropical rainforests. For all we know, they might respond quite differently.

8.8.2 Semi-desert and dry grassland vegetation

Semi-desert and dry grassland vegetation is generally forecast to respond especially strongly to increased CO_2 levels, because it is so limited by water. Since adding CO_2 means that the plants can make use of water more efficiently, this should surely offer a massive boost to them. In one study using CO_2 fertilization models, Jerry Mellilo and colleagues forecast an increase in primary productivity in the world's semi-desert regions of between 50% and 70% if the CO_2 concentration gets to be double what it was 200 years ago. This amount is much greater than the sort of productivity increase forecast for wetter ecosystem types such as the world's forests, which is typically around 20–25%.

How does the experimental evidence match up with this prediction of a big boost for plant growth in deserts? Probably the most realistic study of desert vegetation under increased CO_2 is a FACE experiment that was set up in the Nevada desert of the southwestern USA with a ring of miniature "towers" to match the low vegetation. This experiment increased the CO_2 concentration by 52% above the "background" level across the desert. In some ways the initial response of the CO_2-fertilized plots (compared with the controls at normal CO_2 levels) was dramatic, much as the models would predict. There was a big increase in photosynthesis by 80–100%, and water expenditure through evaporation by the desert plants was only about half of what it would normally be per unit of photosynthetic production. Yet, strangely this did not translate into any increase in shoot or root growth rates of the commonest two desert shrubs creosote bush (*Larrea*) and *Ambrosia*.

However, in contrast to this, closed-chamber experiments with creosote bush and mesquite (*Prosopis*) shrubs grown under doubled CO_2 showed a significant growth response of the shrubs, with an increase in biomass of these species by 69% and 55%, respectively. Quite what is so different between the open air and closed-chamber experiments is not known!

One closed-chamber experiment found that when CO_2 levels were increased there was an increase in survival rates of seedlings under droughty conditions, which is what would be expected since the seedlings would be able to make better use of the water they had available amongst their roots. In a different short-term chamber experiment on various southwestern US semi-desert species, there was a doubling in root nitrogen (N) and phosphorus (P) uptake under high CO_2 by the grass *Bouteloua*, and yet a major decrease in N uptake by the creosote bush *Larrea*—perhaps due to the competition. Because nutrient limitation on plant growth is thought to be important in deserts, this unequal response by different species might tend to bring about longer-term changes in plant communities.

The inconsistency in results between closed-chamber and free air fertilization studies, and between different species, presents a confusing picture for what might happen to semi-desert vegetation in the future. One may regard free air and relatively undisturbed communities at the FACE site as more representative of what will actually happen as ambient CO_2 increases, although some authors have argued that chamber experiments can actually sometimes be more representative than free air studies. The upshot is that it is too early to say with any confidence how even the most intensively studied desert shrub communities of the southwestern USA will respond to rising CO_2, let alone all the other desert areas of the world.

Another interesting observation from the Nevada Desert FACE site is that the non-native invasive grass cheatgrass (*Bromus tectorum*) responds to CO_2 such that it is far more productive than native plants during wet years. Cheatgrass invasion of the southwestern US deserts has been found to greatly increase the frequency of fires, from a 75–100 year cycle to a 4–7 year cycle. These fires are also far more intense than those in native vegetation and usually result in a loss of native shrubs. A further change from shrubs to grasses under increased CO_2 would have a dramatic effect on desert water cycles and wildlife habitat, as well as the suitability of the lands for cattle-ranching.

The results so far from the FACE experiment in Nevada indicate that both desert shrubs and wet-season herbaceous plants such as cheatgrass respond especially strongly to increased CO_2 during the occasional wet years that correspond to El Niño events. There is more year-to-year variation in growth rate at elevated CO_2, suggesting that the whole ecosystem may become even more episodic and thus, in this sense, more desert-like in a future high CO_2 world.

In a study of desert margin species from the semi-desert environment of the Negev Desert (Israel), transplanted into closed chambers, species-rich assemblages of winter annual grasses and herbs showed very little biomass response to doubled CO_2 but significant changes in tissue quality and species dominance. However, these changes were solely the result of the response of a single species of legume (a member of the pea family) which became much more vigorous and abundant. Had this particular species not been included, overall responses would have been minute. The general lack of response to CO_2 for most of the desert species in this system was rather unexpected, since CO_2 fertilization models predict an especially strong effect in arid vegetation.

A FACE experiment on semi-arid Mediterranean-type grassland in California likewise confounded all the expectations of models. Right from its first year at increased CO_2 levels, to the third year when results were reported, there was no significant enhancement of net primary productivity (growth rate) of the plants. This was true across a whole range of treatments, some of which involved increasing nutrient supply and water supply as well as CO_2.

8.8.3 Will C_4 plants lose out in an increased CO_2 world?

It is often expected that plants which use the more water-efficient and CO_2-efficient C_4 photosynthetic system (see Box 8.1) will respond less strongly to raised CO_2 than plants using the conventional C_3 system. Because desert and semi-desert ecosystems contain a high proportion of C_4 species, one might expect those species to decrease as a proportion of the vegetation, relative to increased growth of C_3 species. Closed-chamber experiments with C_4 and C_3 species growing in competition have often supported this view. In a chamber experiment with various southwestern US semi-desert species, the C_4 grass *Bouteloua* responded with only about half as much increase in biomass (a 25% increase) as the C_3 shrubs creosote and mesquite, which is the sort of response that might be expected. However, the grass also greatly increased its nitrogen content, which might seem to suggest that it was also doing better than would be expected from growth rate alone, despite being a C_4 species.

In the semi-arid grasslands of the central US that contain a mixture of C_4 and C_3 plants, the picture of CO_2 response is not at all as models predict. When intact pieces of prairie grassland turf containing both C_4 and C_3 plants were studied in elevated CO_2 in the greenhouse, the greater response forecast for C_3 species was not found and both types responded about equally. A field experiment in open-top chambers on the prairie actually showed the opposite trend: there was no response in the most important C_3 grass (*Poa* spp.) but significant growth stimulation of C_4 prairie grasses! Whether such a situation will "carry over" into other grasslands around the world and into drier environments such as semi-deserts is a moot point, but these results should be considered as a further uncertainty in predicting arid-land vegetation responses to CO_2.

Box 8.1 C_4, C_3 and CAM plants

Many plants in arid environments decrease the problem of water loss through stomata by chemical tricks that help them take up CO_2 with less water loss. These are known as C_4 and CAM plants.

Most plants are known as C_3 plants. They take CO_2 up into leaf cells which handle the whole photosynthetic reaction in the same cells. The CO_2 gets fixed into a three-carbon chain (hence the name C_3), and then in the same cell the water-splitting part of photosynthesis gives the hydrogen needed to tack on to carbon. The hydrogen and the carbon are then combined in that same cell, to make sugars. C_4 and CAM plants do something a bit different.

The most straightforward alternative is in plants that have the Crassulacean Acid Metabolism or CAM system, including the cactus family Cactaceae. These plants open their stomata at night when it's cool (so evaporation loss is low) and soak up CO_2 and store it chemically. The photosynthetic cells then release the stored CO_2 for fixing by photosynthesis during the day. The storage chemical is an organic acid; the plant tissues become acidic during the night as the storage product accumulates, then less so during the day as it's released to yield CO_2. CAM plants tend to live in the most arid environments. They are all succulents: fleshy leaved or with fleshy stems. In addition to occurring in deserts, CAM is often found in plants growing in salt marshes and on seashores, and this shows how significant drought is for these seaside plants, due to salt in the soil exerting an osmotic effect, preventing water from being taken up by their roots.

C_4 is a bit less obvious as a trick. Throughout the day a C_4 plant captures CO_2 in special "CO_2-fixing" cells on the outer parts of its leaf tissue, and concentrates it into the center of its leaves. The outer CO_2-fixing cells are also busy photosynthesizing, but they are only doing part of the photosynthetic reaction, the part that yields oxygen: $2H_2O \Rightarrow 4H + O_2$. The H atoms are stored on special intermediate molecules, ready to help form sugars later on.

At the same time (as I mentioned above) those cells are taking up CO_2, and fixing it into special carrier molecules (which consist of a four-carbon chain, hence the name C_4) that are moved to the innermost part of the leaf where there are other special cells which are also photosynthesising, but using the light energy to combine the stored CO_2 and stored H into sugar molecules, which is how the plant wants them.

So, in summary, there are three parts to this process in a C_4 plant: in the outer photosynthetic cells: (1) H and CO_2 are taken up and fixed (and oxygen produced). Then (2) the H and CO_2 are transported and (3) using more sunlight are made into sugars in other photosynthetic cells deeper within the leaf (Figure 8.7).

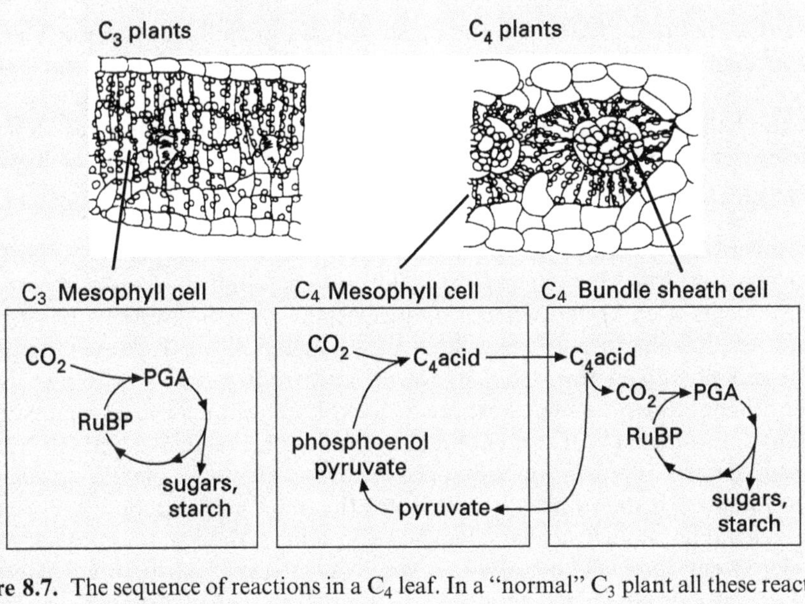

Figure 8.7. The sequence of reactions in a C_4 leaf. In a "normal" C_3 plant all these reactions take place in the same cell.

Why does the C_4 plant do all this? In a "conventional" C_3 plant, something called photorespiration is going on continually in the photosynthesizing cells around the stomata that are also exchanging CO_2 with the atmosphere. Oxygen gets tangled up in the photosynthetic reaction and "spoils" the molecule that has fixed the carbon, which has to be burnt back to CO_2 because it can't be used. This spoiling reaction is known as photorespiration. The burning of the useless by-product of photorespiration spits CO_2 back out within the leaf and many of those CO_2 molecules are lost again as they leak back out of the stomata. To make up for this lost CO_2 the conventional plant has to keep its stomata open for longer. This presents problems: in a dry environment, opening stomata is something that the plant needs to avoid doing because it risks dying from drought.

A C_4 plant, on the other hand, avoids photorespiration because it shuttles the fixed carbon to have the final sugar-making reaction occur in special cells deep within its leaf that aren't producing any oxygen (which is the thing that "spoils" the reaction). And concentrating CO_2 at high levels relative to oxygen also helps suppress photorespiration. Having special "CO_2-gathering" cells that take up CO_2 without producing any CO_2 through photorespiration helps to ensure maximum efficiency in CO_2 uptake (it is like a vacuum cleaner for CO_2), in terms of "stomatal opening time" and water loss. Hence, in a C_4 plant stomata need not be open for as long to take up a unit of carbon, and—for this reason too—water use is more efficient. Thus, the C_4 plant loses less water per unit time per unit of carbon fixed. This should help it to do better in dry environments.

Note that, because photorespiration also occurs especially fast in warm climates and at high light intensity (which causes high temperatures in the leaf), the C_4 system is also directly advantageous for avoiding wasting solar energy, irrespective of water balance. Of course, dry environments also tend to be sunny and hot, so in this respect (avoiding water loss, and allowing effective utilization of high light intensities) they doubly favor C_4 plants.

Not surprisingly, then, C_4 plants tend to be most abundant in warm, fairly dry environments. The C_4 system is especially often found in grasses in semi-arid environments such as the western and especially the southwestern parts of the North American prairies (no trees have the C_4 metabolism). There is a gradient in C_4 abundance going from cool to warm, and from wet to dry. However, the pattern is not always quite as expected. Surely, one should see C_4 plants totally dominating in the most arid environments, the deserts and semi deserts? Yet, in the hot semi-deserts of North America, one or more species of C_3 plants (e.g., creosote bush, *Larrea tridentata*) usually dominates the plant communities. According to their general physiological characteristics, C_3 plants should be the least adapted to hot desert environments because they are less water-efficient than C_4 plants and have a lower optimum temperature and lower rates of photosynthesis, and C_3 plants also reach their maximum response to sunlight at low light intensities. Clearly, there are other factors which we don't quite understand that can contribute to the success of plants which have a "wrong" photosynthetic system for the climate.

Given that C_4 plants are so effective at gathering in CO_2, they do not have so much to gain from an increase in CO_2 concentrations in the atmosphere. So, it is generally expected that they will not respond as much to a higher CO_2 world. In fact, they might end up being pushed out, as C_3 plants do better in response to CO_2 fertilization. It is to be expected that the reason C_4 plants do not occur everywhere is that there is a "cost" in maintaining this complex photosynthetic system, and in places where it is not really needed they lose out in competition to C_3 plants. On the other hand, if climates grow warmer due to the greenhouse effect, photorespiration will tend to increase and C_4 plants might still find themselves favored by this factor, because they suffer less from photorespiration.

CAM plants can be expected to show even less response than C_4 plants to increased CO_2, because they are so good at taking in CO_2 and they also do it at night when they do not lose so much water by evaporation. Experiments show that they are essentially unaffected by doubled or tripled CO_2 levels. As with the C_4 plants, they can be expected to lose their advantage over C_3 plants in at least some situations in nature as CO_2 levels increase.

Among other effects noted in arid-land plants exposed to increased CO_2, it appears that increasing the atmospheric CO_2 concentration can reduce the impact of salinity on plant growth. This could improve crop growth in desert-marginal areas which tend to have salty soils, and perhaps increase productivity and biomass of natural desert vegetation.

From the limited amount of experimental information on responses of desert and arid-land plants to increased CO_2, it seems that most of the preconceptions have to some extent been supported and to some extent challenged. Some experiments suggest that either because of nutrient limitations or their innately low growth rate, desert and semi-desert plants may hardly be able to respond to high CO_2 in terms of growth rate and biomass. Other similar growth studies suggest a strong response by these very same species of drought-tolerators. In certain experiments on mixtures of desert species growing together, there is a disproportionate response to CO_2 by particular plant "types" or even by certain individual species which apparently arbitrarily show a very large response when most others around them barely respond. Overall, the experiments bolster the general expectation that increased CO_2 will favor a stronger response of C_3 plants over C_4 species, but the contradictory results from prairie species add an element of uncertainty to all of this. The amount of idiosyncrasy in responses seen in all of these various experiments seems to make the prediction of CO_2 effects on any particular arid region (or arid regions in general) a rather risky business, for it may vary greatly with the detailed community assemblage and perhaps other local factors such as soil variations and herbivory.

Another factor which should be borne in mind is that many of the free air CO_2 experiments that have been run in moister climate biomes (e.g., tundra and some forest systems) for more than a few years show a decline or even a disappearance of the effects of CO_2 on plant growth rates. It is unclear what this might mean in terms of biomass and species composition as the plant community reaches a rough equilibrium in the longer term. The Nevada desert CO_2 experiment has not been run for as long as some other FACE experiments, and because desert plants tend to be slow-growing, the time taken for the ecosystem to reach a balance in response to higher CO_2 levels may be even longer. Even if growth rates are initially boosted in arid lands with raised CO_2 (as some chamber experiments suggest), there is no certainty that this will translate into greater vegetation biomass beyond a boost in the earliest years, because shortage of nutrients may begin to dominate after a while.

8.9 OTHER FACE EXPERIMENTS

When tundra ecosystems in northern Alaska were exposed to increased CO_2 at nearly 600 ppm (about double the pre-industrial background) in a FACE experiment, there was an initial increase in growth which disappeared after 2–3 years. This particularly short-lived response is thought to occur because tundra ecosystems are strongly limited by nutrients: they cannot respond well to the extra carbon supply by increasing plant tissues because the nutrients they also need to build their tissues are in such short supply.

Other FACE studies have included salt marshes, and tallgrass prairies in the Midwestern USA. These too have tended to show an initial burst of growth, followed by a decline in the advantage of CO_2-fertilized plots.

8.9.1 FACE studies on agricultural systems

Many different experiments have studied how agricultural yields might change as CO_2 increases. Most of these have focused on temperate-zone crops such as wheat, soy, maize and rice. Closed-chamber experiments with the crop plants growing in well-fertilized soils often find a lasting yield increase of around 20–30% (relative to late–20th century levels) for the sort of CO_2 levels that will be reached by the mid–21st century (around 550 ppm). However, the various FACE experiments that have been set up in fields in the mid-latitudes of Europe, Asia and North America (Figure 8.8*) paint a somewhat less optimistic picture of a 10–15% increase in yields for most of these crops. It is not clear what is happening to limit the yield increase in these more realistic open-air systems. Maize (known in North America as corn) actually shows no increase in yields in FACE experiments, presumably due to its C_4 photosynthetic system (see Box 8.1) which already makes very efficient use of CO_2 at lower background levels. While 10–15% is still quite a substantial yield increase for several major crops, it must be balanced against the effects of changing temperature and rainfall while the whole climate system alters with global warming over the coming decades. The best estimates we have, preliminary and imperfect though they are, suggest that the effects of CO_2 fertilization on crop yields will not compensate for the effects of greenhouse gases on climate, and that yields of temperate-zone crops will decline overall during the next few decades (Long *et al.*, 2006). Though there have been no FACE experiments so far on tropical crops, it is reasonable to expect that

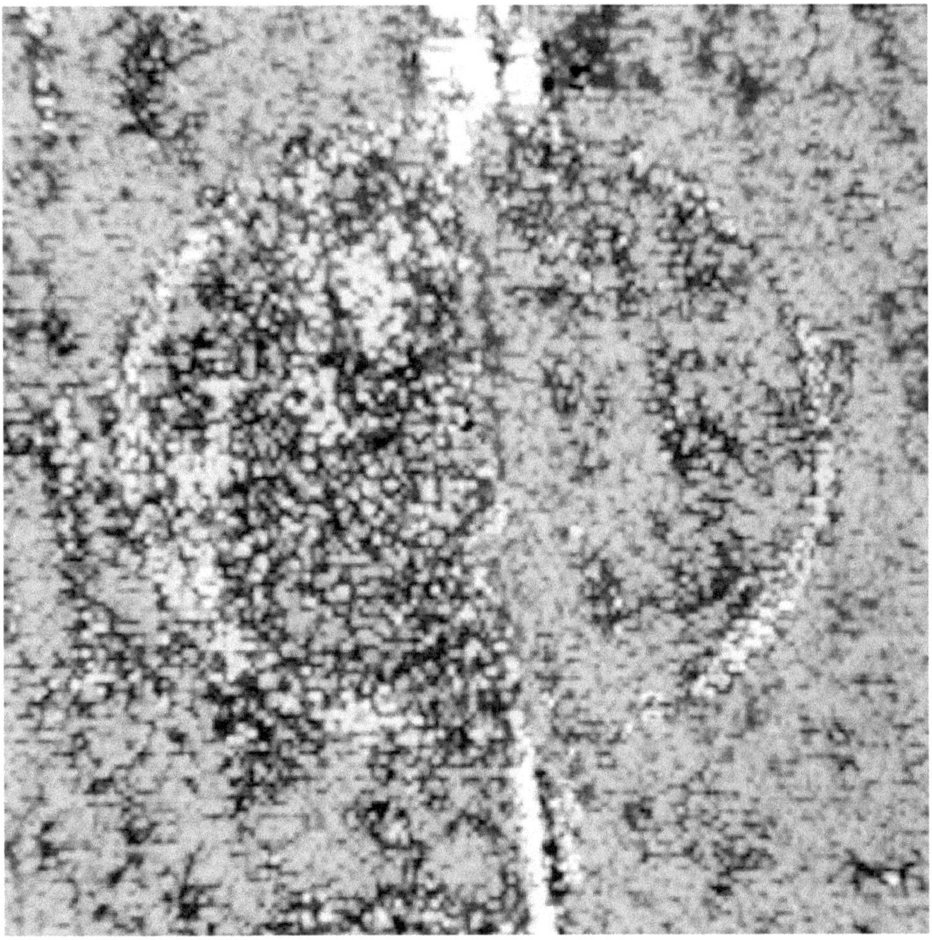

Figure 8.8. This thermal infrared image was taken looking down on a wheat field fertilized by a ring of pipes releasing CO_2 (white and black areas are warmer than gray). The higher temperature on parts of the plot results from less evapotranspiration and decreased latent heat uptake from the leaves, which have their stomata mostly or totally shut. The field across the left-hand part of the ring was not irrigated, and the much stronger effect on temperature shows how CO_2 may make the most difference when plants are water-stressed. Credit: Bruce Kimball. From Kimball *et al.* (1995).

even a slight increase in temperature due to the greenhouse effect will decrease yields due to extra heat stress and photorespiration in these plants, which are already living near the upper limits of their temperature tolerances. Furthermore, many small farmers in tropical countries cannot afford the large amounts of fertilizer that are necessary to ensure a substantial yield increase with additional CO_2. Hence, it seems likely that tropical countries will lose out more heavily than temperate countries in terms of food production over the coming decades.

8.10 SOME CONCLUSIONS ABOUT FACE EXPERIMENTS

The results of CO_2 enrichment experiments are diverse, confusing and sometimes contradictory. Increases in growth rate of plants are often reported during the early part of an experiment, so at the end of the experimental period they are bigger than they would otherwise be (though this relative gain tends to lessen during the experiment). It is important to realise however that this does not necessarily translate into a long-term equilibrium change in biomass. For all we know the plants might just grow more quickly to maturity then fall over and rot, with no change in their long-term biomass. Even less is known about this than the question of whether growth rate of the world's vegetation will be significantly speeded up.

It is scientifically risky to try to make predictions that scale up from such isolated experiments, operated for such a brief period of time, to predict the response of the whole world's vegetation. But, one general lesson that the experiments *do* give us is that CO_2 responses are very complex and are not always what we would expect from a physiological model. Models that extrapolate from cell and leaf-level processes, to forecast global-scale responses in vegetation to increased CO_2, seem to score some successes in predicting the results of individual experiments but also many significant failures. Even the successes are rather ambivalent if one looks at the details of the response. For example, the CO_2-fertilized sweetgum forest in Tennessee showed about the "right" amount of increase in NPP predicted by the models, but the way in which the NPP showed up (in fine-root turnover) is rather bizarre. The expectation of most of the modelers, even if not clearly stated, is that the extra growth put on by plants under increased CO_2 will show up in the form of wood. This will then form a negative feedback in terms of taking up CO_2. In contrast, it is not at all clear what an increase in fine-root turnover would do for overall carbon storage.

8.10.1 Will a high CO_2 world favor C_3 species over C_4 species?

Probably the most generally held principle in forecasting raised CO_2 effects is that plants using the more water-efficient and CO_2-efficient C_4 photosynthetic system will lose out by competition to the "normal" C_3 plants which have more to gain from raised CO_2. When C_4 plants growing alone are fertilized with extra CO_2, they tend to show little gain from it. There is almost no enhancement of photosynthesis, although they do lose a little less water because they can get the CO_2 they need quicker and then shut their stomatal pores. A FACE-type experiment on a corn (maize) field in the USA showed that, much as expected, it did not grow any bigger or faster under doubled CO_2. Various experiments growing wild C_3 and C_4 plants in competition in chambers under raised CO_2 have shown that, as expected, C_4 plants tend to lose out to C_3 species, which benefit much more strongly from the extra CO_2. However, it is rather puzzling that closed-chamber experiments growing apparently realistic combinations of prairie plants under raised CO_2 do not support this (see above). In one case, C_3 and C_4 species responded about equally, and in another experiment C_4 species actually did better than the C_3 species under raised CO_2!

8.10.2 What factors tend to decrease plant responses to CO_2 fertilization?

Some vegetation types respond more strongly to CO_2 than others, but all seem to show at least some decline in CO_2 response over time. Various factors seem to be at work in producing both the variation in response and the decline in response. Probably the most important of these is nutrient supply. The more nutrient-deprived the system is, the less responsive it tends to be to CO_2 fertilization. Also, much of the decline in CO_2 response over time seen in raised CO_2 experiments is thought to be the result of nutrient shortages. Without the mineral nutrients it is impossible for the plant to build more tissues no matter how much extra carbon it has, and so the response to CO_2 diminishes. When a plant undergoes an initial burst of growth under raised CO_2, it is drawing upon a store of nutrients within its own tissues and in the immediate vicinity of its roots. After a while, however, the available nutrients become used up and the plant has difficulty finding more. It is also thought that greater internal shading within the denser growing canopy decreases the gains from CO_2 fertilization as time goes on, by putting a limit on photosynthesis.

Apart from these factors there is also a poorly defined and poorly understood "acclimation" of the photosynthetic response to CO_2 that occurs over time. This is involved in the decline in growth rate following an initial response, even in crop plants that have plenty of mineral nutrients. There seems to be some sort of "switching off" mechanism within the plant over time that causes it to be less responsive to raised CO_2. It is unclear what advantages there are to the plant in doing this. It might perhaps involve a dismantling of part of the photosynthetic apparatus to allow diversion of nitrogen away to other places within the plant where it is needed.

8.11 THERE ARE OTHER EFFECTS OF ENHANCED CO_2 ON PLANTS APART FROM GROWTH RATE

Plants grown at higher CO_2 levels generally have a higher carbon-to-nitrogen ratio. It seems that because they have more carbon to work with, they end up producing more of the carbon-containing structural molecules of cell walls, such as cellulose and lignin. They may also store up excess carbon as starch reserves inside their cells. It is uncertain what implications these changes might have for ecosystem functioning; for example, they might decrease the suitability of the plant as food for herbivores, and the decomposition rate of dead plant tissues when they end up in the soil. There are also concerns that crops might become less nutritious since they have less protein; people could simply fill up on starch.

Plants grown at high CO_2 also tend to have a greater mass of roots relative to shoots. Alternatively, the rate of growth and turnover of the small roots that gather nutrients may increase (as in the sweetgum plots in Tennessee). From the plant's point of view, having more carbon as a result of growing in high CO_2 means that nutrients from the soil are more limiting to its growth, so investing in roots is a good way to gather more nutrients. Getting more nutrients can then mean that it puts on the maximum amount of growth, and produces more offspring.

Another effect, detectable only at the microscopic level, is that plants grown at higher CO_2 levels have fewer stomata on their leaves, presumably because fewer are needed to allow CO_2 in to the plant when CO_2 levels are higher. Perhaps stomata leak water a bit, or use up energy unnecessarily, so that it is advantageous for the plant to have no more stomata than it really needs (Figure 8.9).

One rather strange effect of increased CO_2 levels is that respiration rates of plant tissues tend to be lower. If wasteful burning of carbon was being decreased, then this could be a good thing, allowing the plant to accumulate more carbohydrate for useful tasks. However, much of the respiration that goes on in a plant has a purpose, such as the building and repair of tissues. If this sort of respiration is cut down, plants might not be repairing tissues as thoroughly, with possible long-term consequences.

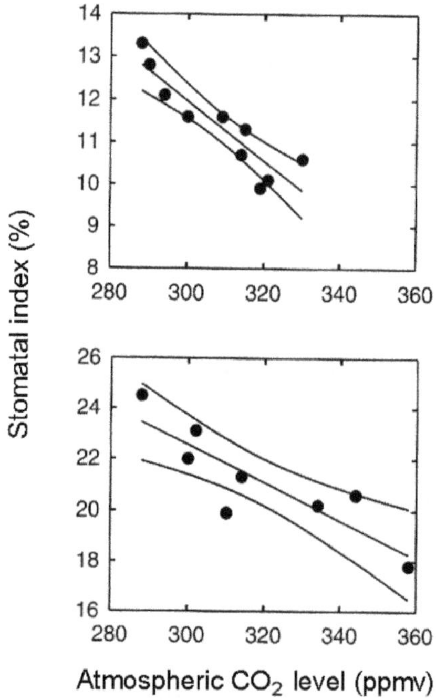

Figure 8.9. Stomatal index change. From Woodward (1987).

8.12 CO_2 FERTILIZATION AND SOILS

If working out what will happen to vegetation due to direct CO_2 fertilization is a challenge, figuring out its effects on soil carbon is even harder to do. Soil organic carbon is the largest reservoir of organic carbon in terrestrial ecosystems, exceeding vegetation two to three-fold. But, it is indirectly a product of the growth of living plants, plus what happens to them after death.

Various CO_2 fertilization studies have found that plants invest proportionately more in roots at high CO_2 levels (which may be explained as being an adaptation to bring in nutrients which become more limiting). Both closed-chamber and FACE experiments (e.g., that Tennessee sweetgum stand, once again) have found a large increase in the rate of turnover of fine roots at higher CO_2 concentrations. Roots supply carbon directly to the soil when they die, and an increase in the size and rate of turnover of root systems seems likely to increase the amount of organic matter in soils.

As well as a change in the rate of supply, the detailed composition of plant materials may alter under increased CO_2, making a difference to decay rates and organic matter in soils. Many studies have shown a decrease in nitrogen content of plant tissues under increased CO_2. Nitrogen generally seems to be a limiting factor in the rate of decay of plant materials by fungi and bacteria, so less nitrogen in these tissues should mean they decay more slowly, perhaps adding to the carbon store in soils. However, one study which looked at the breakdown of the less nitrogen-rich plant parts that had grown at increased CO_2 found no effect on decay rate.

Even if plant materials are lower in nitrogen, it is possible that, because of their greater rate of supply to the soil and litter layer (due to increased growth rates), they will encourage the growth of a specialized microbial and detritivore community. This will be able to break them down quicker overall, because the right organisms are always on hand—the decomposer populations are "primed" with a continual supply of material to feed off. According to this hypothesis, there will be less carbon ending up in soils in a high-CO_2 world.

Where CO_2 fertilization experiments have looked at soil carbon, in some cases they have found that it increased, while in other cases it decreased. One set of chamber experiments with sets of wild tropical plants growing together in artificial communities found a decrease in soil carbon at doubled CO_2, which is what the "priming" hypothesis predicts. Another study comparing soil changes under soy (C_3) and sorghum (C_4) found that soil carbon decreased under soy but increased under sorghum at high CO_2. This latter experiment contrasts two very different photosynthetic metabolisms, and it is not clear how more subtle differences in metabolic and growth characteristics might affect the soil carbon response to raised CO_2. However, this degree of complexity does not bode well for understanding future responses of soil carbon to increased CO_2.

Another effect that might turn out to be important under increased CO_2 is the enhancement of chemical weathering, which itself acts as a CO_2 sink (Chapter 6). If plants produce more roots and more root exudates under increased CO_2 (remember that various studies show they do put more into fine roots when CO_2-fertilized), then this might promote the fungal and bacterial activity that breaks down minerals in the soil. This will act as an increased global sink of CO_2: a negative feedback on CO_2. There is a need for studies which address this question. Some preliminary studies do show that under increased CO_2, chemical weathering is enhanced.

8.13 CO_2 FERTILIZATION EFFECTS ACROSS TROPHIC LEVELS

What sort of effects might increasing CO_2 have on broader community structure? How will the animals and the fungi that feed off living plants respond to changes in the growth and composition of plants that are CO_2-fertilized?

Most of the work on how increased CO_2 can affect such interactions has focused on crop plant systems, although the findings might also apply to more natural communities. Several studies of herbivory on CO_2-fertilized crops have suggested there might be an increase in insect or fungal attack on plants at higher CO_2, which will "take back" part of the gain from CO_2 fertilization. Some work has shown that at raised CO_2 levels insects increase their rate of feeding, perhaps because the leaves have a lower protein content when they are CO_2-fertilized. The insect simply has to eat more leaf in order to get the protein it needs to grow. It has been suggested that this means that in the future high-CO_2 world, insects will cause more damage.

However, it is important to bear in mind that the plants themselves are generally bigger when they are CO_2-fertilized, and the extra amount lost to hungry insects in these experiments often works out to be less as a percentage of the total leaf area. Also, insects which have to eat more leaf material to extract enough protein are generally placed in a difficult situation: it takes a lot of work for the insect to digest the extra material, and the insect may also have to take in extra amounts of poisons the host plant produces in the process of consuming more leaf. The insect may also have to spend more time feeding out on the leaf exposed to enemies when it cannot get enough protein. It seems that insects on CO_2-fertilized plants not only consume a smaller proportion of leaf tissue, they grow more slowly and die more often. In fact, the evidence is that overall with CO_2 fertilization the advantage is tipped in favor of the plant, against the insect. However, not all the experiments have shown this, so there is a definite need for more studies on direct CO_2 effects on herbivore activity.

Most species in the world are herbivorous insects, and it is rather frightening to consider what effects any large change in insect–plant interactions might have on insect biodiversity in the tropics and elsewhere. It is quite possible that a large change in nutrient content will push many insect herbivore species over the edge into extinction. It is widely considered by ecologists that a large part of the reason so many species of tropical trees can coexist in the tropical rainforests is that selective insect herbivores prevent each tree species from becoming too abundant. If we start to see these specialized herbivores dropping out of existence because of a direct CO_2 effect, many tropical trees may go extinct because the most competitive species among them are no longer so closely density-limited and can now push the others out.

Even though plants may benefit from CO_2 fertilization, humans who also want to eat them may suffer from some of the same problems as insects do. Experiments on wheat and rice suggest that with CO_2 fertilization their grain contains proportionally more starch and less protein than when they are grown at background CO_2 levels. This may mean poorer nutrition for human populations in some parts of the world where protein intake is already very limited. Analogous problems might also come up for mammalian herbivores that feed off wild plant materials. If the decline in nutrient content is severe enough, some may go extinct.

8.13.1 Looking for signs of a CO_2 fertilization effect in agriculture

It is always good to back up models and experiments with unfettered observations, showing that what we expect to be happening is actually happening. How about agricultural systems, which we can expect to respond particularly strongly to increasing CO_2? If there is indeed going to be a strong future response of crops to increased CO_2, we might expect that the 40% increase in CO_2 that has occurred over the last

250 years has already had some effect on yields. Is there any direct evidence for this? It is certainly true that there has been a massive increase in crop productivity over that time period. Even in the last several decades in the USA, yields have gone up by 50–100% in many areas. One might take this as indicating that the direct CO_2 effect is at work here. However, such a conclusion would be far too simplistic. Over time, many different factors have changed agricultural yields, including crop-breeding, fertilizer use and pesticide use. Because so many of these other things have changed too, it is basically impossible to "extract" the trend of increasing yields from CO_2 in order to "test" or "prove" models of CO_2 fertilization. It is a reasonable guess that the direct CO_2 fertilization effect is in there somewhere, but we really cannot be sure how large it is.

Crop plants are not the only components of agricultural ecosystems. Weeds are always present too, and they too can be expected to be benefiting from increased CO_2. Some interesting chamber experiments with growing common temperate field weeds in pre-industrial (280 ppm) CO_2 levels showed that they grew 8% slower compared with present-day (around 350 ppm) CO_2 levels. If the weeds in fields are growing faster under increasing CO_2 levels, they can be expected to "take back" some of the gain from increased CO_2 that would otherwise go into more vigorous growth of the crop plants.

8.13.2 Looking for signs of a CO_2 fertilization effect in natural plant communities

Likewise, if there is an increasing CO_2 fertilization effect in natural and semi-natural vegetation, we might expect to see signs of it already. After all, CO_2 has been increasing for many decades now, so if we have good records of how the vegetation was 100 or 200 years ago, we should be able to compare the "before" and "after".

Are there any inexplicable increases in tree growth, for example? Tree rings go back hundreds of years in old trees, so we can compare growth rates now by looking at ring widths: more rapid growth should produce wider rings. In northern and central Europe there has been an increase in tree ring widths—allowing for the growth stage of the tree—over the past couple of hundred years, which shows that trees are growing faster. This is consistent with what might be expected from a CO_2 fertilization effect, but it could equally be due to other factors. Climate has warmed substantially over the last two centuries throughout Europe, perhaps due to natural climate fluctuation and perhaps due to an increasing greenhouse effect. Trees may well be growing faster in response to the warmer climate. Even though climate-warming due to an increase in the greenhouse effect is mostly due to CO_2, this is not the same mechanism as direct CO_2 fertilization that we are talking about here. Another possible explanation for the increase in tree growth is a different form of pollution, from the nitrogen and sulfur oxides produced by power plants, factories and car engines. Although these acidic gases are usually thought of as destroying ecosystems, in small quantities they may act as fertilizers. Many forest soils are very low in nitrogen and sulfur, and experiments suggest that adding traces of sulfate and nitrate salts often promotes tree growth.

There is not even a noticeable positive trend in tree growth in most other parts of the mid-latitudes. For example, a trend towards faster tree growth is not seen in southern Europe. There, growth actually seems to be slowing down despite the ongoing increase in atmospheric CO_2. My friend Gianluca Piovesan of the University of Tuscia has implicated a drying of climate as the main factor behind this trend. In the temperate deciduous forests of the northern USA, Pacala and colleagues have looked for the anomalous increase in ring widths of trees that might indicate a CO_2 fertilization effect over the past century, and found none. In fact, they actually found a decline in growth rate (adjusted for the age of the trees), that might be due to too much air pollution with nitric and sulfuric acids or ozone.

In a wide-ranging tree ring study of high-latitude conifer forests, Schweingruber and colleagues reported that there is no sign of a response of the boreal forest to atmospheric CO_2 growth. In certain parts of the boreal zone, tree ring widths have actually decreased over the past century or so. Northern Siberia is one such region where trees seem to be growing more slowly. This is not at all what we would expect from CO_2 fertilization, and is probably due to a change in climate towards drier conditions.

How about the tropics? In the early 1990s, Oliver Phillips and Alwyn Gentry looked at inventories of the girth of tropical trees taken by generations of foresters, and announced that they had found a clear trend of increasing growth rates throughout the tropics during the previous 60 years. This caused a ripple of excitement through the world of ecology; surely here at last was clear, systematically gathered evidence of a CO_2 fertilization effect. Because it occurred all across the world's tropical forest regions, it was unlikely that any regional climate effect was the cause. Such a widespread trend seemed to leave only CO_2 as the driving factor. However, the trend turned out to be an example of what can go wrong if one does not meticulously check one's sources of data. Someone pointed out that the frequency with which forest managers returned to each tree to check its girth increment had changed over the decades. Nowadays, they were measuring girth less often and this meant that each tree had put on more growth between the measurements. Philips and Gentry had assumed that the interval between measurements had stayed the same throughout, so naturally they found what looked like an increase in tree growth rate. Another "false alert" occurred when a carbon balance study of old-growth tropical forest in the Amazon Basin—using the eddy flux covariance method mentioned in Chapter 7—suggested that the forest was putting on a remarkable burst of growth and was set to double its biomass in another 60 years. This was initially suggested as being a response to direct CO_2 fertilization. This trend turned out to be the result of some combination of problems in the use of the equipment, plus short-term variability in forest processes.

Some years later, there was similar widespread interest when data gathered across the Amazon Basin suggested an ongoing trend in forest turnover—the rate of birth, growth and death of trees—and an increase in the abundance of vines over the last few decades. Most recently, a similar trend has been found in the rainforests of central Africa. In both regions, the forests are also apparently increasing in carbon storage as the trees grow larger, taking up some of the CO_2 that is added to the atmosphere each year by human activities (and agreeing more with those new studies of CO_2 concentration in the atmosphere over the tropics—see Chapter 7—that suggest they are sucking up much of the CO_2 added by human activities). Philips and other authors have suggested this might be evidence of a direct CO_2 effect on the growth of the forests. Whether this is a genuine CO_2 fertilization effect is a matter of dispute, but the increase in vine growth does seem to agree with the expectations of closed-chamber CO_2 fertilization experiments. The trend towards faster turnover and increased carbon accumulation is not occurring on every single sample plot: some of the plots in Africa and the Amazon Basin, plus other forest plots in Central America and Peninsular Malaysia, seem to have been slowing their turnover during the same time. And as Christian Koerner has cautioned (Koerner, 2004) a recent increase in carbon uptake and tree turnover does not necessarily mean that the tropical forest will end up with more carbon overall. For all we know, eventually the new faster growing trees might just fall over and rot, releasing much of the extra carbon they had taken up.

I suspect that a lot of the trend that is seen towards increased carbon storage in the forests of Amazonia and Africa is a result of them recovering from pervasive selective logging by local people over the past centuries and especially the last few decades, rather than a real CO_2 effect. Sandra Brown and her colleagues have found evidence of such a past timber extraction of really big trees by comparing relatively accessible and more inaccessible areas of tropical forest: the more accessible places seem depleted in big trees. There is also a long history of shifting cultivation—now largely abandoned—in many of these forests, that would have involved clearing patches of forest which would then have to slowly recover after being abandoned. Among other possible reasons for the turnover trend in the Amazon is an increase in cloud cover in recent decades, which might be expected to allow faster growth in trees that are exposed to less roasting in direct sunshine. In my opinion, as things stand now, there is still no overwhelming evidence of any direct CO_2 fertilization response in tropical rainforest.

Many firmly established changes in vegetation are being documented in the coldest regions of the world. For example, in the Canadian Arctic islands and in northern Alaska, shrub cover has expanded over the last several decades. On mountains in many parts of the world, the treeline is creeping upwards (Chapter 3). These trends are generally attributed to global warming produced by CO_2 and other greenhouse gases, but it is also possible that the direct CO_2 effect is playing a significant role in promoting the growth of plants. Unfortunately, it is presently impossible to disentangle these two factors, and ecologists rely on their gut instincts (plus the results of experiments in Arctic ecosystems showing that CO_2 fertilization effects are often short-lived and minor) when they attribute these changes to climate rather than CO_2 fertilization.

One indirect effect of increasing CO_2 could be a decrease in the amount of evaporation through stomata, which are expected to stay closed more of the time when the plants get enough CO_2 for their needs. This could have indirect effects on other phenomena in the environment; for example, the amount of water running off land surfaces. With less evaporation, due to CO_2 effects on stomata, a greater proportion of the rain that falls on land surfaces should simply run off. Gedney and colleagues have pointed out that indeed there has been a widespread trend around the world towards increased river runoff in the last few decades, and they suggest that this is best explained as a result of the direct CO_2 effect. These authors find that the pattern and amount of the increase in runoff cannot simply be explained as a result of factors like increased rainfall, or deforestation, leaving it to be best explained as a result of CO_2. However, other authors who work on water runoff have cautioned that this approach is rather simplistic, and that many subtle interacting factors could explain such trends. I also wonder why, if extra CO_2 is letting plants get by with less water, the vegetation doesn't simply get lusher and taller as a result of its improved water supply—and then end up evaporating just as much water as before until the available supply is used up. At least that seems to be how vegetation behaves on a global scale, its leaf area strongly limited by the water available to the vegetation (Chapter 2). Overall, the intruiging trend towards more runoff is worth considering as possible evidence for a direct CO_2 effect showing up in nature, even if we should treat the idea with a certain amount of caution, given the complexity of the systems we are dealing with.

So far, then, there does not seem to be anything in terms of natural vegetation change that is unambiguously a sign of CO_2 fertilization. In such a complex world in which vegetation is buffeted by many different factors, almost anything we see that looks like a direct CO_2 effect could in fact be due to a multiplicity of other causes. Perhaps, as CO_2 levels continue to climb, more striking changes in vegetation that can only be attributed to this effect will begin to show up. Even so, it is very likely that climate will also be warming in parallel with CO_2, always leaving open the possibility that any given change in vegetation will be due to temperature increase, not CO_2 effects on physiology.

8.13.3 The changing seasonal amplitude of CO_2

The CO_2 concentration in the northern hemisphere fluctuates with the seasons: it goes up during the winter when decay dominates (releasing CO_2) and decreases during the summer when photosynthesis is taking up CO_2 and temporarily building it into leaves (Chapter 7). When autumn comes, most leaves in the mid-latitude forests and also the tundra are shed and they decay releasing CO_2.

The seasonal fluctuation in CO_2 in certain places in the northern latitudes has been increasing over the past several decades. This trend towards wider seasonal swings is strongest in northern Alaska, at a CO_2-measuring station located at Barrow. A weaker trend towards more seasonal fluctuation is also found at the Mauna Loa measuring station in Hawaii that ultimately gets a lot of air coming down from the Arctic. However, the trend is absent from other stations around the world.

What is one to make of this trend in the seasonal wiggle in the far north? The first explanation put forward when it was discovered was that it was due to increasing CO_2 fertilization. More CO_2 might be giving greater summer leaf mass in shrubs and herbaceous plants in the far north: more leaves sucked in more CO_2 each growing season, and then this was released by decay after the leaves were dropped at the end of summer. This picture seemed to be reinforced by satellite data showing that the Arctic latitudes had become increasingly greener over the last 10 years. Perhaps this is evidence of an increasingly strong CO_2 fertilization effect?

However, if the trend in the seasonal wiggle is due to CO_2 fertilization, why is it only noticeable in one part of the world? After all, vegetation everywhere should have at least some potential to respond to CO_2 fertilization. And, from what little experimental work has been done on exploring direct CO_2 responses in tundra, it seems to be particularly unresponsive after a few years due to severe nutrient limitation. Also, there are other straightforward explanations as to why the seasonal amplitude of CO_2 in the high latitudes might be increasing. Plants are known to respond strongly to temperature, and greater warmth in the north (where summer temperatures tend to limit the amount of growth that plants can put on) could be allowing the plants to carry more summer leaf mass. Climate records show that indeed there has been a strong warming trend in the Arctic, especially in the parts of Alaska and northeastern Siberia that also show the strongest trend in both CO_2 seasonal fluctuation and in greenness measured from satellites. This neatly explains why the trend in CO_2 seasonality has only occurred in that general part of the world, and there seems no particular need to invoke the poorly understood role of direct CO_2 fertilization. Why temperatures are increasing is another question altogether, and it could be due to CO_2 and other greenhouse gases acting upon climate (Chapter 1).

8.14 CO_2 LEVELS AND STOMATA OUT IN NATURE

Perhaps the only really convincing evidence of a direct CO_2 effect occurring in nature is a change in stomatal indices of leaves (the stomatal index is the abundance of stomatal pores relative to normal epidermal cells in the leaf surface) over the past centuries. The stomatal index has been shown many times to decrease with increasing CO_2 concentration in experimental plants grown at different CO_2 concentrations. Ice core evidence, and old measurements of the CO_2 content of air, show that the atmospheric CO_2 concentration has been increasing since the early 1800s. Ian Woodward of Sheffield University was the first to show that over the last 200 years herbarium specimens of leaves of common trees (e.g., beeches *Fagus*, birches *Betula*) show decreasing stomatal indices that parallel the increase in CO_2. This finding has been repeated many times on herbarium specimens of other species gathered before and after the onset of the main CO_2 increase.

8.15 DIRECT CO_2 EFFECTS AND THE ECOLOGY OF THE PAST

There are some fairly good indications that the CO_2 concentration of the atmosphere has undergone natural variations in the past, before humans began to affect it. The best substantiated changes in CO_2 were those that occurred between glacial and interglacial periods during the last 650,000 years, where bubbles trapped in ice caps preserve samples of the ancient atmosphere that can be analyzed (Chapter 7).

The evidence of such fluctuations in CO_2 has set ecologists wondering what these might have been doing to plants in the past. During glacial phases, with CO_2 concentrations more than 30% lower than at present, plants may have suffered more drought due to the need to open their stomata more to get enough CO_2 to grow. Indeed, the glacial world was much drier with widespread deserts and scrub vegetation, and far less forest than the present-day world (Chapter 3). For instance, large parts of the central African and Asian rainforests were replaced by arid scrub and grassland. The Sahara Desert extended hundreds of kilometers farther south, compressing the vegetation zones on its margin down towards the equator. And Siberia, which is now covered by forests, was a dry cold semi-desert at that time. There are good climatological reasons to expect that the glacial world would have been drier (a more arid world is predicted by GCMs for the glacials, due to less evaporation from colder oceans transporting less water vapor around), but it is possible that the appearance of aridity was intensified by low CO_2 concentrations. Climate models so far seem to have trouble getting the world dry enough to match the indications from the plant fossil record, and it is possible that the additional "missing" factor in the aridity is the direct effect of low CO_2 on the plants.

Some other rather strange aspects of the vegetation of the glacial world might have been due to the effects of CO_2 on plant physiology. For instance, the open, arid steppe–tundra vegetation (see Chapter 3) that covered Siberia and northern Europe during ice ages combined species of plants that do not normally grow together nowadays. There were typical tundra species such as dwarf willows (*Salix*) and sedges (Cyperaceae) alongside steppe species such as Russian thistle (*Kochia*) and wormwoods (*Artemisia*), leading to the name "steppe–tundra" to describe this vegetation type. Also present in the steppe–tundra were plants that are now confined to seashores (e.g. stag's horn plantain, *Plantago maritima*), and genera that are commonly found on builders' rubble (e.g., dock, *Rumex*). Some ecologists who have studied ice age vegetation suggest that the steppe–tundra was a product of low CO_2 levels, bending the niche requirements of plants that cannot normally survive in the same place. However, it could also have been due to the very different climates of the ice age world, with combinations of climatic attributes that do not occur today allowing these plants from different biomes to grow together.

The vegetation on mountains may also have responded particularly strongly to low CO_2 during glacials, because the availability of CO_2 was already limited by the thinner air at high altitude. While ice age climate was too cold for vegetation to grow far up mountains in the high latitudes, tropical mountains did have vegetation growing up to several thousand meters. Even so, vegetation zones on tropical mountains had moved downslope a long way during ice ages. This was certainly in part due to the worldwide cooling of climate at the time, but the vegetation change on mountains seems to correspond to temperature changes that are greater than non-living indicators such as glacier limits would indicate. It has been suggested that this discrepancy is due to the effect of the low CO_2 on mountain plants: the temperature limits at which their growth was viable shifted under reduced CO_2, pushing them still farther down the mountains.

One thing we might expect during the low-CO_2 glacial phases is a greater relative abundance of C_4 plants, because experiments show that they are better at making the most of low concentrations of CO_2. C_4 plant carbon has a characteristic isotopic composition, because the C_4 photosynthetic system distinguishes more strongly between the various stable isotopes of carbon than the C_3 system does. Buried soil carbon—made originally from dead parts of plants—reflects this isotopic composition. If it is composed more of C_4 plants, it has less of the lighter ^{12}C, more of the heavier ^{13}C isotope. It turns out that at least in some places in tropical Africa that are now grasslands, preserved grassland soils from the last glacial have more of the isotopic composition of C_4 plants, suggesting that C_4 plants were doing better in the glacial environment. This has been interpreted as a sign of the influence of low CO_2 levels, favoring C_4 grasses in the tropical grasslands at that time (although a trend towards aridity could produce the same effect as C_4 plants do better in more arid climates). However, the trend is not a consistent one; in other places the soil carbon record shows that C_4 plants were less abundant in grasslands of the glacial times. It could be that in such cases cooling during the ice age is dominating; C_4 plants don't do so well in cooler climates. Because temperature, aridity and CO_2 could all affect the relative abundance of C_4 plants, it is very difficult to disentangle them from one another to say what was really more important during the glacials.

How about an environment which we know stayed wet during the last glacial, because the plants were growing in waterlogged soils throughout? There is a swamp site in the tropical mountains of Burundi that has stayed moist and laying down peat throughout the time since the last glacial. Here at least we can say aridity is removed as a factor, because the plants have had their roots in water all the time. In this site, during the last glacial the peat composition indicates that there was a big shift towards C_4 plants during the glacial, compared with the present. The climate was also cooler at this site during the glacial, and that would be expected to favor C_3 types of plant, and yet the shift was still towards C_4. The only factor left in this case seems to be CO_2 favoring C_4 species that can photosynthesize faster where CO_2 is in very short supply. Perhaps here then, we see one example where the direct CO_2 effect of lower glacial CO_2 levels really does show up unambiguously.

As one would expect, the stomatal index of fossil leaves that grew during the low-CO_2 glacial phases is higher than those of the same species from the interglacials. The lower CO_2 triggered a growth reaction to increase the supply of CO_2 into the leaf, by making more stomatal pores. This again is nice confirmation that the change in CO_2 did affect the plants in at least some way, though how important it actually was in altering vegetation structure and composition is still an unknown. Some clues can be had from chamber experiments which grow plants under lower than present CO_2 levels, similar to those during the last glacial (at 200 ppm). Many herbaceous plants grown under such conditions turned out to have much lower growth rates (often around 50% less) than when they were grown at present-day early 21st century (360 ppm) CO_2 levels. However, it is not clear if evolutionary selection of low-CO_2 tolerant variants would close this gap in a real world low-CO_2 situation.

8.15.1 Direct CO_2 effects on longer geological timescales

It is uncertain how far the CO_2 level has fluctuated on longer geological timescales, over tens and hundreds of millions of years. Some calculations based on balancing the rate at which volcanoes push up CO_2, and the rate at which CO_2 is taken up into weathering reactions with rocks, suggest that there must have been some rather large fluctuations in the CO_2 level of the atmosphere over geological time, perhaps by a factor of 20 or more (Chapter 7). Another way to estimate past CO_2 levels comes from looking at the stomatal indices of leaves that lived during each phase of geological history. By looking at the surface of very well-preserved fossil leaves under a microscope, one can compare the abundance of stomata with similar leaves in the modern-day world, and perhaps deduce the concentration of CO_2 in the ancient atmosphere. It is important to compare the stomatal index of groups of fossil plants that have close present-day relatives for comparison, because the average stomatal index varies quite a lot between different groups of plants. Good candidates include "living fossils" like the maidenhair tree *Ginkgo*, which has existed nearly unchanged for around 150 million years, and the "tulip tree" *Liriodendron* which has existed for 70 million years. When their fossil leaves are very well preserved, it is possible to use a microscope to see the stomata on the leaf surfaces, and to compare their frequency to the abundance of normal epidermal cells in order to work out the stomatal index. By comparing that with the stomatal index of modern specimens grown experimentally under a range of different CO_2 concentrations, it is possible to estimate what the atmospheric CO_2 level was back then, so many millions of years ago. The general picture from stomatal densities seems to agree fairly well with that from balancing volcanoes and rock reactions, supporting the idea that there have been several ten-fold fluctuations in the CO_2 concentration over the last few hundred million years. Some other independent indicators of CO_2 levels in the chemistry of rocks seem to corroborate these estimates, while others disagree, so the picture is perhaps not totally clear overall. However, most geologists who have studied the evidence seem to be convinced that there really were major fluctuations in atmospheric CO_2 levels over geological time.

It is unclear what these fluctuations in CO_2 would have been capable of doing to the ecology of plants. Presumably, the higher CO_2 levels in the distant geological past made plants less drought-susceptible, because they would not have needed to open their stomata quite as much to get the carbon they needed. So, they would be better able to eke out whatever supply of water they had around their roots. It has been suggested that the rise to dominance of the flowering plants (angiosperms) between 120 and 60 million years ago was caused by a large decrease in atmospheric CO_2 levels during the same period. Various features that are common in flowering plants, but rare in other types of plants, seem to be favorable for getting water up from the roots quickly in an environment where leaves are often short of water. For example, the elaborate branching networks of veins in the leaves of flowering plants, and the long open vessels that conduct water up through their stems, may allow better movement of water. It has been suggested, then, that the low CO_2 world exposed leaves to greater drought stress as they had to keep their stomata open for longer to bring in enough CO_2. The flowering plants, having the correct features for keeping leaves supplied with water, were able to flourish under these conditions while other older groups of plants were pushed out. The trouble with such assertions is that the rise of the flowering plants was a one-off event that we cannot re-run under different circumstances: there could in fact be many other reasons why the angiosperms did so well after they first appeared in the lower Cretaceous.

8.15.2 Ancient moist climates or high CO_2 effects?

High CO_2 levels would tend to produce more luxuriant vegetation, for a given level of rainfall, than we would normally see in the present-day world. This does seem to tally in a general way with some aspects of the plant fossil record; for instance, moist climates with tropical and temperate rainforest seem to have dominated the land surfaces around 55 million years ago during the early Tertiary, at a time when geochemical calculations and stomatal indices suggest that CO_2 levels might have been several times higher than at present. It is possible then that the climates were not really as moist as would appear, and that high CO_2 preserving the water balance of plants enabled lush vegetation to thrive under less rainfall than would be needed nowadays. Generally though, indications are that CO_2 was no more than two to three times higher than now during the early Tertiary "moist" world. This is not high enough to fundamentally change the water needs of plants and make it look like rainfall was several times higher over huge areas of the earth's surface. However, further back in time during the Cretaceous and Jurassic periods, rock chemistry calculations suggest that CO_2 levels were as much as 20 times higher than at present, yet semi-arid environments seem to have been fairly widespread.

During the 50 million years that followed the super-moist world of the early Tertiary, plant fossils of drier climate vegetation such as scrub, grassland and semi-desert became progressively more common. From indicators in the rocks, and changing stomatal indices in fossils, geochemists suggest that CO_2 levels were declining during this time. It is certainly tempting to put the shift in vegetation down to lack of CO_2 making it harder for plants to maintain their water balance, so that in many places forests could no longer survive. Grasses that use the C_4 photosynthetic system have become widespread only during the last 7 million years or so (Figure 8.10), leaving a characteristic isotopic trace in the fossilized carbon and soil carbonates they leave behind. Because C_4 grasses are very good at sucking in CO_2 without losing much water, it has been suggested that the progressively lower CO_2 concentrations favored their spread. The spread of grasses in general, and C_4 grasses in particular, during the last 20 million years, has been linked to the evolution of a range of animals adapted mainly to grazing off grasses. Their existence may owe something to the decrease in CO_2 bringing about the dramatic spread of grasses.

However, it is unlikely that CO_2 effects on plant water balance are the entire story in this global shift in vegetation in the last 50 million years. The change in vegetation seems too dramatic for a direct CO_2 effect alone, and surely requires at least some genuine decline in rainfall. C_4 grasses, while using CO_2 more efficiently, also nowadays inhabit arid environments, and any decline in rainfall would also have favored

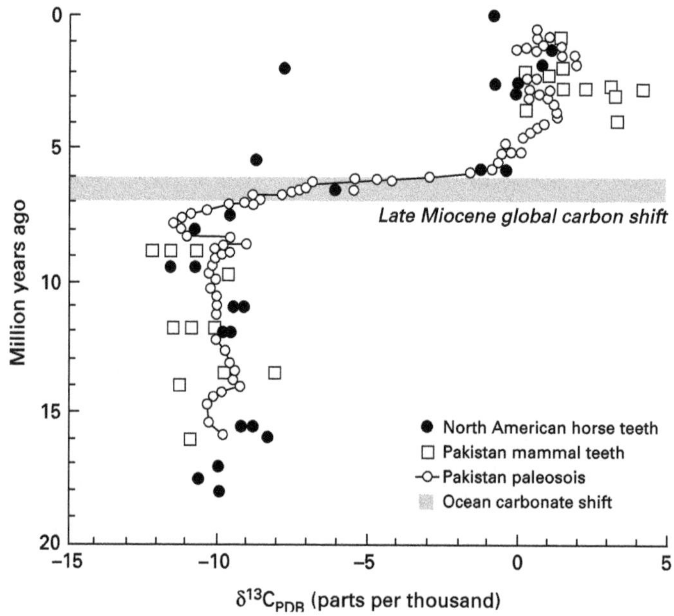

Figure 8.10. The shift in ^{13}C in sediments in North America, South Asia, and in the global ocean indicating a "take-over" by C_4 plants.

them. The moist world 55 million years ago was also very warm (perhaps because of an increased "greenhouse effect", itself partly caused by higher CO_2 or CH_4) and climate models predict that a warm world should produce more rainfall. In addition, the rain-blocking effect of some mountain ranges such as the Himalayas and the Rockies was probably not as strong as it is today. These mountains were just beginning to grow and were still much lower, and this would have allowed rains to reach into the interior of the continents to areas that are now very dry. Once again, it is hard to assign any particular part of the shift in vegetation to CO_2 alone, because so many things changed in parallel.

8.16 OTHER DIRECT CO_2 EFFECTS: IN THE OCEANS

Ocean plankton do not tend to be limited by lack of CO_2, because it is already present in dissolved form in such abundance in the ocean water. In addition, they tend to be severely limited by lack of nutrients, which cuts down how much they could potentially benefit from more CO_2. However, there may be another insidious effect of high CO_2 levels that could have far-reaching consequences for ocean ecosystems and perhaps the whole planetary system.

Some of the most abundant ocean plankton, known as coccolithophores, make shells for themselves out of little calcium carbonate plates. Experiments show that if the CO_2 concentration in the water becomes too high, the calcium carbonate gets eaten away and dissolves as fast as it can be laid down. The result is that the cell collapses in and dies. The studies suggest that at atmospheric CO_2 levels around three times as high as at present (1,000 ppm), coccolithophores will be exposed to so much CO_2 dissolved in ocean waters that they will die. Such CO_2 levels may well be reached in the next century or so as humans continue to burn fossil fuels relentlessly. Corals and seashells, which also lay down calcium carbonate skeletons, seem to be faced with a similar problem. However, it would appear that in the geological past ocean plankton and coral that relied on making calcium carbonate skeletons were able to thrive under CO_2 concentrations higher than this. Either the past CO_2 reconstructions are wrong, or there are ways in which these organisms can eventually evolve to tolerate high CO_2. In the meantime, however, many marine ecosystems could collapse and many extinctions follow from this "other" direct CO_2 effect.

8.17 THE FUTURE DIRECT CO_2 EFFECT: A GOOD OR A BAD THING FOR THE NATURAL WORLD?

If direct CO_2 fertilization turns out to have significant effects on the natural world, will these effects be good or bad? The effects are likely to be complex and multi-faceted, and whether they are, on balance, likely to be good or bad is a subjective issue that depends on one's priorities.

Some scientists, and groups supported by the fossil fuel lobby, have argued that CO_2 fertilization might turn out to be a very good thing for nature in general. By allowing plants to thrive on less water, it might enable tropical rainforests to spread into drier climates. This might help to counterbalance the damaging effects of humans in logging and clearing tropical forests, preserving many of the species that live within them against extinction. There is a general relationship between rainfall and species richness even within tropical forest regions, and rising CO_2 might tend to act rather like an increase in rainfall, maximizing species richness in areas that are already tropical rainforest.

However, the current picture from vegetation–climate modeling in a high-CO_2 world is not encouraging. Models tend to predict that there will either be no substantial increase in tropical rainforest, or a dieback, because of the effects of high temperatures and aridity overwhelming any fertilization effect of CO_2.

Even if rainforest does turn out to grow more lush and cover larger areas, it is not certain that increasing CO_2 will make it easier for large numbers of species to coexist. It is fairly widely established in ecology that if vegetation grows too vigorously, stronger species can triumph and push the weaker ones out. For example, throwing a mineral fertilizer on a grassland will often cause a crash in the species richness of the plant community, as a few fast-growing species that respond particularly well to fertilizer push all the others out. At lower nutrient levels, all species grow relatively slowly but are fairly evenly matched against one another in competition; none can push the others out. The fear is that increasing CO_2 will act as a fertilizer in just the same way, causing plant communities all around the world to undergo a burst of growth that will eliminate many "weaker" species. The result might be massive-scale extinctions of plant species and the insect and fungal life forms which depend upon them. The tropical rainforests could end up far poorer in species, despite conditions that favor their growth.

It is also very difficult to know what "knock-on" effects there will be through the food chain as the primary productivity of terrestrial ecosystems increases. Will food quality for animals increase (with a resulting increase in population densities), or decrease? Will this support more species of animals, or fewer? Also, will forests that have grown under high CO_2 levels burn more easily, or less easily, or be about the same? Will they fall over more easily in wind storms? No-one knows the answer to the many questions that concern the high-CO_2 world. The many nagging uncertainties give us a broad range of possible future scenarios, which frustrate our desire to know exactly what will happen in the future.

8.18 CONCLUSION: THE LIMITS TO WHAT WE CAN KNOW

In terms of painting a picture of the complexity of the living earth, the direct CO_2 effect serves as a final brush stroke, on top of everything else. If there is an overall message from this book, it is that the earth system is immensely subtle and complex, and that life has a key role in just about every aspect of it. And yet, as we marvel at the elegance of the mechanisms that we can see at work, we should not expect to know exactly how they will behave as we tinker with our planet.

The more effort—and the more money—that goes into research on the earth system, the more light will be shed on these processes and how they interact. This can allow us to consider a range of possibilities with more confidence, to spot new possible scenarios that had never been noticed before, and give either more or less credence to each of the various scenarios we already have. But, given the sheer complexity of what we are dealing with here, science can never provide the detailed certainty that politicians crave.

And so we enter a vast experiment, raising greenhouse gas levels beyond anything that has occurred in tens of millions of years. All science can do is offer a range of plausible scenarios, and then try to lay out these possibilities in an accessible way. It is up to people around the world to decide whether to take the gamble on any particular scenario coming true.

Bibliography

Adams, J.M., Constable, J., Guenther, P. and Zimmerman, P. (2001). Estimate of total VOC emission from global vegetation at the Last Glacial Maximum, compared to the Holocene. *Chemosphere – Global Change Science*, **3**, 73–91.

Adams, J.M., Green, W. and Zhang, Y-J. Recalibrating the paleoclimatic thermometer: broad scale analysis of leaf margin trends against temperature in North America. *Global and Planetary Change* (submitted).

Badeck, F., Bondeau, A. and Boettcher, K. (2004). Responses of spring phenology to climate change. *New Phytologist*, **162**, 195–309.

Barnola, J.-M., Raynaud, D., Lorius, C. and Barkov, N.I. (2003). Historical CO_2 record from the Vostok ice core. In: *Trends: A Compendium of Data on Global Change*. Carbon Dioxide Information Analysis Center, Oak Ridge National Laboratory, Oak Ridge, TN. *http://www.cdmc.esd.ornl.gov/trends*

Berner, R.A. (1994). GEOCARB II: A revised model of atmospheric CO_2 over Phanerozoic time. *American Journal of Science*, **294**, 56–91.

Bonan, G. (2002). *Ecological Climatology: Concepts and Applications*. Cambridge University Press, Cambridge, UK.

Bonan, G. (2008). Forests and climate change: Forcings, feedbacks and climate benefits. *Science*, **320**, 1146–1150.

Borchert, R., Robertson, K, Schwartz, M.D. and Williams-Linera, G. (2005). Phenology of temperate trees in tropical climates. *International Journal of Biometeorology*, **50**, 57–65.

Brown, S., Gillespie, A.J.R. and Lugo, A.E. (1991). Biomass of tropical forests of south and southeast Asia. *Canadian Journal of Forest Research*, **2**, 111–117.

Cerling, T.E., Wang, Y. and Quade, J. (1993). Expansion of C_4 ecosystems as an indicator of global ecological change in the late Miocene. *Nature*, **361**, 344–345.

Chase, T.N., Pielke, R.A., Kittel, T.G.F., Nemani, R. and Running, S.W. (1996). Sensitivity of a general circulation model to global changes in leaf area index. *Journal of Geophysical Research*, **101**, 7393–7408.

Copeland, J.H., Pielke. R.A. and Kittel, T.G.F. (1996). Potential climatic impacts of vegetation change: A regional modeling study. *Journal of Geophysical Research*, **101**, 7409–7418.

Costa, M.H. and Foley, J.A. (2000). Combined effects of deforestation and doubled atmospheric CO_2 concentrations on the climate of Amazonia. *Journal of Climate*, **13**, 1834.

Cox, P.M., Betts, R.A., Jones, C.D., Spall, S.A. and Totterdell, I.J. (2000). Acceleration of global warming due to carbon-cycle feedbacks in a coupled climate model. *Nature*, **408**, 184–187.

Cox, P.M., Betts, R.A., Jones, C.D., Spall, S.A. and Totterdall, I.J. (2000). Acceleration of global warming due to carbon cycle feedbacks in a coupled climate model. *Nature*, **409**, 184–187.

Davis, M.B. and Sugita, S. (1995). Re-evaluating the pollen record of Holocene tree migration. In: B. Huntley (eds), *Past and Future Environmental Changes: The Spatial and Evolutionary Responses of Terrestrial Biota*. Proceedings of a NATO workshop. Elsevier, London.

Etheridge, D.M., Steele, L.P., Langenfelds, R.L., Francey, R.J., Barnola, J.-M. and Morgan, V.I. (1998). Historical CO_2 records from the Law Dome DE08, DE08-2 and DSS ice cores. In: *Trends: A Compendium of Data on Global Change*. Carbon Dioxide Information Analysis Center, Oak Ridge National Laboratory, Oak Ridge, TN. *http://www.cdmc.esd.ornl.gov/trends/co2/lawdome.html*

Farquhar, G.D. and Wong, S.C. (1984). An empirical model of stomatal conductance. *Aust. J. Plant Phys*, **11**, 191–210.

Gedney, N., Cox, P.M., Betts, R.A., Huntingford, C. and Stott, P.A. (2006). Detection of a direct carbon dioxie effect in continental river runoff records. *Nature*, **439**, 835–838.

Gibbard, S., Caldeira, K., Govindasamy, B., Phillips, T.J. and Wickett, M. (2005). Climate effects of global land cover change. *Geophysical Research Letters*, **32**, 1029.

Holdridge, L.R. (1967). *Life Zone Ecology*. Tropical Science Center, San Jose, Costa Rica.

Houghton, R.A. and Hackler, J.L. (1995). *Continental Scale Estimates of the Biotic Carbon Flux from Land Cover Change: 1850–1980*. ORNL/CDIAC-79, NDP-050, Carbon Dioxide Information Analysis Center, Oak Ridge National Laboratory, Oak Ridge, TN.

IPCC (2001). *Climate Change 2001*. Intergovernmental Panel on Climate change. UNEP/WMO.

Joseph, L. Eastman, M., Coughenour, B. and Pielke, Sr., R.A. (2001). Does grazing affect regional climate? *Journal of Hydrometeorology*, **2**, 243–253.

Kimball, B., Pinter, P.J. and Garcia, R.L. (1995). Productivity and water use of wheat under free-air CO_2 enrichment. *Global Change Biology*, **1**, 429–442.

Koerner, C. (2004). Through enhanced tree dynamics CO_2 may cause tropical forests to lose carbon. *Philosophical Transactions of the Royal Society of London*, B. **359**, 493–498.

Koerner, C., Asshoff, R., Bignucolo, O., Hättenschwiler, S., Keel, S.G., Peláez-Riedl, S., Pepin, S., Siegwolf, R.T.W. and Zotz, G.. Carbon flux and growth in mature deciduous forest trees exposed to elevated CO_2. *Science*, **309**(5739), 1360–1362.

Lawton, R.O, Nair, U.S., Pielke, Sr., R.A. and Welch, R.M. (2001) Climatic impact of tropical lowland deforestation on nearby montane cloud forests. *Science*, **294**, 584–587.

Levis, S., Foley, J.A. and Pollard, D. (1999). Potential high-latitude vegetation feedbacks on CO_2-induced climate change. *Journal of Geophysical Research*, **26**, 747.

Long, S.P., Ainsworth, E.A., Leakey, A.D.B., Nosburger, J. and Ort, D.R. (2006). Food for thought. *Science*, **312**, 1918–1921.

Machado, L.A., Laurent, T.H. and Lima, A.A. (2002). Diurnal march of the convection observed during TRMM–WETAMC/LBA. *Journal of Geophysical Research*, **107**, 8064.

Melillo, J., Janetos, A., Schimel, D. and Kittel, T. (2001). Vegetation and biogeochemical scenarios. In: National Assessment Synthesis Team (eds), *Climate Change Impacts on the United States: The Potential Consequences of Climate Variability and Change*. Report for the US Global Change Research Program. Cambridge University Press, NY, pp. 73–91.

Menzel, A. *et al.* (2006). Eueopean phenological response to climate change matches the warming pattern. *Global Change Biology*, **12**, 1365–2486.

Mitchell, T. (June 2005). *Sahel Rainfall Index*. Joint Institute for the Study of the Atmosphere and Ocean. Available at *http://earthobservatory.nasa.gov/Features/Desertification/desertification2.php*

Norby, R.J., Long, T.M., Hartz-Rubin, J.S. and O'Neil, E.G. (2000). Nitrogen resorption in senescing leaves in a warmer, CO_2-enriched atmosphere. *Plant and Soil*, **224**, 15–29.

Otterman, J. (1975) Baring high-albedo soils by overgrazing. *Science*, **186**, 531–533.

Overpeck, J.T., Rind, D., Lacis, A. and Healy, R. (1996) Possible role of dust-induced regional warming in abrupt climate change during the last glacial period. *Nature*, **384**, 447–449.

Petit-Maire, N. and Guo, Z. (1996). Mise en evidence de variations climatiques holocenes rapides, en phase dans les deserts actuels de Chine et du Nord de l'Afrique. Comptes Rendus de l'Academie de Sciences, Serie II. *Sciences de la Terre et des Planetes*, **322**, 847–851.

Philips, O.L. and Gentry, A.H. (1994) Increasing turnover time in tropical forests. *Science*, **263**, 954–958.

Pielke, Sr., R.A., Marland, G., Betts, R.A., Chase, T.N., Eastman, J.L., Niles, J.O., Niyogi, D. and Running, S. (2002). The influence of land-use change and landscape dynamics on the climate system. *Philosophical Transactions of the Royal Society*, **360**, 1705–1719.

Raymo, M.E., Ruddiman, W.F. and Froelich, P.N. (1988) Influence of late Cenozoic mountain building on ocean geochemical cycles. *Geology*, **16**, 649–653.

Renssen, H., Brovkin, V., Fichefet, T. and Goosse, H. (2003) Holocene climate instability during the termination of the African Humid Period. *Geophysical Research Letters*, **30**, 1184.

Schwartzman, D. (1999). *Life, Temperature and the Earth: The Self-Organizing Biosphere*. Columbia University Press, USA.

Schweingruber, F.H., Briffa, K.R. and Jones, P.D. (1993). A tree-ring densiometrric transect from Alaska to Labrador. Comparison of ring width and latewood density. *International Journal of Biometerology*, **37**, 151–169.

Sitch, S., Huntingford, C. and Givney, N. (2008). Evaluation of the terrestrial carbon cycle, future plant geography and climate–carbon cycle feedbacks using five dynamic global vegetation models (DVGMs). *Global Change Biology*, **14**, 1–28.

Stowe, D.A. *et al.* (2004). Remote sensing of vegetation and land cover change in Arctic tundra ecosystems. *Remote Sensing of Environment*. **89**, 281–308.

Thuiller, W., Lavorel, S., Arajo, M.B., Sykes, M.T. and Prentice, I.C. (2005). Climate change threats to plant diversity in Europe. *Proc. Natl. Academy Sci.*, **102**, 8245–8250.

Woodward, F.I. (1987) Stomatal numbers are sensitive to increases in atmospheric CO_2. *Nature*, **327**, 617–618.

Yoshioka, M., Mahowald, N.M., Conley, A.J., Collins, W.D., Fillmore, D.W., Zender, C.S., Coleman, D.B. (2006) Impact of desert dust radiative forcing on Sahel precipitation: Relative importance of dust compared to sea surface temperature variations, vegetation changes and greenhouse gas warming. *Journal of Climate*, **19**, 2661–2672.

Zeng, H. and Neelin, D. (2000). The role of vegetation-climate interaction and interannual variability in shaping the African savanna. *Journal of Climate*, **13**, 2665–2670.

Index

Acer saccharum 88
Air masses 8
Albedo 123, 155, 161, 173
Alkalinity (ocean) 184, 190
Amazonia 93, 200–201, 217, 226, 235
Aspect 110

Betula 76, 79–80
Biological pump 193
Boundary layer 99–102
Bowen ratio 157

Canopy microclimates 103–105
Carbon dioxide, history of 186, 188, 193, 201, 257
Carbon isotopes 207–209
Cloud forest 105, 167
Continental climate 17
Coriolis effect 7, 8
Corylus avelana 79
Cushion plants 101

Deciduousness 43–52
Deforestation 161–170, 201, 202, 204
Desert vegetation 54, 122
Desert climates 21–25
Dust 145–149

Eddy flux covariance 215–217
El Niño 210, 218–219
Ekman spiral 7
Endolithic life forms 114
Eocene 13
Epiphytes 104–105
Erica tetralix 60
Eucalyptus 110

FACE experiments 228–245
Farquahar Model 223
Frost hollows 113

Gaia hypothesis 117–119
GCMs 133–136
Global warming 81–96
Grasslands 53
Great Plains 136
Green Sahara 132, 139–143, 150–151
Gulf Stream 15
Gyres 9–10

Heat sum 111
Hedera helix 60
Humus 183
Hyacinthoides non-scripta 94

Interglacials 68, 183, 193–198
ITCZ 6, 18–21
Inversion layer, canopy 107

Latent heat 15, 155, 170
Lichens 190
Little Ice Age 171–172

Marine snow 183
Mediterranean climate 22–23
Methane 187
Missing sink 213–215
Mountain mass effect 42
Mountains 4

Ocean circulation 9–14
Oceanic distribution 60

Peat 183
Pinus sylvestris 79–80
Pollen record 76
Positive feedback 124–127

Quaternary 67–74, 173, 183, 193–198

Rain shadow 24
Rhododendron ponticum 60
Roman era 173
Roughness 102–103, 130, 160, 161, 173

Sahel 127–132
Scrub 53
Selaginella willendowi 108
Sequoia sempervirens 105
Soil microclimates 98
Species richness 60–64
Stomatal index 246
Sun angle 2, 112
Sun leaves 108–109

Thermohaline circulation 10–12, 15
Tillandsia 105
Tree growth habit 39–43
Tree line 42–43
Trillium 94
Tropical rainforest biome 28–31

Understory environment 106

Vegetation–climate schemes 55–59
Volatile organic compounds 116, 176–177

Weathering (Rock) 189–192

Younger Dryas 77

Lightning Source UK Ltd.
Milton Keynes UK
UKHW031136131021
391862UK00012B/390

9 783642 269059